Operator Theory
Advances and Applications
Vol. 82

Editor
I. Gohberg

Schur Parameters, Factorization and Dilation Problems

Tiberiu Constantinescu

Birkhäuser Verlag
Basel · Boston · Berlin

Authors' address:

Tiberiu Constantinescu
Programs in Mathematical Sciences
University of Texas at Dallas
Richardson, TX 75083
USA

1991 Mathematics Subject Classification 46B70, 46M35, 47A40, 93C55

A CIP catalogue record for this book is available from the Library of Congress, Washington D.C., USA

Deutsche Bibliothek Cataloging-in-Publication Data
Constantinescu, Tiberiu:
Schur parameters, factorization and dilation problems / Tiberiu
Constantinescu. - Basel; Boston; Berlin: Birkhäuser, 1996
 (Operator theory; Vol. 82)
ISBN-13: 978-3-0348-9910-9   e-ISBN-13: 978-3-0348-9108-0
DOI: 10.1007/978-3-0348-9108-0
NE: GT

© 1996 Birkhäuser Verlag, P.O. Box 133, CH-4010 Basel, Switzerland
Softcover reprint of the hardcover 1st edition 1996

Printed on acid-free paper produced from chlorine-free pulp. TCF ∞
Cover design: Heinz Hiltbrunner, Basel

9 8 7 6 5 4 3 2 1

Contents

Preface

The subject of this book is about the ubiquity of the Schur parameters, whose introduction goes back to a paper of I. Schur in 1917 concerning an interpolation problem of C. Carathéodory. What followed there appears to be a truly fascinating story which, however, should be told by a professional historian. Here we provide the reader with a simplified version, mostly related to the contents of the book.

In the twenties, the theory of orthogonal polynomials on the unit circle was developed by G. Szegö and the formulae relating these polynomials involved numbers (usually called Szegö parameters) similar to the Schur parameters. Meanwhile, R. Nevanlinna and G. Pick studied the theory of another interpolation problem, known since then as the Nevanlinna-Pick problem, and an algorithm similar to Schur's one was obtained by Nevanlinna. In 1957, Z. Nehari solved an L^∞ problem which contained both Carathéodory-Schur and Nevannlina-Pick problems as particular cases. Apparently unrelated work of H. Weyl, J. von Neumann and K. Friedericks concerning selfadjoint extensions of symmetric operators was connected to interpolation by M.A. Naimark and M.G Krein using some general dilation theoretic ideas. Classical moment problems, like the trigonometric moment and Hamburger moment problems, were also related to these topics and a comprehensive account of what can be called the classical period has appeared in the monograph of M.G. Krein and A.A. Nudelman, [KN].

The fifties and sixties witnessed the first interference of this field with engineering applications. In connection with the Wiener filtering problem in discrete time, N. Levinson developed an efficient algorithm for solving normal equations, which is strongly related to the Szegö formulae. The algorithm was rediscovered in 1960 by Durbin, while Schur's algorithm was rediscovered about the same time in seismic oil prospecting. A transmission-line (or lattice) model is related to these algorithms and similar structures turned out to be of common interest in circuit synthesis and linear estimation. The Burg technique in spectral analysis of stationary time series clarified the connections between the Szegö theory and the maximum entropy method. In the last two decades, the Nevanlinna-Pick problem turned into a main tool in robust control.

These applications motivated an abundance of new approaches to the interpolation problems mentioned before. The main ideas came with the new dilation theoretic achievements in the work of M.S. Livsic and M.S. Brodskii, B. Sz-Nagy and C. Foias and of L. de Branges and J. Rovnyak concerning modeling of non-selfadjoint operators. A major breakthrough was realized in 1967, when D. Sarason obtained the solution of the Carathéodory-Schur and Nevanlinna-Pick problems as a consequence of a representation theorem for operators commuting with special contractions. One year later, B. Sz.-Nagy and C. Foias proved by dilation theoretic methods a vast generalization of Sarason's result, referred to since then as the commutant lifting theorem. At about the same time, V.M. Adamjan, D.Z. Arov and M.G. Krein related the solution of the Nehari problem with the theory of Hankel operators. This work was later generalized to the framework of the commutant lifting (see [FoF], [RR1]). An approach to the commutant lifting based on the Beurling-Lax-Halmos theorem for Krein spaces was developed by J.A. Ball and J.W. Helton and is presented in [Helt5], [BGR]. Another method, known as the band method, was initiated by H. Dym and I. Gohberg (see [GGK]). One more method is based on the results concerning de Branges reproducing kernel spaces, and was employed by H. Dym (see [Dy1]). State-space computations were used by K. Glover and the monograph [BGR] contains a comprehensive account on various interpolation problems from this point of view. Yet another method was developed by V.P. Potapov and his school, and its presentation may be found in [KP] or [DFK]. Finally, a method which encodes a generalization of the Schur algorithm and, therefore, can be used in interpolation, was developed over the years by T. Kailath and his school (see [KKM]).

However, in all these developments, the Schur parameters play a rather modest role, mostly in connection with the parametrization of the solutions of the interpolation problems. This situation prompted the present attempt of making the Schur parameters the main character of a book. We want to emphasize that they might play a distinguished role not only in interpolation, but also in some other related ares, such as dilation and factorization theory.

In Chapter 1 we describe the structure of positive definite kernels on the set of integers in terms of Schur parameters. This construction explores the interplay between some dilation theoretic ideas (Kolmogorov decomposition) and (Cholesky) factorization. A multiplicative structure of the Kolmogorov decomposition is also emphasized as a key technical tool. Chapter 2 explores the role of the Schur parameters in the structure of the triangular contractions and in the realization theory of unitary systems. Then, for another perspective on these results, the models for families of contractions given by Sz.-Nagy and Foias and, respectively, de Branges and Rovnyak are presented in some details.

Chapter 3 deals with interpolation and moment problems. A method employed by M.A. Naimark and M.G. Krein is further developed, allowing us to view all these problems in a general framework of extending certain "structured matrices". The positivity of these matrices turns into the solvability criterion and the Schur parameters can be easily used to provide a description of all the solutions. The commutant lifting theorem is also analyzed in this chapter. The term of "structured matrix" is explained in Chapter 4 by the introduction of the concept of matrix with displacement structure. Here, the main achievement is a generalized Schur algorithm that encompasses all the algorithms mentioned above, starting with the algorithms of Schur and Nevanlinna and ending with least squares algorithms in adaptive filtering. Sometimes, it is possible to associate to this generalized Schur algorithm a generalized transmission-line model. The consideration of so-called transmission zeros of the line leads to the connection with the interpolation problems analyzed in Chapter 3. We also show that this connection can be explained as a consequence of the method of the commutant lifting. In Chapter 5 we unify the various factorization results used in the previous chapters by proving the existence of a spectral factor of a positive definite kernel. Chapter 6 contains an application to the study of nonstationary processes. Schur parameters are used to compute various angles between parts of past and future, as well as to discuss the Szegö limit theorems.

In the last two chapters we pass from structured matrices to arbitrary patterns. It is shown that the Schur parameters still play a role if we stick to patterns associated to chordal graphs. In Chapter 7 we discuss completion problems, while in Chapter 8 we focus on determinantal formulae. We prove a general determinantal formula using the Schur parameters, we discuss a maximum determinant principle which is the right analogue of a maximum entropy principle analyzed in Chapter 5 and we show an inheritance property of the maximum determinant solution. And, of course, we give some hints about the limitations of the use of the Schur parameters.

We must also say that there are some other applications and developments involving Schur parameters. For instance, we can go beyond positive matrices, or we can study integral equations and still be able to introduce some sort of Schur parameters in a natural way. However, the presentation of the positive definite matrix case gives the general ideas about the capabilities (and the limitations) of the analysis based on this type of parameters.

Acknowledgement   The writing of this book came to an end mostly due to the constant encouragements of Professors I. Gohberg and T. Kailath. The project was ready in 1989 and it was a joint work with M. Bakonyi, based on lectures given in the Department of Mathematics at INCREST, Bucharest, and following the ideas in some of the work of Professor C. Foias in dilation theory. The discussions with A. Gheondea and M. Putinar played a major role in understanding most of these ideas. Also discussions with Professors C.R. Johnson and J. Rovnyak were useful for the preparation of some parts of the book. The first step of this project, which was viewed as an introduction to this volume, was the booklet on Schur's algorithm and its applications, [BC2]. Since 1991, Professor Kailath's influence became direct and A. Sayed's thesis written under his guidance played a significant role in the development of this book.

January 9, 1995

Chapter 1

Schur Parameters and Positive Block Matrices

Our aim in this chapter is to emphasize the role played by the Schur parameters in connection with the structure of the positive definite kernels on the set of integers. These parameters appear as operator angles describing the geometry of a certain space associated to the given kernel by a simple renorming construction, usually referred to as the Kolmogorov decomposition. In this setting we make the connection with the dilation theory by emphasizing a key multiplicative structure of the Kolmogorov decomposition. This explains the structure of the Naimark and Sz.-Nagy dilations as elementary rotations of certain contractions. A few immediate applications are also presented. Thus, we discuss Cholesky factorizations, determinantal formulae for positive block matrices and orthogonal polynomials on the unit circle.

1.1 Preliminaries

In this section we describe some elementary properties of Hilbert spaces and their linear operators. Throughout, the conjugate of a complex number z in $\mathbb{C}$ is denoted by z^*.

An *inner product* on a complex linear space $\mathcal{H}$ is a map $\langle \cdot, \cdot \rangle : \mathcal{H} \times \mathcal{H} \longrightarrow \mathbb{C}$ such that

(1) $\langle \alpha_1 h_1 + \alpha_2 h_2, g \rangle = \alpha_1 \langle h_1, g \rangle + \alpha_2 \langle h_2, g \rangle$ for α_1, α_2 in $\mathbb{C}$ and h_1, h_2, g in $\mathcal{H}$.

(2) $\langle h, g \rangle = \langle g, h \rangle^*$ for h, g in $\mathcal{H}$.

(3) $\langle h, h \rangle \geq 0$ for h in $\mathcal{H}$ and $\langle h, h \rangle = 0$ if and only if $h = 0$.

Using the *Cauchy-Schwarz inequality*, $|\langle h, g \rangle|^2 \leq \langle h, h \rangle \langle g, g \rangle$, which is valid for all h and g in $\mathcal{H}$, it follows that the formula $\|h\| = \langle h, h \rangle^{1/2}$ defines a norm on $\mathcal{H}$. The complex linear space $\mathcal{H}$ equipped with an inner product $\langle \cdot, \cdot \rangle$ is called a *Hilbert space* if the metric induced by the above norm is complete. This means that if $\{h_n\}_{n \geq 1}$ is a sequence of elements h_n in $\mathcal{H}$ such that $\|h_n - h_m\| \to 0$ as m and n tend to ∞, then there exists h in $\mathcal{H}$ such that $\|h_n - h\| \to 0$ as n tends to ∞. Several examples will be frequently used.

1.1 Example   Let $\mathbb{C}^n$ denote the linear space of n-tuples $x = \{x_k\}_{k=0}^{n-1}$ of complex numbers and define an inner product on $\mathbb{C}^n$ by the formula

$$\langle x, y \rangle = \sum_{k=0}^{n-1} x_k y_k^*.$$

Then $\mathbb{C}^n$ is a Hilbert space with respect to this inner product and the associated norm is the usual Euclidean norm, $\|x\| = (\sum_{k=0}^{n-1} |x_k|^2)^{1/2}$. This construction can be extended in the following way. Let l^2 denote the linear space of the square-summable sequences $\{x_n\}_{n \in \mathbb{Z}}$ of complex numbers. This means that $x = \{x_n\}_{n \in \mathbb{Z}}$ belongs to l^2 if and only if the series $\sum_{n \in \mathbb{Z}} |x_n|^2$ converges. It follows that if $x = \{x_n\}_{n \in \mathbb{Z}}$ and $y = \{y_n\}_{n \in \mathbb{Z}}$ are elements in l^2, then

$$\langle x, y \rangle = \sum_{n \in \mathbb{Z}} x_n y_n^*$$

is well-defined (*i.e.* the series in the right side converges) and the map $\langle \cdot, \cdot \rangle$ is an inner product. It is readily checked that l^2 is a Hilbert space with respect to this inner product. The space $\mathbb{C}^n$ may be viewed as a subspace of l^2 by identifying an element $x = \{x_0, x_1, \ldots, x_{n-1}\}$ of $\mathbb{C}^n$ with the element $x = \{\ldots, 0, x_0, x_1, \ldots, x_{n-1}, 0, \ldots, \}$ of l^2.   □

1.2 Example   Let μ be a probability measure on the interval $[0, 2\pi)$. Let $\mathcal{L}^2$ be the set of the measurable complex functions f on the unit circle $\mathbb{T} = \{\zeta \in \mathbb{C} \mid |\zeta| = 1\}$ which are square-integrable with respect to μ, *i.e.*

$$\int_0^{2\pi} |f(e^{it})|^2 d\mu(t) < \infty,$$

and define an inner product on $\mathcal{L}^2$ by the formula:

$$\langle f, g \rangle = \frac{1}{2\pi} \int_0^{2\pi} f(e^{it}) g(e^{it})^* d\mu(t).$$

Since the subspace $\mathcal{N}^2 = \mathcal{L}^2 \cap \{f \mid \int_0^{2\pi} |f(e^{it})|^2 d\mu(t) = 0\}$ may be nontrivial ($\neq \{0\}$), we define $L^2(\mu) = \mathcal{L}^2/\mathcal{N}^2$. Denote by $[f]$ the class in $L^2(\mu)$ of the element f of $\mathcal{L}^2$ and it is easy to check that the formula

$$\langle [f], [g] \rangle = \frac{1}{2\pi} \int_0^{2\pi} f(e^{it}) g(e^{it})^* d\mu(t)$$

is independent of the choice of the representatives f and g of $[f]$ and, respectively, $[g]$. Moreover, this formula defines an inner product on $L^2(\mu)$. Equipped with this inner product, $L^2(\mu)$ becomes a Hilbert space. We maintain the usual convention of speaking about the elements of $L^2(\mu)$ as functions instead of classes of functions which are equal almost everywhere. If μ is the Lebesgue measure on $[0, 2\pi)$, then we denote the corresponding space $L^2(\mu)$ by L^2.

We also mention that according to the Riesz-Markov representation theorem, every probability measure μ on $[0, 2\pi)$ can be viewed as a positive bounded linear functional on $C(\mathbb{T})$, the Banach space of the complex valued continuous functions on the unit circle. The space $C(\mathbb{T})$ is equipped with the norm $\|f\|_\infty = \sup\{|f(\zeta)| \mid |\zeta| = 1\}$ and the linear functional associated to μ is given by the formula

$$\mu(f) = \frac{1}{2\pi} \int_0^{2\pi} f(e^{it}) d\mu(t)$$

for f in $C(\mathbb{T})$. $\qquad\square$

Two vectors h, g of the Hilbert space $\mathcal{H}$ are said to be orthogonal if $\langle h, g \rangle = 0$. If $\mathcal{G}$ is a closed subspace of $\mathcal{H}$, then the *orthogonal complement* $\mathcal{G}^\perp$ (or $\mathcal{H} \ominus \mathcal{G}$) of $\mathcal{G}$ is defined by

$$\mathcal{G}^\perp (= \mathcal{H} \ominus \mathcal{G}) = \{h \in \mathcal{H} \mid \langle h, g \rangle = 0 \text{ for all } g \in \mathcal{G}\}.$$

Two closed subspaces $\mathcal{G}_1$ and $\mathcal{G}_2$ of $\mathcal{H}$ are called *orthogonal subspaces*, in symbols $\mathcal{G}_1 \perp \mathcal{G}_2$, if $\langle h, g \rangle = 0$ for every h in $\mathcal{G}_1$ and g in $\mathcal{G}_2$. We introduce now some other examples of Hilbert spaces.

1.3 Example (a) Suppose $\{\mathcal{H}_n\}_{n\in\mathbb{Z}}$ is a family of closed subspaces of a Hilbert space $\mathcal{H}$. Then, the closure of the linear span of these spaces is denoted by $\bigvee_{n\in\mathbb{Z}} \mathcal{H}_n$. If the subspaces $\mathcal{H}_n$ are pairwise orthogonal, i.e. $\mathcal{H}_i \perp \mathcal{H}_j$ for $i \neq j$, then the notation $\oplus_{n\in\mathbb{Z}}\mathcal{H}_n$ is used instead of $\bigvee_{n\in\mathbb{Z}} \mathcal{H}_n$. This space $\oplus_{n\in\mathbb{Z}}\mathcal{H}_n$ will be called *the orthogonal sum of the pairwise orthogonal subspaces* $\mathcal{H}_n$. Since for every vector f in $\mathcal{H}$ there exist unique vectors g in $\mathcal{G}$ and h in $\mathcal{G}^\perp$ such that $f = g + h$, it follows that $\mathcal{G} \oplus \mathcal{G}^\perp = \mathcal{H}$.

(b) Suppose $\{\mathcal{H}_n\}_{n\in\mathbb{Z}}$ is a family of Hilbert spaces. The symbol $\oplus_{n\in\mathbb{Z}}\mathcal{H}_n$ denotes the Hilbert space (called the *orthogonal sum of the spaces* $\mathcal{H}_n$) consisting of the sequences $\{h_n\}_{n\in\mathbb{Z}}$ such that h_n belongs to $\mathcal{H}_n$ for $n \in \mathbb{Z}$ and $\sum_{n\in\mathbb{Z}} \|h_n\|^2 < \infty$, equipped with the inner product

$$\langle \{h_n\}_{n\in\mathbb{Z}}, \{g_n\}_{n\in\mathbb{Z}} \rangle = \sum_{n\in\mathbb{Z}} \langle h_n, g_n \rangle_{\mathcal{H}_n},$$

where $\langle \cdot, \cdot \rangle_{\mathcal{H}_n}$ is the inner product on $\mathcal{H}_n$. If the spaces $\mathcal{H}_n$ are all equal to a given Hilbert space $\mathcal{H}$, then the notation $l^2(\mathcal{H})$ is used to denote their orthogonal sum. We also remark that if $\mathcal{H}_n = \mathbb{C}$ for all $n \in \mathbb{Z}$, then $\oplus_{n\in\mathbb{Z}}\mathcal{H}_n = l^2(\mathbb{C}) = l^2$.

We mention that every space $\mathcal{H}_n$ may be identified with a closed subspace of $\oplus_{n\in\mathbb{Z}}\mathcal{H}_n$ and then, $\oplus_{n\in\mathbb{Z}}\mathcal{H}_n$ is the orthogonal sum of these pairwise orthogonal subspaces. This is the reason for our use of the same notation for the orthogonal sum of a family of pairwise subspaces of a certain Hilbert space, as well as for the orthogonal sum of an arbitrary family of Hilbert spaces. In any case, we will also use the notation $\oplus_{n\in\mathbb{Z}} h_n$ to denote the element $\{h_n\}_{n\in\mathbb{Z}}$ of $\oplus_{n\in\mathbb{Z}}\mathcal{H}_n$. $\qquad\square$

A subset $\{e_i\}_{i\in\mathbb{I}}$ of a Hilbert space $\mathcal{H}$ is said to be *orthonormal* if $\langle e_i, e_j\rangle = 0$ for $i \neq j$, and $\|e_i\| = 1$ for all $i \in \mathbb{I}$. If, in addition, the subspace spanned by the set $\{e_i\}_{i\in\mathbb{I}}$ is dense in $\mathcal{H}$, then we say that $\{e_i\}_{i\in\mathbb{I}}$ is an *orthonormal basis* of $\mathcal{H}$. It is known that every Hilbert space ($\neq \{0\}$) has an orthonormal basis. It is also well known that any two orthonormal basis of a Hilbert space have the same cardinality, which is the *dimension* of $\mathcal{H}$. Moreover, two Hilbert spaces are isomorphic if and only if their dimensions are equal. We will restrict our attention to *separable* Hilbert spaces, *i.e.* to those Hilbert spaces possessing a countable orthonormal basis. For this type of spaces there is a method, called the *Gram-Schmidt orthonormalization procedure*, to construct an orthonormal basis starting from a given *total* sequence (*i.e.* a sequence with dense linear span). Indeed, let $\{f_n\}_{n\geq 1}$ be such a sequence in $\mathcal{H}$. Without loss of generality, we can assume that the vectors $f_n, n \geq 1$, are linearly independent and the Gram-Schmidt procedure sets:

$$e_1 = \frac{f_1}{\|f_1\|},$$

$$e_{n+1} = \frac{f_{n+1} - \sum_{k=1}^{n}\langle f_{n+1}, e_k\rangle e_k}{\|f_{n+1} - \sum_{k=1}^{n}\langle f_{n+1}, e_k\rangle e_k\|}.$$

Then $\{e_n\}_{n\geq 1}$ is an orthonormal basis for $\mathcal{H}$ such that the space spanned by e_1, e_2, ..., e_n coincides with the space spanned by f_1, f_2, ..., f_n.

1.4 Example   Some of the Hilbert spaces introduced so far possess orthonormal basis which are very easy to describe. Thus, the set $\{E_n\}_{n\in\mathbb{Z}}$, where E_n is the vector in l^2 with the n-th entry equal to one and all the other entries equal to zero, is an orthonormal basis for l^2. As a consequence, $\{E_n\}_{k=0}^{n-1}$ is an orthonormal basis for $\mathbb{C}^n$. If we define $\nu_n(\zeta) = \zeta^n$ for $n \in \mathbb{Z}$ and $\zeta = e^{it}, t \in [0, 2\pi)$, then $\{\nu_n\}_{n\in\mathbb{Z}}$ is an orthonormal basis for L^2. $\square$

We now review a few facts about linear operators on Hilbert spaces. If $\mathcal{H}$ and $\mathcal{H}'$ are two given Hilbert spaces, then $\mathcal{L}(\mathcal{H}, \mathcal{H}')$ denotes the set of all the linear bounded operators from $\mathcal{H}$ to $\mathcal{H}'$; we write $\mathcal{L}(\mathcal{H})$ instead of $\mathcal{L}(\mathcal{H}, \mathcal{H})$. The norm of an operator T in $\mathcal{L}(\mathcal{H}, \mathcal{H}')$ is defined by the formula

$$\|T\| = \sup\{\|Th\| \mid h \in \mathcal{H}, \|h\| \leq 1\}.$$

We will use the symbol 0 (or $0_{\mathcal{H},\mathcal{H}'}$ when necessary) to denote the zero operator in $\mathcal{L}(\mathcal{H}, \mathcal{H}')$. The identity operator on the Hilbert space $\mathcal{H}$ is denoted by I (or $I_{\mathcal{H}}$ when necessary). The *spectrum* of an operator T on the Hilbert space $\mathcal{H}$ is the set

$$\sigma(T) = \{\lambda \in \mathbb{C} \mid T - \lambda I \quad \text{is not invertible}\}.$$

The complex number λ is an eigenvalue of T if there exists a vector $h \neq 0$ such that $Th = \lambda h$. If T is an operator in $\mathcal{L}(\mathcal{H}, \mathcal{H}')$, then we denote by $\mathcal{R}(T)$ its *range*, *i.e.* the set $\{Th \mid h \in \mathcal{H}\}$, and we denote by $\text{cl}\mathcal{R}(T)$ the closure of $\mathcal{R}(T)$ in $\mathcal{H}'$. Moreover, the *kernel* of T is the set $\ker T = \{h \in \mathcal{H} \mid Th = 0\}$.

The *adjoint* T^* of an operator T in $\mathcal{L}(\mathcal{H}, \mathcal{H}')$ is the operator mapping $\mathcal{H}'$ into $\mathcal{H}$, defined by the equality $\langle h, T^*h' \rangle = \langle Th, h' \rangle$ for all h in $\mathcal{H}$ and h' in $\mathcal{H}'$. A convention that will be frequently employed consists in denoting the adjoint of an operator $T(x)$ which depends on a parameter x by $T^*(x)$ instead of $(T(x))^*$. An operator A in $\mathcal{L}(\mathcal{H})$ is called *selfadjoint* if $A = A^*$ and A in $\mathcal{L}(\mathcal{H})$ is called *positive* (and we write $A \geq 0$) if $\langle Ah, h \rangle \geq 0$ for all h in $\mathcal{H}$. If A is positive and invertible, then we say that A is *strictly positive* and occasionally write $A > 0$. For a positive operator A on the Hilbert space $\mathcal{H}$ there exists a unique positive operator B such that $B^2 = A$. We write $B = A^{1/2}$ and the operator $A^{1/2}$ is called the *square root* of A. An operator T in $\mathcal{L}(\mathcal{H}, \mathcal{H}')$ is a *contraction* if $I - T^*T \geq 0$ (or, equivalently, $\|T\| \leq 1$). If $I - T^*T > 0$ (or, equivalently, $\|T\| < 1$), then the operator T is called *strict contraction*.

An operator V in $\mathcal{L}(\mathcal{H}, \mathcal{H}')$ is a *partial isometry* if $\|Vh\| = \|h\|$ for every h in $\mathcal{H}$ which is orthogonal to $\ker V$. If, in addition, $\ker V = \{0\}$, then V is an *isometry*. It can be verified that V is an isometry if and only if $V^*V = I$. An operator U is *unitary* if both U and U^* are isometries.

An operator P is a *projection* if $P^2 = P = P^*$. Suppose $\mathcal{G}$ is a closed subspace of the Hilbert space $\mathcal{H}$. Define $P_\mathcal{G}$ to be the map $P_\mathcal{G} f = g$, where $f = g + h$ with g in $\mathcal{G}$ and h in $\mathcal{G}^\perp$. Then $P_\mathcal{G}$ is a projection and, conversely, for every projection P in $\mathcal{L}(\mathcal{H})$, there exists a closed subspace $\mathcal{G}(= \mathcal{R}(P))$ such that $P = P_\mathcal{G}$ (whenever necessary to emphasize the space $\mathcal{H}$, we write $P = P_\mathcal{G}^\mathcal{H}$).

We next consider the block matrix representation of an operator T in $\mathcal{L}(\mathcal{H}, \mathcal{H}')$, where direct sum decompositions $\mathcal{H} = \oplus_{n \in \mathbb{Z}} \mathcal{H}_n$ and $\mathcal{H}' = \oplus_{n \in \mathbb{Z}} \mathcal{H}'_n$ are taken into account. Thus, the *matrix elements* (or *entries*) of T with respect to the given direct sum decompositions of $\mathcal{H}$ and $\mathcal{H}'$ are defined by

$$T_{ij} = P_{\mathcal{H}'_i} T / \mathcal{H}_j, \qquad i, j \in \mathbb{Z}.$$

Therefore, T_{ij} belongs to $\mathcal{L}(\mathcal{H}_j, \mathcal{H}'_i)$ and the corresponding *block matrix representation* of T is $T = [T_{ij}]_{i,j \in \mathbb{Z}}$. If $T_{ij} = 0$ for $i \neq j$, then we will also use the notation $T = \oplus_{n \in \mathbb{Z}} T_{nn}$, saying that T is the *direct sum* of the operators T_{nn}, $n \in \mathbb{Z}$.

In addition to the *norm topology* induced on $\mathcal{L}(\mathcal{H}, \mathcal{H}')$ by the norm of operators, we mention here the following two topologies which will arise most frequently later. The *strong operator topology* is the weakest topology on $\mathcal{L}(\mathcal{H}, \mathcal{H}')$ such that the maps $T \to \|Th\|$ are continuous for all h in $\mathcal{H}$. We denote the strong limit by the symbol *s*-lim. The *weak operator topology* is the weakest topology on $\mathcal{L}(\mathcal{H}, \mathcal{H}')$ such that the maps $T \to \langle Th, h' \rangle$ are continuous for every h in $\mathcal{H}$ and every h' in $\mathcal{H}'$.

Several times we will use the spectral theory for unitary or selfadjoint operators. For instance, for unitary operators this will be stated as follows: if U is a unitary operator on $\mathcal{H}$, then there exists a spectral measure (*i.e.* a linear multiplicative map) E from $C(\mathbb{T})$ into $\mathcal{L}(\mathcal{H})$ such that $E(\nu_0) = I_\mathcal{H}$ and $E(\nu_1) = U$.

1.2   Renorming Hilbert Spaces and Elementary Rotations

Let A be a positive operator on the Hilbert space $\mathcal{H}$. Define the map

$$\langle \cdot, \cdot \rangle : \mathcal{H} \times \mathcal{H} \longrightarrow \mathbb{C} \tag{2.1}$$

$$\langle h, g \rangle_A = \langle Ah, g \rangle, \qquad h, g \in \mathcal{H},$$

and remark that it satisfies all the properties of an inner product, except for the fact that the set $\mathcal{N}_A = \{ h \in \mathcal{H} \mid \langle h, h \rangle_A = 0 \}$ may be nontrivial. The equality $\mathcal{N}_A = \{ h \in \mathcal{H} \mid \langle h, g \rangle_A = 0 \text{ for all } g \in \mathcal{H} \}$ is a consequence of the Cauchy-Schwarz inequality. It follows that $\mathcal{N}_A$ is a linear subspace of $\mathcal{H}$. Then, the quotient space $\mathcal{H}/\mathcal{N}_A$ is also a linear space. If $[h]$ denotes the class in $\mathcal{H}/\mathcal{N}_A$ of the element h, then the map given by the formula $\langle [h], [g] \rangle_A = \langle h, g \rangle_A$, $h, g \in \mathcal{H}$, is well defined and it is easily checked that $\langle \cdot, \cdot \rangle_A$ is an inner product. The completion of $\mathcal{H}/\mathcal{N}_A$ with respect to the norm induced by this inner product is a Hilbert space denoted by $\mathcal{H}_A$.

This construction is useful especially if we can render it explicit in terms of a given factorization of the positive operator A. More precisely, assume

$$A = F^* B F, \tag{2.2}$$

where B is a positive operator on the Hilbert space $\mathcal{H}'$ and F belongs to $\mathcal{L}(\mathcal{H}, \mathcal{H}')$. Define the linear space $\mathcal{F} = \mathcal{R}(F)$ and by the same construction as above, we obtain the Hilbert space $\mathcal{F}_B$. We introduce

$$\omega : \mathcal{H}_A \longrightarrow \mathcal{F}_B \tag{2.3}$$

$$\omega[h] = [Fh], \qquad h \in \mathcal{H},$$

and remark that this map is a well defined isometry which can be extended by continuity to a unitary operator, also denoted by ω.

As a first example, we may consider the factorization $A = A^{1/2} A^{1/2}$ which is of type (2.2) with $F = A^{1/2}$ and $B = I$, hence the space $\mathcal{H}_A$ may be identified with the space $\mathrm{cl}\mathcal{R}(A^{1/2})(= \mathrm{cl}\mathcal{R}(A))$. In this case, the unitary operator in $\mathcal{L}(\mathcal{H}_A, \mathrm{cl}\mathcal{R}(A))$ defined by formula (2.3) is denoted by ω_A. A more interesting example can be obtained by considering the positive operator $A = \begin{bmatrix} X & Y \\ Y^* & Z \end{bmatrix}$. The structure of this operator is described by the following result.

2.1 Lemma   *Let X and Z be positive operators in $\mathcal{L}(\mathcal{H}')$ and, respectively, $\mathcal{L}(\mathcal{H})$. The following are equivalent:*

(a)   *The operator $A = \begin{bmatrix} X & Y \\ Y^* & Z \end{bmatrix}$ is positive.*

(b)   *There exists a unique contraction Γ in $\mathcal{L}(\mathrm{cl}\mathcal{R}(Z), \mathrm{cl}\mathcal{R}(X))$ such that $Y = X^{1/2} \Gamma Z^{1/2}$.*

Proof Assume (a). Suppose first that X and Z are both invertible operators. In this case, the following *Frobenius-Schur identities* hold for every fixed operator S in $\mathcal{L}(\mathcal{H}', \mathcal{H})$:

$$\begin{bmatrix} X & Y \\ S & Z \end{bmatrix} = \begin{bmatrix} I & 0 \\ SX^{-1} & I \end{bmatrix} \begin{bmatrix} X & 0 \\ 0 & Z - SX^{-1}Y \end{bmatrix} \begin{bmatrix} I & X^{-1}Y \\ 0 & I \end{bmatrix} \quad (2.4)$$

and

$$\begin{bmatrix} X & Y \\ S & Z \end{bmatrix} = \begin{bmatrix} I & YZ^{-1} \\ 0 & I \end{bmatrix} \begin{bmatrix} X - YZ^{-1}S & 0 \\ 0 & Z \end{bmatrix} \begin{bmatrix} I & 0 \\ Z^{-1}S & I \end{bmatrix}. \quad (2.5)$$

The two operators $Z - SX^{-1}Y$ and $X - YZ^{-1}S$ are called the *Schur complements* associated to A. An immediate consequence of the equality (2.4) is that the operator A is positive if and only if the Schur complement $Z - Y^*X^{-1}Y$ is positive. Therefore, $I - Z^{-1/2}Y^*X^{-1/2}X^{-1/2}YZ^{-1/2} \geq 0$ and if one defines $\Gamma = X^{-1/2}YZ^{-1/2}$, then it is concluded that Γ is a contraction and $Y = X^{1/2}\Gamma Z^{1/2}$.

Dropping the assumption that X and Z are invertible operators, define for $n \geq 1$ the operators $X_n = X + \frac{1}{n}I$ and $Z_n = Z + \frac{1}{n}I$. It turns out that these operators are strictly positive and according to the previous considerations, there exist contractions Γ_n such that $Y = X_n^{1/2}\Gamma_n Z_n^{1/2}$ for $n \geq 1$. Using the weak-compactness of the unit ball of $\mathcal{L}(\mathcal{H}', \mathcal{H})$, a limit point of the sequence $\{\Gamma_n\}_{n \geq 1}$ in the weak topology will be a contraction Γ that satisfies $Y = X^{1/2}\Gamma Z^{1/2}$. Finally, note that if Γ' is another contraction in $\mathcal{L}(\mathrm{cl}\mathcal{R}(Z), \mathrm{cl}\mathcal{R}(X))$ such that $Y = X^{1/2}\Gamma'Z^{1/2}$, then $X^{1/2}\Gamma Z^{1/2}h = X^{1/2}\Gamma'Z^{1/2}h$ for all h in $\mathcal{H}$. Since the operator X is one-to-one on the closure of its range, it follows that $\Gamma Z^{1/2}h = \Gamma'Z^{1/2}h$ for all h in $\mathcal{H}$. That is, there exists a unique contraction Γ in $\mathcal{L}(\mathrm{cl}\mathcal{R}(Z), \mathrm{cl}\mathcal{R}(X))$ such that $Y = X^{1/2}\Gamma Z^{1/2}$. The other direction is straightforward. $\square$

As a consequence of Lemma 2.1, we obtain that the operator $A = \begin{bmatrix} I & T \\ T^* & I \end{bmatrix}$ is positive if and only if T is a contraction. The two associated Schur complements are $I - T^*T$ and, respectively, $I - TT^*$. It is convenient to introduce the *defect operator* $D_T = (I - T^*T)^{1/2}$ and the *defect space* $\mathcal{D}_T = \mathrm{cl}\mathcal{R}(D_T)$ of the contraction T. Using the relation $T(I - T^*T) = (I - TT^*)T$ and the continuous functional calculus for selfadjoint operators, we deduce that

$$TD_T = D_{T^*}T. \quad (2.6)$$

Using Lemma 2.1 and the Frobenius-Schur identities, we obtain the relations

$$A = \begin{bmatrix} X & Y \\ Y^* & Z \end{bmatrix} = \begin{bmatrix} X^{1/2} & 0 \\ Z^{1/2}\Gamma^* & Z^{1/2}D_\Gamma \end{bmatrix} \begin{bmatrix} X^{1/2} & \Gamma Z^{1/2} \\ 0 & D_\Gamma Z^{1/2} \end{bmatrix}$$
$$= \begin{bmatrix} X^{1/2}D_{\Gamma^*} & X^{1/2}\Gamma \\ 0 & Z^{1/2} \end{bmatrix} \begin{bmatrix} D_{\Gamma^*}X^{1/2} & 0 \\ \Gamma^*X^{1/2} & Z^{1/2} \end{bmatrix},$$

which are factorizations of the form (2.2). Two natural identifications of the Hilbert space $(\mathcal{H}' \oplus \mathcal{H})_A$ can be obtained using these factorizations. Thus, consider the contraction $G = \omega_X^*\Gamma\omega_Z$ in $\mathcal{L}(\mathcal{H}_Z, \mathcal{H}_X)$ and two maps Ω and Ω' which are defined by setting:

$$\Omega : (\mathcal{H}' \oplus \mathcal{H})_A \longrightarrow \mathcal{H}_X \oplus \mathcal{D}_G \quad (2.7)$$

$$\Omega[k] = ([h'] + G[h]) \oplus D_G[h]$$

and

$$\widetilde{\Omega} : (\mathcal{H}' \oplus \mathcal{H})_A \longrightarrow \mathcal{H}_Z \oplus \mathcal{D}_{G^*} \tag{2.8}$$

$$\widetilde{\Omega}[k] = ([h] + G^*[h']) \oplus D_{G^*}[h'],$$

where $[k]$ denotes an element in $(\mathcal{H}' \oplus \mathcal{H})/\mathcal{N}_A$ such that $h' \oplus h$ is one of its representatives, while $[h]$ and $[h']$ denote the classes of h and h' in $\mathcal{H}_Z$ and, respectively, $\mathcal{H}_X$. These maps are well defined isometries and can be extended by continuity to unitary operators, denoted by the same symbols Ω and, respectively, $\widetilde{\Omega}$. Then it is also natural to introduce the unitary operator

$$R(G) : \mathcal{H}_Z \oplus \mathcal{D}_{G^*} \longrightarrow \mathcal{H}_X \oplus \mathcal{D}_G \tag{2.9}$$

$$R(G) = \Omega \widetilde{\Omega}^*.$$

We can easily describe the structure of this operator.

2.2 Proposition   *The unitary operator $R(G)$ has the following block matrix representation:*

$$R(G) = \begin{bmatrix} G & D_{G^*} \\ D_G & -G^* \end{bmatrix}.$$

Proof The set $\{(G^*[h'] + [h]) \oplus D_{G^*}[h'] \mid h \in \mathcal{H}, \ h' \in \mathcal{H}'\}$ is a dense part of the space $\mathcal{H}_Z \oplus \mathcal{D}_{G^*}$ and by direct computations using the relation (2.6), one obtains:

$$\Omega \widetilde{\Omega}^*((G^*[h'] + [h]) \oplus D_{G^*}[h']) = ([h'] + G[h]) \oplus D_G[h]$$

$$= \begin{bmatrix} G & D_{G^*} \\ D_G & -G^* \end{bmatrix} ((G^*[h'] + [h]) \oplus D_{G^*}[h']). \quad \square$$

We are led by Proposition 2.2 to introduce, for an arbitrary contraction T in $\mathcal{L}(\mathcal{H}, \mathcal{H}')$, the unitary operator

$$R(T) = \begin{bmatrix} T & D_{T^*} \\ D_T & -T^* \end{bmatrix} \tag{2.10}$$

mapping $\mathcal{H} \oplus \mathcal{D}_{T^*}$ onto $\mathcal{H}' \oplus \mathcal{D}_T$. The operator $R(T)$ is called the *elementary rotation* or the *Julia operator* of T and it plays a key role in most of our developments. The elementary rotation can be alternatively characterized as follows.

2.3 Proposition   *If T is a contraction in $\mathcal{L}(\mathcal{H}, \mathcal{H}')$, then $R(T)$ has the properties:*

(a) $P_{\mathcal{H}'} R(T)/\mathcal{H} = T$.

(b) $\mathcal{H}' \oplus \mathcal{D}_T = \mathcal{H}' \vee R(T)\mathcal{H}$ *(equivalently,* $\mathcal{H} \oplus \mathcal{D}_{T^*} = \mathcal{H} \vee R^*(T)\mathcal{H}'$*).*

(c)   *For any other unitary operator R in $\mathcal{L}(\mathcal{K}_1, \mathcal{K}_0)$ satisfying the properties (a) and (b) (consequently, $\mathcal{H} \subset \mathcal{K}_1$ and $\mathcal{H}' \subset \mathcal{K}_0$), there exist unitary operators Φ_0 in $\mathcal{L}(\mathcal{H}' \oplus \mathcal{D}_T, \mathcal{K}_0)$ and Φ_1 in $\mathcal{L}(\mathcal{H} \oplus \mathcal{D}_{T^*}, \mathcal{K}_1)$ such that $\Phi_0/\mathcal{H}' = I$, $\Phi_1/\mathcal{H} = I$ and $\Phi_0 R(T) = R\Phi_1$.*

Proof The first two properties (a) and (b) are obvious by the definition of $R(T)$. Then, given any other unitary operator R in $\mathcal{L}(\mathcal{K}_1, \mathcal{K}_0)$ satisfing the properties (a) and (b), define $\Phi_0(h' + R(T)h) = h' + Rh$ and $\Phi_1(h + R^*(T)h') = h + R^*h'$ for arbitrary h in $\mathcal{H}$ and h' in $\mathcal{H}'$. These definitions are motivated by (b), and taking into account the property (a) gives

$$
\begin{aligned}
\|\Phi_0(h' + R(T)h)\|^2 &= \|h' + Rh\|^2 = \|h'\|^2 + \langle h', Rh \rangle + \langle Rh, h' \rangle + \|Rh\|^2 \\
&= \|h'\|^2 + \langle h', R(T)h \rangle + \langle R(T)h, h' \rangle + \|R(T)h\|^2 \\
&= \|h' + R(T)h\|^2.
\end{aligned}
$$

Similarly, $\|\Phi_1(h + R^*(T)h')\| = \|h + R^*(T)h'\|$. Consequently, Φ_0 and Φ_1 are well defined isometries which can be extended by continuity to unitary operators, also denoted by Φ_0 and, respectively, Φ_1, and it is easily checked that $\Phi_0 R(T) = R\Phi_1$. $\qquad \square$

2.4 Remark As a consequence of the construction of the elementary rotation, we may see that every contraction can be interpreted as an operator angle between two closed subspaces of a certain Hilbert space. Thus, for two closed subspaces $\mathcal{H}_1$ and $\mathcal{H}_2$ of a Hilbert space $\mathcal{H}$, the operator $c(\mathcal{H}_1, \mathcal{H}_2) = P_{\mathcal{H}_1}/\mathcal{H}_2$ in $\mathcal{L}(\mathcal{H}_2, \mathcal{H}_1)$ is called the *operator angle* between $\mathcal{H}_1$ and $\mathcal{H}_2$. Then, let T be a contraction in $\mathcal{L}(\mathcal{H}, \mathcal{H}')$. Consider the positive operator $A = \begin{bmatrix} I & T \\ T^* & I \end{bmatrix}$ and the operators Ω, $\widetilde{\Omega}$ are defined by (2.7) and, respectively, (2.8). Define the spaces $\mathcal{G} = \widetilde{\Omega}^* \mathcal{H}$, $\mathcal{G}' = \Omega^* \mathcal{H}'$, and the unitary operators $\omega = \Omega^*/\mathcal{H}'$ in $\mathcal{L}(\mathcal{H}', \mathcal{G}')$ and $\widetilde{\omega} = \widetilde{\Omega}^*/\mathcal{H}$ in $\mathcal{L}(\mathcal{H}, \mathcal{G})$. We can check that $c(\mathcal{G}', \mathcal{G})\widetilde{\omega} = \omega T$. Thus, for h in $\mathcal{H}$ and g' in $\mathcal{H}'$,

$$
\begin{aligned}
\langle \omega T h, g' \rangle &= \langle Th, \omega^* g' \rangle = \langle P_{\mathcal{H}'} \Omega \widetilde{\Omega}^* h, \omega^* g' \rangle \\
&= \langle \Omega \widetilde{\Omega}^* h, \Omega g' \rangle = \langle c(\mathcal{G}', \mathcal{G})\widetilde{\omega} h, g' \rangle.
\end{aligned}
$$

Finally, let us remark that if T is a real number with $|T| \leq 1$, then its representation as an operator angle is nothing but the relation $T = \cos \vartheta$ for a certain $\vartheta \in [0, \pi]$. We notice that, in this case, the elementary rotation of T has the familiar form

$$
R(T) = \begin{bmatrix} \cos \vartheta & \sin \vartheta \\ \sin \vartheta & -\cos \vartheta \end{bmatrix}. \qquad \square
$$

1.3 Kolmogorov Decompositions. I

In this section we extend the construction of the Hilbert space $\mathcal{H}_A$ by replacing the positive operator A with a positive definite kernel on the set $\mathbb{Z}$ of integers.

Consider a family $\mathbf{H} = \{\mathcal{H}_n\}_{n \in \mathbb{Z}}$ of Hilbert spaces. A map A on $\mathbb{Z} \times \mathbb{Z}$ such that $A(i, j)$ belongs to $\mathcal{L}(\mathcal{H}_j, \mathcal{H}_i)$ for $i, j \in \mathbb{Z}$, is called a *positive definite kernel* if

$$
\sum_{i, j \in \mathbb{Z}} \langle A(i, j)h_j, h_i \rangle \geq 0
$$

for all sequences $\{h_n\}_{n \in \mathbb{Z}}$ in $\oplus_{n \in \mathbb{Z}} \mathcal{H}_n$ with finite support (*i.e.* $h_n = 0$ except for finitely many n's). The main result of this section is the following.

3.1 Theorem *Let A be a positive definite kernel. Then there exists a Hilbert space $\mathcal{H}_A$ and a map V defined on $\mathbb{Z}$ such that $V(n)$ belongs to $\mathcal{L}(\mathcal{H}_n, \mathcal{H}_A)$ for each $n \in \mathbb{Z}$ and*

(a) $A(i,j) = V^*(i)V(j)$, $i, j \in \mathbb{Z}$.

(b) $\mathcal{H}_A = \bigvee_{n \in \mathbb{Z}} V(n)\mathcal{H}_n$.

(c) *If there exist another Hilbert space $\mathcal{H}'$ and a map V' on $\mathbb{Z}$ such that $V'(n)$ belongs to $\mathcal{L}(\mathcal{H}_n, \mathcal{H}')$ for $n \in \mathbb{Z}$ and (a), (b) hold, then there exists a unitary operator $\Phi : \mathcal{H}_A \longrightarrow \mathcal{H}'$ such that $\Phi V(n) = V'(n)$ for all $n \in \mathbb{Z}$.*

Proof Let $\mathcal{F}$ be the linear space of the elements of $\oplus_{n \in \mathbb{Z}} \mathcal{H}_n$ with finite support. Define the map $\langle \cdot, \cdot \rangle_A$ on $\mathcal{F}$ by the formula

$$\langle \varphi, \psi \rangle_A = \sum_{i,j \in \mathbb{Z}} \langle A(i,j)\varphi(j), \psi(i) \rangle \tag{3.1}$$

for φ and ψ in $\mathcal{F}$. If $\mathcal{N}_A = \{\varphi \in \mathcal{F} \mid \langle \varphi, \varphi \rangle_A = 0\}$, then an application of the Cauchy-Schwarz inequality shows that $\mathcal{N}_A$ is a linear subspace of $\mathcal{F}$. Factoring out this subspace, we get a linear space $\mathcal{F}/\mathcal{N}_A$ endowed with an inner product, also denoted by $\langle \cdot, \cdot \rangle_A$, and which is defined by the formula: for φ, ψ in $\mathcal{F}$,

$$\langle [\varphi], [\psi] \rangle_A = \langle \varphi, \psi \rangle_A, \tag{3.2}$$

where $[\varphi], [\psi]$ denote the classes in $\mathcal{F}/\mathcal{N}_A$ of φ and, respectively, ψ. The definition of this inner product is independent of the chosen representatives and let $\mathcal{H}_A$ denote the completion of $\mathcal{F}/\mathcal{N}_A$ with respect to the norm induced by the inner product (3.2). Further, if h belongs to $\mathcal{H}_i$, then the element $h_{(i)}$ in $\mathcal{F}$ is defined as follows:

$$h_{(i)}(j) = \begin{cases} h & \text{if } j = i \\ 0 & \text{if } j \neq i. \end{cases}$$

The map V can be defined by the formula:

$$V(n) : \mathcal{H}_n \longrightarrow \mathcal{H}_A \tag{3.3}$$

$$V(n)h = [h_{(n)}], \qquad h \in \mathcal{H}_n.$$

If the element h belongs to $\mathcal{H}_n$, then we have

$$\begin{aligned}\|V(n)h\|_A^2 &= \|[h_{(n)}]\|_A^2 = \langle [h_{(n)}], [h_{(n)}] \rangle_A \\ &= \langle A(n,n)h, h \rangle \leq \|A(n,n)\|\|h\|^2,\end{aligned}$$

which shows that $V(n)$ belongs to $\mathcal{L}(\mathcal{H}_n, \mathcal{H}_A)$. For h in $\mathcal{H}_j$ and g in $\mathcal{H}_i$, one obtains that

$$\langle A(i,j)h, g \rangle = \langle [h_{(j)}], [g_{(i)}] \rangle_A = \langle V(j)h, V(i)g \rangle = \langle V^*(i)V(j)h, g \rangle,$$

hence the assertion (a) holds.

By the definition of $V(n)$, it follows that the span of the spaces $V(n)\mathcal{H}_n$, $n \in \mathbb{Z}$, is contained in $\mathcal{F}/\mathcal{N}_A$. Conversely, pick $[\varphi]$ in $\mathcal{F}/\mathcal{N}_A$ and its representative φ in $\mathcal{F}$ will have the form $\varphi = \sum_{j \in \mathbb{Z}} \varphi(j)_{(j)}$, where only a finite number of the vectors $\varphi(j)$ are different from zero. Then

$$[\varphi] = \sum_{j \in \mathbb{Z}} [\varphi(j)_{(j)}] = \sum_{j \in \mathbb{Z}} V(j)\varphi(j),$$

which shows that the span of the spaces $V(n)\mathcal{H}_n$, $n \in \mathbb{Z}$, is $\mathcal{F}/\mathcal{N}_A$. Consequently, the assertion (b) is proved.

Finally, consider a Hilbert space $\mathcal{H}'$ and a map V' on $\mathbb{Z}$ such that $V'(n)$ belongs to $\mathcal{L}(\mathcal{H}_n, \mathcal{H}')$ for $n \in \mathbb{Z}$ and (a), (b) hold. Define, as suggested by (b), the map

$$\Phi : \mathcal{H}_A \longrightarrow \mathcal{H}' \tag{3.4}$$

$$\Phi\Big(\sum_{n \in \mathbb{Z}} V(n)h_n\Big) = \sum_{n \in \mathbb{Z}} V'(n)h_n,$$

where $h_n = 0$ except for finitely many n's. Then, one has

$$\Big\|\Phi\Big(\sum_{n \in \mathbb{Z}} V(n)h_n\Big)\Big\|^2 = \Big\|\sum_{n \in \mathbb{Z}} V'(n)h_n\Big\|^2 = \sum_{i,j \in \mathbb{Z}} \langle V'(j)h_j, V'(i)h_i \rangle$$

$$= \sum_{i,j \in \mathbb{Z}} \langle A(i,j)h_j, h_i \rangle = \Big\|\sum_{n \in \mathbb{Z}} V(n)h_n\Big\|_A^2,$$

which shows that the map Φ can be extended by continuity to a unitary operator from $\mathcal{H}_A$ onto $\mathcal{H}'$ and, moreover, $\Phi V(n) = V'(n)$ for all $n \in \mathbb{Z}$. The assertion (c) is proved. $\qquad\square$

A map V satisfying the property (a) in Theorem 3.1 will be called a *Kolmogorov decomposition* of the kernel A. The property (b) is referred to as the *minimality property* of the Kolmogorov decomposition. The meaning of the property (c) is that, under the minimality condition (b), the Kolmogorov decomposition is essentially unique.

There is an important particular case when the construction in Theorem 3.1 can be significantly improved. Thus, suppose the family $\mathbf{H} = \{\mathcal{H}_n\}_{n \in \mathbb{Z}}$ reduces to a single Hilbert space, *i.e.* $\mathcal{H}_n = \mathcal{H}$ for all $n \in \mathbb{Z}$, and the positive definite kernel A has the property that $A(i,j) = T(j - i)$ for a certain map T from $\mathbb{Z}$ to $\mathcal{L}(\mathcal{H})$. In this case, the kernel A is called a *positive definite Toeplitz kernel*.

3.2 Theorem *Let A be a positive definite Toeplitz kernel. Then there exist a Hilbert space $\mathcal{H}_A$, a unitary operator S in $\mathcal{L}(\mathcal{H}_A)$ and an operator Q in $\mathcal{L}(\mathcal{H}, \mathcal{H}_A)$ such that*

(a) $A(i,j) = Q^* S^{j-i} Q$, $\quad i, j \in \mathbb{Z}$.

(b) $\mathcal{H}_A = \bigvee_{n \in \mathbb{Z}} S^n Q \mathcal{H}$.

(c) *If there exist another Hilbert space $\mathcal{H}'$, a unitary operator S' in $\mathcal{L}(\mathcal{H}')$ and an operator Q' in $\mathcal{L}(\mathcal{H}, \mathcal{H}')$ such that (a), (b) hold, then there exists a unitary operator Φ mapping $\mathcal{H}_A$ onto $\mathcal{H}'$ such that $\Phi Q h = Q' h$ for h in $\mathcal{H}$ and $S'\Phi = \Phi S$.*

Proof By Theorem 3.1, there exist a Hilbert space $\mathcal{H}_A$ and a map V from $\mathbb{Z}$ to $\mathcal{L}(\mathcal{H}, \mathcal{H}_A)$ such that the following assertions hold: (a') $A(i,j) = V^*(i)V(j)$, $i, j \in \mathbb{Z}$ and (b') $\mathcal{H}_A = \bigvee_{n \in \mathbb{Z}} V(n)\mathcal{H}$. Using (b'), one defines the operator

$$S : \mathcal{H}_A \longrightarrow \mathcal{H}_A \qquad\qquad (3.5)$$

$$S\left(\sum_{n \in \mathbb{Z}} V(n)h_n\right) = \sum_{n \in \mathbb{Z}} V(n+1)h_n,$$

where h_n belongs to $\mathcal{H}$ for all $n \in \mathbb{Z}$ and $h_n \neq 0$ only for a finite number of indices $n \in \mathbb{Z}$. The assumption that A is a positive definite Toeplitz kernel implies:

$$
\begin{aligned}
\left\| S\left(\sum_{n \in \mathbb{Z}} V(n)h_n\right) \right\|_A^2 &= \left\| \sum_{n \in \mathbb{Z}} V(n+1)h_n \right\|_A^2 = \sum_{i,j \in \mathbb{Z}} \langle V(j+1)h_j, V(i+1)h_i \rangle_A \\
&= \sum_{i,j \in \mathbb{Z}} \langle V^*(i+1)V(j+1)h_j, h_i \rangle = \sum_{i,j \in \mathbb{Z}} \langle A(i+1,j+1)h_j, h_i \rangle \\
&= \sum_{i,j \in \mathbb{Z}} \langle A(i,j)h_j, h_i \rangle = \left\| \sum_{n \in \mathbb{Z}} V(n)h_n \right\|_A^2.
\end{aligned}
$$

Then, it follows by the definition of S that $SV(n) = V(n+1)$ for all $n \in \mathbb{Z}$. Hence, for $i, j \in \mathbb{Z}$ and $j > i$, the equality $S^{j-i}V(0) = V(j-i)$ holds and then

$$A(i,j) = A(0, j-i) = V^*(0)V(j-i) = V^*(0)S^{j-i}V(0).$$

In order to bring to an end the proof of (a), one takes $Q = V(0)$. Then, the assertion (b) follows immediately from (b') and (3.6), and the proof of (c) is a repetition of the corresponding part in Theorem 3.1. $\square$

The operator S defined by (3.5) plays a similar role with the elementary rotation and it will be referred to as the *Naimark dilation* of the considered positive definite Toeplitz kernel. The interpretation of the Naimark dilation as an elementary rotation, as well as the structure of the Kolmogorov decomposition, will be clarified after the developments in the next two sections. We conclude this section with an example.

3.3 Example Let μ be a probability measure on the interval $[0, 2\pi)$. $L^2(\mu)$ is the Hilbert space introduced in Example 1.2. For each $n \in \mathbb{Z}$, ν_n denotes the function $\nu_n(e^{it}) = e^{int}$ and the *Fourier coefficients* of μ are defined by $A_n = \frac{1}{2\pi} \int_0^{2\pi} e^{-int} d\mu(t)$, $n \in \mathbb{Z}$. The map A from $\mathbb{Z} \times \mathbb{Z}$ into $\mathbb{C}$, defined by $A(j,k) = A_{k-j}$ for $j, k \in \mathbb{Z}$, is a positive definite Toeplitz kernel since

$$
\sum_{j,k=0}^{n} A(j,k)\lambda_k \lambda_j^* = \sum_{j,k=0}^{n} \left(\frac{1}{2\pi} \int_0^{2\pi} e^{-i(k-j)t} d\mu(t) \right) \lambda_k \lambda_j^*
$$

$$
= \frac{1}{2\pi} \int_0^{2\pi} \left| \sum_{j=0}^{n} \lambda_j e^{-ijt} \right|^2 d\mu(t) \geq 0
$$

for all $n \geq 0$ and all families $\{\lambda_0, \lambda_1, \dots \lambda_n\}$ of complex numbers. The Hilbert space l^2 was introduced in Example 1.1 and its orthonormal basis $\{E_n\}_{n \in \mathbb{Z}}$ was introduced in Example 1.4. Our task is to show the explicit connection between $(l^2)_A$ and $L^2(\mu)$. Let us remark that, by the property (b) in Theorem 3.1, the set $\{[E_n]\}_{n \in \mathbb{Z}}$ is total in $(l^2)_A$ (we denote by $[E_n]$ the class of E_n in $(l^2)_A$). Since $\{\nu_n\}_{n \in \mathbb{Z}}$ is a total set in $L^2(\mu)$, we define:

$$\Psi : L^2(\mu) \longrightarrow (l^2)_A \qquad (3.6)$$

$$\Psi(\sum_{n \in \mathbb{Z}} \lambda_n \nu_{-n}) = \sum_{n \in \mathbb{Z}} \lambda_n [E_n],$$

where only finitely many of the complex numbers λ_n are different from zero. Then

$$\|\Psi(\sum_{n \in \mathbb{Z}} \lambda_n \nu_{-n})\|_A^2 = \|\sum_{n \in \mathbb{Z}} \lambda_n [E_n]\|_A^2 = \sum_{j,k \in \mathbb{Z}} \langle \lambda_k [E_k], \lambda_j [E_j] \rangle_A$$

$$= \sum_{j,k \in \mathbb{Z}} \lambda_k \lambda_j^* \langle [E_k], [E_j] \rangle_A = \sum_{j,k \in \mathbb{Z}} A(j,k) \lambda_k \lambda_j^*$$

$$= \frac{1}{2\pi} \int_0^{2\pi} |\sum_{n \in \mathbb{Z}} \lambda_n e^{-int}|^2 d\mu(t),$$

hence Ψ extends to a unitary operator from $L^2(\mu)$ to $(l^2)_A$. Finally, let us note that if μ is the Lebesgue measure, then we obtain the equality $(l^2)_A = l^2$. $\quad\square$

1.4 Row and Column Contractions

Strictly speaking, in this section we describe the structure of the row and column contractions. But we do this as another illustration of the constructions in Section 2 and in such a way as to bring to the light the idea that there are some free operator angles holding the whole structure of the object of interest (row or column contraction, in our case). These operator angles can already be viewed as a first example of Schur parameters.

We consider the contraction T with the following block matrix representation:

$$T = [T_1 \quad T_2 \quad T_3 \quad \ldots] \in \mathcal{L}(\oplus_{k=1}^{\infty} \mathcal{H}_k, \mathcal{H}'). \qquad (4.1)$$

Denote by P_n, $n \geq 1$, the projection of $\oplus_{k=1}^{\infty} \mathcal{H}_k$ onto $\oplus_{k=1}^{n} \mathcal{H}_k$ and notice the relation $T = s\text{-}\lim_{n \to \infty} TP_n$. This shows that in order to describe the structure of a row contraction T with the matrix representation (4.1), it is sufficient to describe the structure of the row contractions of finite length, $T^{(n)} = [T_1 \quad T_2 \quad \ldots \quad T_n]$, in $\mathcal{L}(\oplus_{k=1}^{n} \mathcal{H}_k, \mathcal{H}')$.

4.1 Lemma *Let X and Y be operators in $\mathcal{L}(\mathcal{H}', \mathcal{H})$. The following are equivalent:*

(a) *There exists a contraction Γ in $\mathcal{L}(\mathcal{H})$ such that $X = \Gamma Y$.*

(b) *$X^*X \leq Y^*Y$.*

Proof Assume (b). For h in $\mathcal{H}'$, one defines $\Gamma_0 Y h = Xh$ and it follows that

$$\|\Gamma_0 Y h\|^2 = \langle X^* Xh, h \rangle \leq \langle Y^* Y h, h \rangle = \|Yh\|^2,$$

hence Γ_0 is a well defined contraction. Extending Γ_0 with zero on $\ker Y^*$, a contraction Γ in $\mathcal{L}(\mathcal{H})$ satisfying (a) is obtained. The other direction is also straightforward. $\quad\square$

The next result describes the structure of the row contractions of finite length.

4.2 Proposition   *The following are equivalent:*

(a)   *The operator $T^{(n)} = [\,T_1 \quad T_2 \quad \ldots \quad T_n\,]$ in $\mathcal{L}(\oplus_{k=1}^n \mathcal{H}_k, \mathcal{H}')$ is a contraction.*

(b) $T_1 = \Gamma_1$ *is a contraction and, for $k \geq 2$, there exist uniquely determined contractions Γ_k in $\mathcal{L}(\mathcal{H}_k, \mathcal{D}_{\Gamma_{k-1}^*})$ such that $T_k = D_{\Gamma_1^*} D_{\Gamma_2^*} \ldots D_{\Gamma_{k-1}^*} \Gamma_k$.*

Moreover, if $T^{(n)}$ is a contraction, then there exist unitary operators $\alpha_{T^{(n)}}$ and $\beta_{T^{(n)}}$ identifying the defect spaces of $T^{(n)}$ as follows:

$$\alpha_{T^{(n)}} : \mathcal{D}_{T^{(n)}} \longrightarrow \oplus_{k=1}^n \mathcal{D}_{\Gamma_k} \tag{4.2}$$

$$\alpha_{T^{(n)}} D_{T^{(n)}} = \begin{bmatrix} D_{\Gamma_1} & -\Gamma_1^* \Gamma_2 & \ldots & -\Gamma_1^* D_{\Gamma_2^*} \ldots D_{\Gamma_{n-1}^*} \Gamma_n \\ 0 & D_{\Gamma_2} & \ldots & -\Gamma_2^* D_{\Gamma_3^*} \ldots D_{\Gamma_{n-1}^*} \Gamma_n \\ \vdots & \vdots & \ddots & \vdots \\ 0 & 0 & \ldots & D_{\Gamma_n} \end{bmatrix}$$

and

$$\beta_{T^{(n)}} : \mathcal{D}_{(T^{(n)})^*} \longrightarrow \mathcal{D}_{\Gamma_n^*} \tag{4.3}$$

$$\beta_{T^{(n)}} D_{(T^{(n)})^*} = D_{\Gamma_n^*} D_{\Gamma_{n-1}^*} \ldots D_{\Gamma_1^*}.$$

Proof   Assume (a) holds for $T^{(2)} = [\,T_1 \quad T_2\,]$. The fact that $T_2 = D_{\Gamma_1^*} \Gamma_2$ is an immediate consequence of Lemma 4.1. However, it is convenient for our purposes to examine this case from a "geometric" point of view, based on Remark 2.4. Consider the positive block matrix

$$A = \begin{bmatrix} I & T_1 & T_2 \\ T_1^* & I & 0 \\ T_2^* & 0 & I \end{bmatrix}$$

and define a unitary operator Ω from $(\mathcal{H}' \oplus \mathcal{H}_1 \oplus \mathcal{H}_2)_A$ onto $(\mathcal{H}' \oplus \mathcal{H}_1)_B \oplus \mathcal{D}_G$ according to the formula (2.7). The operator G essentially appears as the operator angle between the spaces $\mathcal{H}_2$ and $(\mathcal{H}' \oplus \mathcal{H}_1)_B$, where $B = \begin{bmatrix} I & T_1 \\ T_1^* & I \end{bmatrix}$. In its turn, the space $(\mathcal{H}' \oplus \mathcal{H}_1)_B$ is identified with $\mathcal{D}_{T_1^*} \oplus \mathcal{H}_1$. Since it is easy to remark that $\mathcal{H}_1$ and $\mathcal{H}_2$ are orthogonal inside $(\mathcal{H}' \oplus \mathcal{H}_1 \oplus \mathcal{H}_2)_A$, then G will determine only the angle between the spaces $\mathcal{D}_{T_1^*}$ and $\mathcal{H}_2$.

The algebraic counterpart of these remarks can be easily followed. Thus, define $\Gamma_1 = T_1$ and note the identity

$$\begin{bmatrix} T_2 \\ 0 \end{bmatrix} = \begin{bmatrix} I & \Gamma_1 \\ \Gamma_1^* & I \end{bmatrix}^{1/2} G$$

which is a consequence of Lemma 2.1. The identification of the space $(\mathcal{H}' \oplus \mathcal{H}_1)_B$ with the space $\mathcal{D}_{T_1^*} \oplus \mathcal{H}_1$ means the replacement of G with the contraction $\begin{bmatrix} \Gamma_1' \\ \Gamma_2' \end{bmatrix}$

in $\mathcal{L}(\mathcal{H}_2, \mathcal{D}_{T_1^*} \oplus \mathcal{H}_1)$ such that

$$\begin{bmatrix} T_2 \\ 0 \end{bmatrix} = \begin{bmatrix} D_{\Gamma_1^*} & \Gamma_1 \\ 0 & I \end{bmatrix} \begin{bmatrix} \Gamma_1' \\ \Gamma_2' \end{bmatrix} = \begin{bmatrix} D_{\Gamma_1^*} \Gamma_1' + \Gamma_1 \Gamma_2' \\ \Gamma_2' \end{bmatrix}.$$

So, $\Gamma_2' = 0$ must hold and defining $\Gamma_2 = \Gamma_1' \in \mathcal{L}(\mathcal{H}_2, \mathcal{D}_{\Gamma_1^*})$, one obtains $T_2 = D_{\Gamma_1^*} \Gamma_2$. The unitary operators $\alpha_{T^{(2)}}$ and $\beta_{T^{(2)}}$ identifying the defect spaces of $T^{(2)}$ can be determined by taking into account the Frobenius-Schur factorizations for the operator A.

The general case is proved by induction. Assume the proposition is true for all contractions with at most $n-1$ entries. Then $T^{(n)} = [\,T_1 \;\; T_2 \;\; \ldots \;\; T_n\,]$ is a contraction if and only if $S = [\,T_1 \;\; T_2 \;\; \ldots \;\; T_{n-1}\,]$ is a contraction and $T_n = D_{S^*} \Gamma'$, where Γ' is a uniquely determined contraction in $\mathcal{L}(\mathcal{H}_n, \mathcal{D}_{S^*})$. But, by the induction hypothesis, there exist uniquely determined contractions Γ_k in $\mathcal{L}(\mathcal{H}_k, \mathcal{D}_{\Gamma_{k-1}^*})$ such that $T_k = D_{\Gamma_1^*} D_{\Gamma_2^*} \ldots D_{\Gamma_{k-1}^*} \Gamma_k$ for $k = 1, 2, \ldots, n-1$, and define $\Gamma_n = \beta_S \Gamma'$. Consequently, $T_n = D_{\Gamma_1^*} D_{\Gamma_2^*} \ldots D_{\Gamma_{n-1}^*} \Gamma_n$. It remains to obtain the identifications of the defect spaces of $T^{(n)}$. For this purpose, it is remarked that $T^{(n)} = [\,\Gamma_1 \;\; D_{\Gamma_1^*} T'\,]$, where T' is the row contraction with $n-1$ entries determined by $\Gamma_2, \Gamma_3, \ldots \Gamma_n$. Therefore, $\alpha_{T^{(n)}}$ is defined according to the rule $\alpha_{T^{(n)}} = (I_{\mathcal{D}_{\Gamma_1}} \oplus \alpha_{T'})\alpha$, where α is the unitary operator that identifies $\mathcal{D}_{T^{(n)}}$ with $\mathcal{D}_{\Gamma_1} \oplus \mathcal{D}_{T'}$, when $T^{(n)}$ is regarded as a row contraction with the two entries Γ_1 and $D_{\Gamma_1^*} T'$, while $\alpha_{T'}$ is available by the induction hypothesis. The identification of the defect space $\mathcal{D}_{(T^{(n)})^*}$ with $\mathcal{D}_{\Gamma_n^*}$ is immediate. $\square$

Once the structure of the contraction $T^{(n)}$ is determined, it is useful to clarify the structure of its elementary rotation. For this purpose, it is convenient to consider the unitary operator:

$$R_{T^{(n)}} : \oplus_{k=1}^n \mathcal{H}_k \oplus \mathcal{D}_{\Gamma_n^*} \longrightarrow \mathcal{H}' \oplus \oplus_{k=1}^n \mathcal{D}_{\Gamma_k} \tag{4.4}$$

$$R_{T^{(n)}} = (I \oplus \alpha_{T^{(n)}}) R(T^{(n)})(I \oplus \beta_{T^{(n)}}^*),$$

where $R(T^{(n)})$ is the elementary rotation of $T^{(n)}$. Given the family $\{\Gamma_k\}_{k=1}^n$ of contractions associated to $T^{(n)}$ by Proposition 4.2, we define the following unitary operators: for $1 \le k \le n$,

$$R_n(\Gamma_k) : \oplus_{j=1}^{k-1} \mathcal{H}_j \oplus (\mathcal{H}_k \oplus \mathcal{D}_{\Gamma_k^*}) \oplus \oplus_{j=k+1}^n \mathcal{D}_{\Gamma_j} \longrightarrow \tag{4.5}$$

$$\longrightarrow \oplus_{j=1}^{k-1} \mathcal{H}_j \oplus (\mathcal{D}_{\Gamma_{k-1}^*} \oplus \mathcal{D}_{\Gamma_k}) \oplus \oplus_{j=k+1}^n \mathcal{D}_{\Gamma_j}$$

$$R_n(\Gamma_k) = I \oplus R(\Gamma_k) \oplus I,$$

where $\mathcal{H}_0 = 0$ and $\Gamma_0 = 0$.

4.3 Proposition *The unitary operator $R_{T^{(n)}}$ has the multiplicative structure*

$$R_{T^{(n)}} = R_n(\Gamma_1) R_n(\Gamma_2) \ldots R_n(\Gamma_n).$$

Proof In order to prove this result, it will be first shown that

$$K_{T^{(n)}} = [\,\Gamma_1^* D_{\Gamma_2^*} \dots D_{\Gamma_n^*} \quad \Gamma_2^* D_{\Gamma_3^*} \dots D_{\Gamma_n^*} \quad \Gamma_{n-1}^* D_{\Gamma_n^*} \quad \Gamma_n^*\,]^{\mathsf{T}}, \qquad (4.6)$$

where $K_{T^{(n)}} = \alpha_{T^{(n)}} (T^{(n)})^* \beta_{T^{(n)}}^*$ and the symbol T denotes the matrix transpose. Since the set $\{D_{\Gamma_n^*} \dots D_{\Gamma_1^*} h \mid h \in \mathcal{H}_1\}$ is a dense part of the space $\mathcal{D}_{\Gamma_1^*}$, the equality (2.6) and Proposition 4.2 give that $K_{T^{(n)}} D_{\Gamma_n^*} \dots D_{\Gamma_1^*} = \alpha_{T^{(n)}} (T^{(n)})^* D_{(T^{(n)})^*} = \alpha_{T^{(n)}} D_{T^{(n)}} (T^{(n)})^*$. Using the definition of $\alpha_{T^{(n)}} D_{T^{(n)}}$, one obtains that the matrix elements of $\alpha_{T^{(n)}} D_{T^{(n)}} (T^{(n)})^*$ are the following: for $n > k \geq 1$,

$$(\alpha_{T^{(n)}} D_{T^{(n)}} (T^{(n)})^*)_k = \Gamma_k^* D_{\Gamma_{k+1}^*} \dots D_{\Gamma_{n-1}^*} D_{\Gamma_n^*}^2 D_{\Gamma_{n-1}^*} \dots D_{\Gamma_1^*}$$

and

$$(\alpha_{T^{(n)}} D_{T^{(n)}} (T^{(n)})^*)_n = \Gamma_n^* D_{\Gamma_n^*} D_{\Gamma_{n-1}^*} \dots D_{\Gamma_1^*},$$

thereby concluding the proof of (4.6). Now, the following identities are simple consequences of (4.6):

$$\alpha_{T^{(n)}} D_{T^{(n)}} = \begin{bmatrix} \alpha_{T^{(n-1)}} D_{T^{(n-1)}} & -K_{T^{(n-1)}} \Gamma_n \\ 0 & D_{\Gamma_n} \end{bmatrix} \qquad (4.7)$$

and

$$K_{T^{(n)}} = [\,K_{T^{(n-1)}} D_{\Gamma_n^*} \quad \Gamma_n^*\,]. \qquad (4.8)$$

Finally, the multiplicative structure of $R_{T^{(n)}}$ is proven by induction on n. The case $n = 1$ is obvious and we assume that the proposition holds for all row contractions with at most $n - 1$ matrix elements. By (4.6), (4.7), (4.8) and the induction hypothesis, it follows that $R_n(\Gamma_1) R_n(\Gamma_2) \dots R_n(\Gamma_n) = (R_{T^{(n-1)}} \oplus I)(I \oplus R_n(\Gamma_n)) = R_{T^{(n)}}$ and the proof is complete. $\qquad \square$

We now set about the task of describing contractions of the form (4.1). Let T be such a contraction. Then, we deduce by Proposition 4.2 that T is uniquely determined by a family $\{\Gamma_n\}_{n=1}^{\infty}$ of associated parameters, consisting of the contractions Γ_1 in $\mathcal{L}(\mathcal{H}_1, \mathcal{H}')$ and Γ_n in $\mathcal{L}(\mathcal{H}_n, \mathcal{D}_{\Gamma_{n-1}^*})$, $n \geq 2$, such that $T_1 = \Gamma_1$ and $T_n = D_{\Gamma_1^*} D_{\Gamma_2^*} \dots D_{\Gamma_{n-1}^*} \Gamma_n$ for $n \geq 2$. Since the description of the elementary rotation of T will play a distinguished role in some of our further developments, we tackle this question more carefully.

4.4 Lemma *Let T be a contraction of the form (4.1) and let $\{\Gamma_n\}_{n=1}^{\infty}$ be the set of its associated parameters. Then, there exist as bounded operators, the following:*

$$D_\infty(T) : \oplus_{n=1}^{\infty} \mathcal{H}_n \longrightarrow \oplus_{n=1}^{\infty} \mathcal{D}_{\Gamma_n} \qquad (4.9)$$

$$D_\infty(T) = s\text{-} \lim_{n \to \infty} \alpha_{T^{(n)}} D_{T^{(n)}} P_n$$

and

$$H_\infty(T) : \mathcal{H}' \longrightarrow \mathcal{H}' \qquad (4.10)$$

$$H_\infty(T) = (s\text{-} \lim_{n \to \infty} D_{\Gamma_1^*} D_{\Gamma_2^*} \dots D_{\Gamma_{n-1}^*} D_{\Gamma_n^*}^2 D_{\Gamma_{n-1}^*} \dots D_{\Gamma_1^*})^{1/2}.$$

Proof For h in $\cup_{n=1}^{\infty} \oplus_{k=1}^{n} \mathcal{H}_k$ there exists $q \in \mathbb{N}$ such that $h \in \oplus_{k=1}^{q} \mathcal{H}_k$. Note that $\alpha_{T^{(n)}} D_{T^{(n)}} P_n h = \alpha_{T^{(q)}} D_{T^{(q)}} P_q h$ for $n \geq q$. Hence, $\{\alpha_{T^{(n)}} D_{T^{(n)}} P_n h\}_{n=1}^{\infty}$ is a Cauchy sequence. But $\|\alpha_{T^{(n)}} D_{T^{(n)}} P_n\| \leq 1$ and $\cup_{n=1}^{\infty} \oplus_{k=1}^{n} \mathcal{H}_k$ is a dense subset of $\oplus_{n=1}^{\infty} \mathcal{H}_n$, so that there exists the bounded operator

$$D_\infty(T) = s\text{-}\lim_{n \to \infty} \alpha_{T^{(n)}} D_{T^{(n)}} P_n$$

which is a contraction. Furthermore, $\{D_{\Gamma_1^*} D_{\Gamma_2^*} \dots D_{\Gamma_{n-1}^*} D_{\Gamma_n^*}^2 D_{\Gamma_{n-1}^*} \dots D_{\Gamma_1^*}\}_{n=1}^{\infty}$ is a monotone bounded sequence of positive operators and its strong limit exists and it is a positive contraction. Therefore, $H_\infty(T)$ also exists and is a positive contraction. $\qquad\square$

4.5 Proposition   *Let T be a contraction with matrix representation as in (4.1) and let $\{\Gamma_n\}_{n=1}^{\infty}$ be the set of its associated parameters. Then, there exist the unitary operators*

$$\alpha_T : \mathcal{D}_T \longrightarrow \oplus_{n=1}^{\infty} \mathcal{D}_{\Gamma_n} \qquad (4.11)$$
$$\alpha_T D_T = D_\infty(T)$$

and

$$\beta_T : \mathcal{D}_{T^*} \longrightarrow \mathrm{cl}\mathcal{R}(H_\infty(T)) \qquad (4.12)$$
$$\beta_T D_{T^*} = H_\infty(T).$$

Proof Noting that $s\text{-}\lim_{n \to \infty} T^{(n)} P_n = T$ and $s\text{-}\lim_{n \to \infty} (T^{(n)})^* = T^*$, it simply follows that $\lim_{n \to \infty} \|D_{T^{(n)}} P_n h\| = \|D_T h\|$ for all h in $\oplus_{n=1}^{\infty} \mathcal{H}_n$. On the other hand, Lemma 4.4 gives $\|\alpha_{T^{(n)}} D_{T^{(n)}} P_n h\| = \|D_\infty(T) h\|$ for all h in $\oplus_{n=1}^{\infty} \mathcal{H}_n$, hence α_T is a unitary operator. Finally, it is easy to check that $T^{(n)} (T^{(n)})^* + D_{\Gamma_1^*} D_{\Gamma_2^*} \dots D_{\Gamma_{n-1}^*} D_{\Gamma_n^*}^2 D_{\Gamma_{n-1}^*} \dots D_{\Gamma_1^*} = I$ for all $n \geq 1$, and taking the strong limit in both sides when $n \to \infty$, one obtains the equality $TT^* + H_\infty(T)^2 = I$. Consequently, β_T is a unitary operator. $\qquad\square$

As it was the case with the row contractions of finite length, the existence of the unitary operators α_T and β_T makes it more convenient to replace the elementary rotation $R(T)$ of T with the following unitary operator:

$$R_T : \oplus_{n=1}^{\infty} \mathcal{H}_n \oplus \mathrm{cl}\mathcal{R}(H_\infty(T)) \longrightarrow \mathcal{H}' \oplus \oplus_{n=1}^{n} \mathcal{D}_{\Gamma_n} \qquad (4.13)$$
$$R_T = (I \oplus \alpha_T) R(T)(I \oplus \beta_T^*).$$

4.6 Remark The structure of the yet unknown part of R_T, which is $K_T = \alpha_T T^* \beta_T^*$, can be elucidated in much the same way as in the proof of Proposition 4.3. The conclusion is that, considering the matrix representation $K_T = [K_1 \quad K_2 \quad K_3 \quad \dots]^\top$, then $K_n = \Gamma_n^* D_{*,n+1}$ for all $n \geq 1$, where $D_{*,n}$ are operators which can be easily described. $\qquad\square$

4.7 Remark Similar results can be established for the column contractions with matrix representation $T = [T_1 \quad T_2 \quad T_3 \quad \dots]^\top$ in $\mathcal{L}(\mathcal{H}, \oplus_{n=1}^{\infty} \mathcal{H}_n')$. In this case, we obtain that T is a contraction if and only if $T_1 = \Gamma_1$ is a contraction and $T_n = \Gamma_n D_{\Gamma_{n-1}} D_{\Gamma_{n-2}} \dots D_{\Gamma_1}$ for $n \geq 2$, where $\Gamma_n \in \mathcal{L}(\mathcal{D}_{\Gamma_{n-1}}, \mathcal{H}_n')$ are uniquely determined contractions. $\qquad\square$

1.5   The Structure of Positive Definite Kernels

We have introduced an operation of renorming Hilbert spaces in the presence of a positive definite kernel. For two particular cases, described in Section 2 and Section 4, we have seen that this operation can be made explicit in terms of certain operator angles. In this section we prove a similar result for arbitrary $n \times n$ positive block matrices. The associated angle operators will be the Schur parameters of the block matrix. We illustrate the result by presenting some particular cases and by showing two immediate applications. First, it is shown the role of the Schur parameters in the triangular factorization of positive block matrices. Then, a formula for the computation of the determinant of a positive block matrix is obtained.

It is useful to begin with the analysis of 3×3 positive block matrices. This case is still simple and motivates the approach we intend to follow in the general case.

5.1 Example   Let us consider a positive block matrix

$$A = \begin{bmatrix} A_{11} & A_{12} & A_{13} \\ A_{12}^* & A_{22} & A_{23} \\ A_{13}^* & A_{23}^* & A_{33} \end{bmatrix}, \tag{5.1}$$

where $A_{ij} \in \mathcal{L}(\mathcal{H}_j, \mathcal{H}_i)$ for $i, j = 1, 2, 3$ and $\mathcal{H}_1$, $\mathcal{H}_2$ and $\mathcal{H}_3$ are Hilbert spaces. Since A is positive, it follows that the block matrices

$$A^{(12)} = \begin{bmatrix} A_{11} & A_{12} \\ A_{12}^* & A_{22} \end{bmatrix} \quad \text{and} \quad A^{(23)} = \begin{bmatrix} A_{22} & A_{23} \\ A_{23}^* & A_{33} \end{bmatrix}$$

are also positive. It follows by Lemma 2.1 that there exist uniquely determined contractions Γ_{12} in $\mathcal{L}(\text{cl}\mathcal{R}(A_{22}), \text{cl}\mathcal{R}(A_{11}))$ and Γ_{23} in $\mathcal{L}(\text{cl}\mathcal{R}(A_{33}), \text{cl}\mathcal{R}(A_{22}))$ such that $A_{12} = A_{11}^{1/2} \Gamma_{12} A_{22}^{1/2}$ and $A_{23} = A_{22}^{1/2} \Gamma_{23} A_{33}^{1/2}$. Moreover, by the same Lemma 2.1, there exists a contraction G from $\text{cl}\mathcal{R}(A^{(23)})$ into $\text{cl}\mathcal{R}(A_{11})$ such that $\begin{bmatrix} A_{12} & A_{13} \end{bmatrix} = A_{11}^{1/2} G(A^{(23)})^{1/2}$. Using a unitary operator of type (2.7), it follows that G may be replaced by another contraction $\Gamma' = \begin{bmatrix} \Gamma_1' & \Gamma_2' \end{bmatrix}$ in $\mathcal{L}(\text{cl}\mathcal{R}(A_{22}) \oplus \mathcal{D}_{\Gamma_{23}}, \text{cl}\mathcal{R}(A_{11}))$ such that

$$\begin{bmatrix} A_{12} & A_{13} \end{bmatrix} = A_{11}^{1/2} \begin{bmatrix} \Gamma_1' & \Gamma_2' \end{bmatrix} \begin{bmatrix} A_{22}^{1/2} & \Gamma_{23} A_{33}^{1/2} \\ 0 & D_{\Gamma_{23}} A_{33}^{1/2} \end{bmatrix}$$
$$= \begin{bmatrix} A_{11}^{1/2} \Gamma_1' A_{22}^{1/2} & A_{11}^{1/2}(\Gamma_1' \Gamma_{23} + \Gamma_2' D_{\Gamma_{23}}) A_{33}^{1/2} \end{bmatrix}.$$

Consequently, $\Gamma_1' = \Gamma_{12}$ and then, Proposition 4.2 gives $\Gamma_2' = D_{\Gamma_{12}^*} \Gamma_{13}$, where Γ_{13} is a contraction in $\mathcal{L}(\mathcal{D}_{\Gamma_{23}}, \mathcal{D}_{\Gamma_{12}^*})$. Summing up, the block matrix A in (5.1) is positive if and only if $A_{12} = A_{11}^{1/2} \Gamma_{12} A_{22}^{1/2}$, $A_{23} = A_{22}^{1/2} \Gamma_{23} A_{33}^{1/2}$ and $A_{13} = A_{11}^{1/2}(\Gamma_{12}\Gamma_{23} + D_{\Gamma_{12}^*} \Gamma_{13} D_{\Gamma_{23}}) A_{33}^{1/2}$, where Γ_{12}, Γ_{23} and Γ_{13} are contractions, uniquely determined by A.

This structure of the entries of A shows that it is convenient, without being an essential loss of generality, to suppose that $A_{11} = I_{\mathcal{H}_1}$, $A_{22} = I_{\mathcal{H}_2}$ and $A_{33} =$

$I_{\mathcal{H}_3}$. Moreover, it is worth mentioning a "geometric" counterpart of the previous algebraic computations. Set $\mathcal{H} = \oplus_{i=1}^{3}\mathcal{H}_i$ and remark that the unitary operator defined by (2.7) gives an identification of $\mathcal{H}_A$ with the space $\mathcal{D}_{G^*} \oplus (\mathcal{H}_2 \oplus \mathcal{H}_3)_{A^{(23)}}$. In its turn, $(\mathcal{H}_2 \oplus \mathcal{H}_3)_{A^{(23)}}$ may be identified with $\mathcal{H}_2 \oplus \mathcal{D}_{\Gamma_{23}^*}$. It follows that $\mathcal{D}_{G^*} = \mathcal{D}_{\Gamma_{12}^*}$ and Γ_{13} measures the angle between $\mathcal{D}_{\Gamma_{12}^*}$ and $\mathcal{D}_{\Gamma_{23}}$ inside $\mathcal{H}_A$. Thus, an identification of the space $\mathcal{H}_A$ with the space $\mathcal{D}_{\Gamma_{12}^*} \oplus \mathcal{H}_2 \oplus \mathcal{D}_{\Gamma_{23}}$ is obtained and the positions of the key spaces inside $\mathcal{H}_A$ are illustrated by Figure 1.1. $\qquad\square$

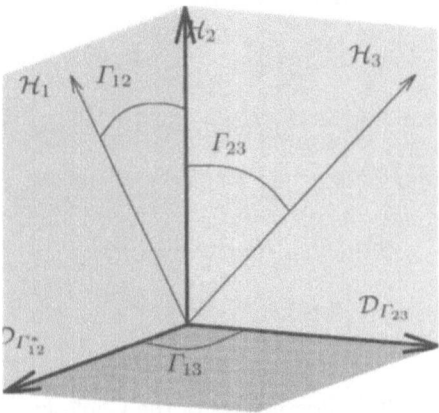

Figure 1.1: The geometry of $\mathcal{H}_A$

We now set about the task of describing the structure of a positive definite kernel following the path suggested in Example 5.1. It is expected to have a large family of operator angles as main parameters and let us introduce some necessary elements. Consider two families $\mathbf{E} = \{\mathcal{E}_n\}_{n\in\mathbb{Z}}$ and $\mathbf{F} = \{\mathcal{F}_n\}_{n\in\mathbb{Z}}$ of Hilbert spaces. Then $\Pi(\mathbf{E}, \mathbf{F}) = \Pi(\{\mathcal{E}_n\}_{n\in\mathbb{Z}}, \{\mathcal{F}_n\}_{n\in\mathbb{Z}})$ is the family of sets $\Gamma = \{\Gamma_{ij} \mid i, j \in \mathbb{Z}, i \le j\}$ of contractions such that $\Gamma_{nn} = 0_{\mathcal{E}_n, \mathcal{F}_n}$ for all $n \in \mathbb{Z}$ and the following *compatibility condition* holds: for $i < j$,

$$\Gamma_{ij} \in \mathcal{L}(\mathcal{D}_{\Gamma_{i+1,j}}, \mathcal{D}_{\Gamma_{i,j-1}^*}). \tag{5.2}$$

The elements of the set Γ are called *Schur parameters*. If $\mathcal{E}_n = \mathcal{F}_n$ for all $n \in \mathbb{Z}$, then we write $\Pi(\mathbf{E}, \mathbf{E}) = \Pi(\mathbf{E})$.

Let $\mathbf{H} = \{\mathcal{H}_n\}_{n\in\mathbb{Z}}$ be a family of Hilbert spaces and let Γ in $\Pi(\mathbf{H})$ be a set of Schur parameters. It is convenient to associate to Γ several objects suggested by the developments in Section 4. First, we consider row and column contractions as follows: for $i, j \in \mathbb{Z}$, $i < j$,

$$L_i^{(j)} : \oplus_{k=i+1}^{j} \mathcal{D}_{\Gamma_{i+1,k}} \longrightarrow \mathcal{H}_i \tag{5.3}$$

$$L_i^{(j)} = \quad \text{the row contraction associated by Proposition 4.2}$$
$$\text{to the parameters} \quad \{\Gamma_{ik} \mid i < k \le j\}.$$

and

$$C_j^{(i)} : \mathcal{H}_j \longrightarrow \oplus_{k=-(j-1)}^{-i} \mathcal{D}_{\Gamma_{-k,j-1}^*} \tag{5.4}$$

$C_j^{(i)} = $ the column contraction associated by Remark 4.7
to the parameters $\{\Gamma_{-k,j} \mid -j < k \leq -i\}$.

Then, the unitary operators $R_{L_i^{(j)}}$ defined by the formula (4.4) are denoted by R_{ij} and, in addition, we introduce unitary operators called *generalized rotations*, as follows: $U_{nn} = I_{\mathcal{H}_n}$ for $n \in \mathbb{Z}$ and for $i < j$,

$$U_{ij} : \oplus_{k=-j}^{-i} \mathcal{D}_{\Gamma_{-k,j}^*} \longrightarrow \oplus_{k=i}^{j} \mathcal{D}_{\Gamma_{ik}} \tag{5.5}$$

$$U_{ij} = R_{ij}(U_{i+1,j} \oplus I_{\mathcal{D}_{\Gamma_{ij}^*}}).$$

Finally, we define the upper triangular *Cholesky operators*: $F_{nn} = I_{\mathcal{H}_n}$ for $n \in \mathbb{Z}$ and for $i < j$,

$$F_{ij} : \oplus_{k=i}^{j} \mathcal{H}_k \longrightarrow \oplus_{k=i}^{j} \mathcal{D}_{\Gamma_{ik}} \tag{5.6}$$

$$F_{ij} = \begin{bmatrix} F_{i,j-1} & U_{i,j-1}C_j^{(i)} \\ 0 & \mathcal{D}_{\Gamma_{ij}} \dots \mathcal{D}_{\Gamma_{j-1,j}} \end{bmatrix}.$$

It is useful to note some other descriptions of the Cholesky operators F_{ij}.

5.2 Lemma   *The Cholesky operators satisfy the following relations: for $i < j$,*

$$F_{ij} = \begin{bmatrix} I & L_i^{(j)} F_{i+1,j} \\ 0 & \alpha_{L_i^{(j)}} D_{L_i^{(j)}} F_{i+1,j} \end{bmatrix} \tag{5.7}$$

and

$$F_{ij} = R_{ij} \begin{bmatrix} (L_i^{(j)})^* & F_{i+1,j} \\ D_{\Gamma_{ij}^*} \dots D_{\Gamma_{i,i+1}^*} & 0 \end{bmatrix}. \tag{5.8}$$

Proof First of all remember that the operator $\alpha_{L_i^{(j)}} D_{L_i^{(j)}}$ is the one introduced in Proposition 4.2. The equality (5.7) will be proven by induction on j. For $j = i+1$, this equality is obvious and suppose it is also true up to a certain $j > i$. It follows that:

$$F_{i,j+1} = \begin{bmatrix} F_{ij} & U_{ij}C_{j+1}^{(i)} \\ 0 & D_{\Gamma_{i,j+1}} \dots D_{\Gamma_{j,j+1}} \end{bmatrix} = \begin{bmatrix} I & L_i^{(j)} F_{i+1,j} & U_{ij}C_{j+1}^{(i)} \\ 0 & \alpha_{L_i^{(j)}} D_{L_i^{(j)}} F_{i+1,j} & \\ 0 & 0 & D_{\Gamma_{i,j+1}} \dots D_{\Gamma_{j,j+1}} \end{bmatrix}.$$

Now, the definition of the generalized rotation U_{ij} and the structure of the operator R_{ij} may be used in order to conclude the proof of (5.7). The equality (5.8) is a direct consequence of (5.7), as shown by the following computations:

$$
R_{ij}
\begin{bmatrix}
(L_i^{(j)})^* & F_{i+1,j} \\
D_{\Gamma_{ij}^*} \dots D_{\Gamma_{i,i+1}^*} & 0
\end{bmatrix}
=
\begin{bmatrix}
L_i^{(j)} & D_{\Gamma_{i,i+1}^*} \dots D_{\Gamma_{ij}^*} \\
\alpha_{L_i^{(j)}} D_{L_i^{(j)}} & -K_{L_i^{(j)}}
\end{bmatrix}
$$

$$
\times
\begin{bmatrix}
(L_i^{(j)})^* & F_{i+1,j} \\
D_{\Gamma_{ij}^*} \dots D_{\Gamma_{i,i+1}^*} & 0
\end{bmatrix}
=
\begin{bmatrix}
I & L_i^{(j)} F_{i+1,j} \\
0 & \alpha_{L_i^{(j)}} D_{L_i^{(j)}} F_{i+1,j}
\end{bmatrix}
= F_{ij}. \qquad \square
$$

After these preliminaries we can prove the main result of this section. It states that a positive definite kernel is uniquely determined by a family of Schur parameters. Moreover, the exact dependence on parameters shows that each entry of the kernel belongs to a set of the form $\{C_{ij} + S_{ij}\Gamma_{ij}D_{ij} \mid \|\Gamma_{ij}\| \le 1\}$, which can be viewed as an analogue of a disc in the complex plane. We also obtain exact descriptions of the center C_{ij} and of the both radii S_{ij} and D_{ij}.

Consider a family $\mathbf{H} = \{\mathcal{H}_n\}_{n\in\mathbb{Z}}$ of Hilbert spaces and a positive definite kernel A such that $A(i,j) = A_{ij}$ belongs to $\mathcal{L}(\mathcal{H}_j, \mathcal{H}_i)$ for all $i,j \in \mathbb{Z}$. For $m \le n$, we define the positive block matrices

$$
A^{(mn)} : \oplus_{k=m}^n \mathcal{H}_k \longrightarrow \oplus_{k=m}^n \mathcal{H}_k \tag{5.9}
$$

$$
A^{(mn)} = [\, A_{ij} \mid m \le i,j \le n \,].
$$

5.3 Theorem *Let A be a positive definite kernel and suppose that $A_{nn} = I_{\mathcal{H}_n}$ for all $n \in \mathbb{Z}$. Then there exists a uniquely determined set $\Gamma = \{\Gamma_{ij} \mid i,j \in \mathbb{Z}, i \le j\}$ of Schur parameters in $\Pi(\mathbf{H})$ such that $A_{i,i+1} = \Gamma_{i,i+1}$ for $i \in \mathbb{Z}$ and for $i,j \in \mathbb{Z}, j > i+1$,*

$$
A_{ij} = L_i^{(j-1)} U_{i+1,j-1} C_j^{(i+1)} + D_{\Gamma_{i,i+1}^*} \dots D_{\Gamma_{i,j-1}^*} \Gamma_{ij} D_{\Gamma_{i+1,j}} \dots D_{\Gamma_{j-1,j}}. \tag{5.10}
$$

Proof In order to prove this result, the same idea already employed in Example 5.1 will be used. To that end, it will be shown by induction on $k = j - i > 0$ that for an arbitrary $i \in \mathbb{Z}$, the following statements hold:

$(\alpha)_k \quad [\, A_{i,i+1} \quad \dots \quad A_{i,i+k} \,] = L_i^{(i+k)} F_{i+1,i+k}.$

$(\beta)_k \quad [\, A_{i,i+k} \quad \dots \quad A_{i+k-1,i+k} \,]^\top = F_{i,i+k-1}^* U_{i,i+k-1} C_{i+k}^{(i)}.$

$(\gamma)_k \quad A^{(i,i+k)} = F_{i,i+k}^* F_{i,i+k}.$

$(\delta)_k \quad$ There exists a uniquely determined contraction $\Gamma_{i,i+k+1}$ from $\mathcal{D}_{\Gamma_{i+1,i+k+1}}$ into $\mathcal{D}_{\Gamma_{i,i+k}^*}$ satisfying (5.10) for $j = i + k + 1$.

The case $k = 1$ was already dealt with in Example 5.1 and suppose that these statements are true for all p, $1 \leq p < k$. It follows from $(\alpha)_{k-1}$, $(\delta)_{k-1}$ and (5.6) that

$$
[A_{i,i+1} \ \cdots \ A_{i,i+k-1} \ A_{i,i+k}] = L_i^{(i+k)} \begin{bmatrix} F_{i+1,i+k-1} & U_{i+1,i+k-1}C_{i+k}^{(i+1)} \\ 0 & D_{\Gamma_{i+1,i+k}} \cdots D_{\Gamma_{i+k-1,i+k}} \end{bmatrix}
$$
$$
= L_i^{(i+k)} F_{i+1,i+k}.
$$

and $(\alpha)_k$ is proven. Next, it is remarked that

$$
\begin{bmatrix} A_{i,i+k} & A_{i+1,i+k} & \cdots & A_{i+k-1,i+k} \end{bmatrix}^\top
$$
$$
= \begin{bmatrix} L_i^{(i+k-1)}U_{i+1,i+k-1} & D_{\Gamma_{i,i+1}^*} \cdots D_{\Gamma_{i,i+k-1}^*} \\ F_{i+1,i+k-1}^* U_{i+1,i+k-1} & 0 \end{bmatrix} C_{i+k}^{(i)}.
$$

The proof of the statement $(\beta)_k$ is now concluded by the induction hypothesis and the equality (5.8) in Lemma 5.2. In order to obtain the required factorization of $A^{(i,i+k)}$, the statement $(\beta)_k$ and the induction hypothesis are used in the following way:

$$
A^{(i,i+k)} = \begin{bmatrix} A^{(i,i+k-1)} & \begin{bmatrix} A_{i,i+k} & \cdots & A_{i+k-1,i+k} \end{bmatrix}^\top \\ \begin{bmatrix} A_{i,i+k}^* & \cdots & A_{i+k-1,i+k}^* \end{bmatrix} & I \end{bmatrix}
$$
$$
= \begin{bmatrix} F_{i,i+k-1}^* F_{i,i+k-1} & F_{i,i+k-1}^* U_{i,i+k-1} C_{i+k}^{(i)} \\ (C_{i+k}^{(i)})^* U_{i,i+k-1}^* F_{i,i+k-1} & I \end{bmatrix} = F_{i,i+k}^* F_{i,i+k}.
$$

Finally, since A is a positive definite kernel, it follows that the block matrices $A^{(mn)}$ are positive operators for all $m \leq n$. In particular, $A^{(i,i+k+1)}$ is positive and, by Lemma 2.1, there exists a contraction G mapping $\mathrm{cl}\mathcal{R}(A^{(i+1,i+k+1)})$ into $\mathcal{H}_i$ such that

$$
[A_{i,i+1} \ \cdots \ A_{i,i+k} \ A_{i,i+k+1}] = G(A^{(i+1,i+k+1)})^{1/2}.
$$

Using a unitary operator of type (2.7), one obtains that G may be replaced by another contraction $G' = [\Gamma_i' \ \cdots \ \Gamma_{i+k+1}']$, such that

$$
[A_{i,i+1} \ \cdots \ A_{i,i+k} \ A_{i,i+k+1}] = G' \begin{bmatrix} F_{i+1,i+k} & U_{i+1,i+k}C_{i+k+1}^{(i+1)} \\ 0 & D_{\Gamma_{i+1,i+k+1}} \cdots D_{\Gamma_{i+k,i+k+1}} \end{bmatrix}.
$$

Therefore, $[\Gamma_i' \ \cdots \ \Gamma_{i+k}'] = L_i^{(i+k)}$ and, by Proposition 4.2, there exists a contraction $\Gamma_{i,i+k+1}$ in $\mathcal{L}(\mathcal{D}_{\Gamma_{i+1,i+k+1}}, \mathcal{D}_{\Gamma_{i,i+k}^*})$ such that

$$
\Gamma_{i+k+1}' = D_{\Gamma_{i,i+1}^*} \cdots D_{\Gamma_{i,i+k}^*} \Gamma_{i,i+k+1}.
$$

The proof is complete. $\square$

5.4 Remark   The assumption that $A_{nn} = I_{\mathcal{H}_n}$ for all $n \in \mathbb{Z}$, made in the statement of Theorem 5.3, is not a serious restriction. It is easy to remark that a slight modification of the preceding proof shows that a positive definite kernel A is uniquely determined by the pair $(\{A_{nn}\}_{n\in\mathbb{Z}}, \Gamma)$, where $\Gamma = \{\Gamma_{ij} \mid i,j \in \mathbb{Z}, i \le j\}$ belongs to $\Pi(\{\mathrm{cl}\mathcal{R}(A_{nn})\}_{n\in\mathbb{Z}})$. This correspondence is described by the following formulae: for $i \in \mathbb{Z}$, $A_{i,i+1} = A_{ii}^{1/2}\Gamma_{i,i+1}A_{i+1,i+1}^{1/2}$ and for $i,j \in \mathbb{Z}$, $j > i+1$,

$$A_{ij} = A_{ii}^{1/2}(L_i^{(j-1)}U_{i+1,j-1}C_j^{(i+1)} + D_{\Gamma_{i,i+1}^*}\cdots D_{\Gamma_{i,j-1}^*}\Gamma_{ij}D_{\Gamma_{i+1,j}}\cdots D_{\Gamma_{j-1,j}})A_{jj}^{1/2}. \quad \square$$

We now consider several particular cases of Theorem 5.3.

5.5 Example   Let $A = A^{(0n)} = [A_{ij}]_{i,j=0}^n$ be a positive block matrix and suppose $A_{ii} = I_{\mathcal{H}_i}$ for $i = 0,\dots,n$. Then, the entries of A are uniquely determined by a finite set $\{\Gamma_{ij} \mid 0 \le i \le j \le n\}$ of Schur parameters. The correspondence between A and the set of Schur parameters is described by the formula (5.10).

Using the factorization $A = A^{(0n)} = F_{0n}^*F_{0n}$, where F_{0n} is the associated Cholesky operator, we obtain a natural identification of $\mathcal{H}_A$ in terms of the Schur parameters of A. Indeed, the above mentioned factorization is of type (2.2), with $F = F_{0n}$ and $B = I$. Since F_{0n} is an upper triangular operator, it follows that the corresponding map defined by (2.3) is a unitary operator from $\mathcal{H}_A$ onto $\oplus_{k=0}^n \mathcal{D}_{\Gamma_{0k}}$.
$$\square$$

It is a nice feature of the Schur parameters that they characterize the positive definite Toeplitz kernels in a natural way, as shown by the following result.

5.6 Proposition   *Let A be a positive definite kernel and let Γ be the set of its Schur parameters. Suppose $A_{nn} = I_{\mathcal{H}_n}$ for all $n \in \mathbb{Z}$. The following are equivalent:*

(a)   *A is a positive definite Toeplitz kernel.*

(b)   *$\Gamma_{ij} = \Gamma_{i+k,j+k}$ for $i,j \in \mathbb{Z}$, $i \le j$ and $k \in \mathbb{Z}$.*

Proof   Assume (b) and define for $i \le j$ the contractions $\Gamma_{j-i} = \Gamma_{ij}$. It follows that $\Gamma_0 = 0_\mathcal{H}$, where $\mathcal{H}_n = \mathcal{H}$ for all $n \in \mathbb{Z}$ and, for $n > 0$, Γ_n are contractions from $\mathcal{D}_{\Gamma_{n-1}}$ into $\mathcal{D}_{\Gamma_{n-1}^*}$. Taking into account the formula (5.10) gives

$$A_{i+k,j+k} = L^{(j-i-1)}U_{j-i-2}C^{(j-i-1)} + D_{\Gamma_1^*}\cdots D_{\Gamma_{j-i-1}^*}\Gamma_{j-i}D_{\Gamma_{j-i-1}}\cdots D_{\Gamma_1}, \tag{5.11}$$

where $L^{(j-i-1)}$ denotes the row contraction associated to the parameters $\{\Gamma_{k-i} \mid i < k \le j\}$ by Proposition 4.2 and $C^{(j-i-1)}$ denotes the column contraction associated to the parameters $\{\Gamma_{j+k} \mid -j < k \le -i\}$ by Remark 4.7. Moreover, $\{U_n\}_{n\ge 0}$ is the corresponding family of generalized rotations, defined as follows: $U_0 = I_\mathcal{H}$ and for $n > 0$,

$$U_n : \oplus_{k=0}^n \mathcal{D}_{\Gamma_k^*} \longrightarrow \oplus_{k=0}^n \mathcal{D}_{\Gamma_k} \tag{5.12}$$

$$U_n = R_{L^{(n)}}(U_{n-1} \oplus I_{\mathcal{D}_{\Gamma_n^*}}).$$

The equality (5.11) shows that A is a positive definite Toeplitz kernel. Conversely, suppose A is a positive definite Toeplitz kernel. An induction argument based on the formula (5.10) and the compatibility conditions (5.2) may conclude the proof. $\qquad \square$

5.7 Remark If A is a positive definite Toeplitz kernel whose entries are complex numbers, then its Schur parameters are complex numbers satisfying $|\Gamma_n| \leq 1$ for all $n \geq 0$ and the compatibility condition (5.2). This condition means that whenever $|\Gamma_{n_0}| = 1$ for a certain $n_0 \geq 0$, then $\Gamma_n = 0$ for all $n > n_0$.

It is also convenient to introduce the following notation. For two Hilbert spaces $\mathcal{E}$ and $\mathcal{F}$, we denote by $\Pi^0(\mathcal{E},\mathcal{F})$ the set of families $\{\Gamma_n\}_{n\geq 0}$ of contractions such that $\Gamma_0 = 0_{\mathcal{E},\mathcal{F}}$ and Γ_n belongs to $\mathcal{L}(\mathcal{D}_{\Gamma_{n-1}^*}, \mathcal{D}_{\Gamma_{n-1}})$ for $n > 0$. If $\mathcal{E} = \mathcal{F}$, then we write $\Pi^0(\mathcal{E},\mathcal{E}) = \Pi^0(\mathcal{E})$ and if $\mathcal{E} = \mathcal{F} = \mathbb{C}$, then we write $\Pi^0(\mathbb{C}) = \Pi^0$. We also use the notation $\widetilde{\Pi}^0$ to denote the set of families $\{\Gamma_n\}_{n\geq 0}$ of complex numbers with the properties: $|\Gamma_n| \leq 1$ for all $n \geq 0$ and if $|\Gamma_{n_0}| = 1$ for a certain $n_0 \geq 0$, then $\Gamma_n = 0$ for all $n \geq n_0$. Hence, $\Pi^0 \subset \widetilde{\Pi}^0$ and $\{\Gamma_n\}_{n\geq 0}$ in $\widetilde{\Pi}^0$ belongs to Π^0 if and only if $\Gamma_0 = 0$. $\qquad \square$

We now mention another particular case of Theorem 5.3.

5.8 Proposition *Let A be a positive definite kernel and let Γ be the set of its Schur parameters. Suppose $A_{nn} = I_{\mathcal{H}_n}$ for all $n \in \mathbb{Z}$. The following are equivalent:*

(a) $A_{ij} = T_i T_{i+1} \ldots T_{j-1}$ *for $j > i$, where $\{T_n\}_{n\in\mathbb{Z}}$ is a given set of contractions T_n from $\mathcal{H}_{n+1}$ into $\mathcal{H}_n$.*

(b) $\Gamma_{i,i+1} = T_i$ *for $i \in \mathbb{Z}$ and $\Gamma_{ij} = 0$ for $j > i+1$.*

Proof Assume (b). Then, it is easily seen that $L_i^{(j-1)} = [\,\Gamma_{i,i+1} \quad 0 \quad \ldots \quad 0\,]$ and $C_j^{(i+1)} = [\,\Gamma_{j-1,j} \quad 0 \quad \ldots \quad 0\,]^\top$ for $j > i+1$. Moreover, $P_{\mathcal{H}_{j-1}} U_{i+1,j-1}/\mathcal{H}_{i+1} = \Gamma_{i+1,i+2} \ldots \Gamma_{j-2,j-1}$. Using the formula (5.10), it follows that

$$A_{ij} = \Gamma_{i,i+1}\Gamma_{i+1,i+2} \ldots \Gamma_{j-1,j}.$$

Conversely, suppose that $A_{ij} = T_i T_{i+1} \ldots T_{j-1}$ for $j > i$. Therefore, $\Gamma_{i,i+1} = T_i$ for all $i \in \mathbb{Z}$. An induction argument based on formula (5.10) and the compatibility condition (5.2) concludes the proof. $\qquad \square$

5.9 Corollary *The map defined by*

$$A_{ij} = \begin{cases} T_i T_{i+1} \ldots T_{j-1} & \text{if } j > i \\ I_{\mathcal{H}_i} & \text{if } j = i \\ A_{ji}^* & \text{if } j < i, \end{cases}$$

where $\{T_n\}_{n\in\mathbb{Z}}$ is a given set of operators T_n in $\mathcal{L}(\mathcal{H}_{n+1}, \mathcal{H}_n)$ is a positive definite kernel if and only if T_n are contractions for all $n \in \mathbb{Z}$. $\qquad \square$

We conclude this section with the presentation of two applications of Theorem 5.3. First, let $A^{(ij)} = [A_{mk}]_{i \leq m,k \leq j}$ be a positive block matrix with $A_{pp} = I_{\mathcal{H}_p}$ for $p = i, i+1, \ldots, j$. The factorization

$$A^{(ij)} = F_{ij}^* F_{ij} \tag{5.13}$$

obtained in the proof of Theorem 5.3 will be referred to as the *(lower-upper) Cholesky factorization* of $A^{(ij)}$. It is useful to introduce another Cholesky factorization in the following way. Define the unitary block matrix of appropriate dimension,

$$\mathcal{I} = \begin{bmatrix} 0 & 0 & \cdots & 0 & I \\ 0 & 0 & & I & 0 \\ \vdots & & \ddots & & \\ 0 & I & & & 0 \\ I & 0 & & 0 & 0 \end{bmatrix} \tag{5.14}$$

and the positive block matrix $B^{(ij)} = \mathcal{I} A^{(ij)} \mathcal{I}$. Let $\widetilde{F}_{ij}$ be the upper triangular Cholesky operator of $B^{(ij)}$ and define $G_{ij} = \mathcal{I} \widetilde{F}_{ij} \mathcal{I}$. Then,

$$A^{(ij)} = \mathcal{I} B^{(ij)} \mathcal{I} = \mathcal{I} \widetilde{F}_{ij}^* \mathcal{I}^2 \widetilde{F}_{ij} \mathcal{I} = G_{ij}^* G_{ij} \tag{5.15}$$

and remark that G_{ij} is a lower triangular matrix. The factorization in (5.15) will be referred to as the *(upper-lower) Cholesky factorization* of $A^{(ij)}$ and it is easy to describe the structure of G_{ij} in terms of the Schur parameters of $A^{(ij)}$.

The second application of Theorem 5.3 which is presented in this section consists of a formula for the computation of the determinant of a positive matrix in terms of its Schur parameters.

5.10 Theorem   *Let $A = [A_{ij}]_{i,j=0}^n$ be a positive block matrix such that A_{ij} belongs to $\mathcal{L}(\mathcal{H}_j, \mathcal{H}_i)$ and suppose that, for $i = 0, \ldots n$, $\mathcal{H}_i$ are Hilbert spaces of finite dimension. Let Γ_{ij}, $0 \leq i \leq j \leq n$, be the Schur parameters of A. Then the following formula holds:*

$$\det A = \prod_{i=0}^n \det A_{ii} \times \prod_{0 \leq i < j \leq n} \det D_{\Gamma_{ij}}^2.$$

Proof   The Cholesky factorization $A = A^{(0n)} = (\oplus_{k=0}^n A_{kk}^{1/2}) F_{0n} F_{0n}^* (\oplus_{k=0}^n A_{kk}^{1/2})$ gives

$$\det A = \prod_{i=0}^n \det A_{ii} \times (\det F_{0n})^2.$$

Then, it follows by the definition of the Cholesky operator F_{0n} that

$$\det F_{0n} = \det F_{0,n-1} \times \prod_{k=0}^{n-1} \det D_{\Gamma_{kn}}.$$

This equality leads immediately to the required formula for $\det A$.   □

1.6   Kolmogorov Decompositions. II

In this section we use the Schur parameters in order to describe the Kolmogorov decomposition of a positive definite kernel. Thus, a multiplicative structure involving elementary rotations is pointed out for the map V in Theorem 3.1. Besides, the Naimark dilation of a positive definite Toeplitz kernel appears to be essentially the elementary rotation of a well defined row contraction of infinite length. We also discuss a number of particular cases. We show the connection between the Kolmogorov decomposition, the Cholesky factorization and the Gram-Schmidt orthonormalization procedure. We illustrate these connections by introducing the orthogonal polynomials on the unit circle. Finally, we discuss a so-called Toeplitz embedding.

Consider a positive definite kernel A such that $A(i,j) = A_{ij}$ belongs to $\mathcal{L}(\mathcal{H}_j, \mathcal{H}_i)$ and $\mathbf{H} = \{\mathcal{H}_n\}_{n \in \mathbb{Z}}$ is a family of Hilbert spaces. We continue to assume, without loss of generality, that $A_{nn} = I_{\mathcal{H}_n}$ for all $n \in \mathbb{Z}$ and let $\Gamma = \{\Gamma_{ij} \mid i, j \in \mathbb{Z}, i \leq j\}$ in $\Pi(\mathbf{H})$ be the set of the Schur parameters of A. For $i \in \mathbb{Z}$, we introduce the row contractions of infinite length

$$L_i : \oplus_{k=i+1}^{\infty} \mathcal{D}_{\Gamma_{i+1,k}} \longrightarrow \mathcal{H}_i \tag{6.1}$$

$$L_i = \quad \text{the row contraction associated to} \\ \text{the parameters} \quad \{\Gamma_{ik} \mid i < k\}.$$

Denote by $\mathcal{D}_i$ the space $\oplus_{k=i+1}^{\infty} \mathcal{D}_{\Gamma_{ik}}$ introduced in (4.11) as an identification of the defect space $\mathcal{D}_{L_i}$ and denote by $\mathcal{D}_{i,*}$ the space $\mathrm{cl}\mathcal{R}(H_\infty(L_i))$ introduced in (4.12) as an identification of the defect space $\mathcal{D}_{L_i^*}$. Consider the Hilbert spaces

$$\mathcal{K}_i = \oplus_{j=-\infty}^{i-1} \mathcal{D}_{j,*} \oplus \mathcal{H}_i \oplus \mathcal{D}_i \tag{6.2}$$

for $i \in \mathbb{Z}$ and define the unitary operators

$$W_i : \mathcal{K}_{i+1} \longrightarrow \mathcal{K}_i \tag{6.3}$$

$$W_i = I \oplus \begin{bmatrix} I & 0 \\ 0 & \alpha_{L_i} \end{bmatrix} R(L_i) \begin{bmatrix} 0 & I \\ \beta_{L_i}^* & 0 \end{bmatrix},$$

with respect to the decompositions $\mathcal{K}_{i+1} = (\oplus_{j=-\infty}^{i-1} \mathcal{D}_{j,*}) \oplus (\mathcal{D}_{i,*} \oplus \mathcal{H}_{i+1} \oplus \mathcal{D}_{i+1})$ and, respectively, $\mathcal{K}_i = (\oplus_{j=-\infty}^{i-1} \mathcal{D}_{j,*}) \oplus (\mathcal{H}_i \oplus \mathcal{D}_i)$. The symbol $R(L_i)$ denotes the elementary rotation of L_i, while α_{L_i} and β_{L_i} are the unitary operators defined by (4.11) and (4.12). Consequently, the operators W_i are essentially the elementary rotations of the contractions L_i. With these elements, we can describe the multiplicative structure of the Kolmogorov decomposition of a positive definite kernel.

6.1 Theorem   *Let A be a positive definite kernel and suppose that $A_{nn} = I_{\mathcal{H}_n}$ for $n \in \mathbb{Z}$. Then the Kolmogorov decomposition of A is given by the map V_0 defined* on $\mathbb{Z}$ such that $V_0(n)$ belongs to $\mathcal{L}(\mathcal{H}_n, \mathcal{K}_0)$ for $n \in \mathbb{Z}$, $\mathcal{K}_0$ is defined according to (6.2) and

$$V_0(n) = \begin{cases} W_{-1}^* W_{-2}^* \ldots W_n^*/\mathcal{H}_n & \text{if } n < 0 \\ P_{\mathcal{H}_0}^{\mathcal{K}_0}/\mathcal{H}_0 & \text{if } n = 0 \\ W_0 W_1 \ldots W_{n-1}/\mathcal{H}_n & \text{if } n > 0. \end{cases}$$

Proof In order to prove this result, it will be shown that V_0 defined as above satisfies the properties (a) and (b) in Theorem 3.1. First, we claim that

$$A_{ij} = P_{\mathcal{H}_i} U_{ij}/\mathcal{H}_j \tag{6.4}$$

for $j > i$. Taking into account the multiplicative structure of the operators R_{ij} as established in Proposition 4.3, it follows:

$$\begin{aligned} P_{\mathcal{H}_i} U_{ij}/\mathcal{H}_j &= P_{\mathcal{H}_i} \begin{bmatrix} U_{i,j-1} & 0 \\ 0 & I \end{bmatrix} \begin{bmatrix} C_j^{(i)} \\ D_{\Gamma_{ij}} \ldots D_{\Gamma_{j-1,j}} \end{bmatrix} = P_{\mathcal{H}_i} U_{i,j-1} C_j^{(i)} \\ &= P_{\mathcal{H}_i} R_{i,j-1} \begin{bmatrix} U_{i+1,j-1} & 0 \\ 0 & I \end{bmatrix} \begin{bmatrix} C_j^{(i+1)} \\ \Gamma_{ij} D_{\Gamma_{i+1,j}} \ldots D_{\Gamma_{j-1,j}} \end{bmatrix} \\ &= \begin{bmatrix} L_{i,j-1} & D_{\Gamma_{i,i+1}^*} \ldots D_{\Gamma_{i,j-1}^*} \end{bmatrix} \begin{bmatrix} U_{i+1,j-1} & 0 \\ 0 & I \end{bmatrix} \begin{bmatrix} C_j^{(i+1)} \\ \Gamma_{ij} D_{\Gamma_{i+1,j}} \ldots D_{\Gamma_{j-1,j}} \end{bmatrix}. \end{aligned}$$

By Theorem 5.3, the last expression gives exactly A_{ij} and the equality (6.4) is proved. Next, we show that for $j > i$,

$$V_0^*(i) V_0(j) = P_{\mathcal{H}_i} W_i W_{i-1} \ldots W_{j-1}/\mathcal{H}_j = A_{ij}. \tag{6.5}$$

Denote by $L_i^{(j)}$ the finite sections of the row contractions L_i. Then, it is transparent that

$$P_{\oplus_{k=i}^j \mathcal{D}_{\Gamma_{ik}}} W_i W_{i-1} \ldots W_{j-1}/\mathcal{H}_j = U_{ij}/\mathcal{H}_j. \tag{6.6}$$

Hence, the equality (6.5) is a consequence of (6.4) and (6.6) and the map V_0 has the property (a) of Theorem 3.1. The fact that V_0 obeys (b) in Theorem 3.1 is an easy consequence of the definitions of $\mathcal{K}_0$ and $V_0(n)$. The proof is now complete. $\square$

6.2 Remark The preceding construction leads to a natural identification of the space $\mathcal{H}_A$ in terms of the Schur parameters of the positive definite kernel A. Together with $\mathcal{H}_A$, a map V satisfying $A(i,j) = V^*(i)V(j)$ for all $i, j \in \mathbb{Z}$ was introduced in Theorem 3.1. Taking into account the elements used in the proof of Theorem 6.1, we define the following maps: for $i \in \mathbb{Z}$,

$$\Phi_i : \mathcal{H}_A \longrightarrow \mathcal{K}_i \tag{6.7}$$

$$\Phi_i\left(\sum_{n \in \mathbb{Z}} V(n) h_n\right) = \ldots + W_{i-1}^* W_{i-2}^* h_{i-2} + W_{i-1}^* h_{i-1} + h_i + W_i h_{i+1} + \ldots,$$

where only a finite number of the vectors h_n in $\mathcal{H}_n$ are different from zero. Obviously, Φ_i extends to a unitary operator from $\mathcal{H}_A$ onto $\mathcal{K}_i$, also denoted by Φ_i. We may notice that

$$\Phi_{i+1} = W_i^* \Phi_i \tag{6.8}$$

for all $i \in \mathbb{Z}$. Indeed, we have

$$\Phi_{i+1}(\sum_{n \in \mathbb{Z}} V(n)h_n) = \ldots + W_i^* W_{i-1}^* h_{i-1} + W_i^* h_i + h_{i+1} + W_{i+1}h_{i+2} + \ldots$$

$$= W_i^*(\ldots + W_{i-1}^* h_{i-1} + h_i + W_i h_{i+1} + W_i W_{i+1} h_{i+2} + \ldots)$$

$$= W_i^* \Phi_i(\sum_{n \in \mathbb{Z}} V(n)h_n).$$

Consequently, the equality (6.8) holds.   □

Several examples are now presented in order to illustrate Theorem 6.1.

6.3 Example   Assume that A is a positive definite Toeplitz kernel. Then, we have by Proposition 5.8 that $\Gamma_{ij} = \Gamma_{i+k,j+k} = \Gamma_{j-i}$ for $i, j \in \mathbb{Z}$, $i \leq j$ and $k \in \mathbb{Z}$. Consequently, $\mathcal{K}_i = \mathcal{K}$ and $W_i = W$ for all $i \in \mathbb{Z}$, and we may notice that in this situation, the equality (6.8) is replaced by the following one:

$$\Phi S = W\Phi, \tag{6.9}$$

where $\Phi = \Phi_n$ for all $n \in \mathbb{Z}$ and S is the operator defined by (3.5). This equality explains a previous statement that the operator S defined by (3.5) plays a similar role with the elementary rotation. Precisely, S is unitarily equivalent to the elementary rotation of the block matrix $\begin{bmatrix} 0 & L \end{bmatrix}^{\top}$ representing a contraction from $\oplus_{n=0}^{\infty} \mathcal{D}_{\Gamma_n}$ into $\oplus_{n=-\infty}^{0} \mathcal{D}_{n,*}$ ($\mathcal{D}_{n,*} = \mathcal{D}_* = \mathrm{cl}\mathcal{R}(H_{\infty}(L))$), where $L = \begin{bmatrix} \Gamma_1 & D_{\Gamma_1^*}\Gamma_2 & D_{\Gamma_1^*}D_{\Gamma_2^*}\Gamma_3 & \ldots \end{bmatrix}$ is the row contraction of infinite length associated to the Schur parameters of A. With a slight abuse of language, we will call W the *Naimark dilation* of the considered positive definite Toeplitz kernel.   □

6.4 Example   Suppose that A is a positive definite Toeplitz kernel as in Corollary 5.9. Then, $T_i = T$ and $\mathcal{K}_i = \mathcal{K}$ for all $i \in \mathbb{Z}$, where $\mathcal{K} = \ldots \oplus \mathcal{D}_{T^*} \oplus \mathcal{D}_{T^*} \oplus \mathcal{H} \oplus \mathcal{D}_T \oplus \mathcal{D}_T \ldots$ and W has the following matrix representation with respect to this direct sum decomposition of $\mathcal{K}$:

$$W = \begin{bmatrix} \ddots & & & \vdots & & \\ & I & 0 & 0 & 0 & \\ \cdots & 0 & D_{T^*} & T & 0 & \cdots \\ & 0 & -T^* & D_T & 0 & \\ & 0 & 0 & 0 & I & \\ & & & \vdots & & \ddots \end{bmatrix}.$$

This matrix representation of W is known as the *Schäffer form of the Sz.-Nagy minimal unitary dilation* of the contraction T. By Theorem 6.1, W has the properties: $T^n = P_{\mathcal{H}} W^n / \mathcal{H}$ and $\mathcal{K} = \vee_{n=-\infty}^{\infty} W^n \mathcal{H}$.   □

6.5 Example   Suppose that $A = A^{(0n)} = [A_{ij}]_{i,j=0}^n$ is a positive block matrix. We explain now the connection between the Kolmogorov decomposition and the

Cholesky factorization. Thus, we deduce from (6.6) that

$$V_0(i) = U_{0i}/\mathcal{H}_i = \begin{bmatrix} U_{0,i-1}C_i^{(0)} \\ D_{\Gamma_{0i}} \cdots D_{\Gamma_{i-1,i}} \end{bmatrix}$$

for $0 \leq i \leq n$, and taking into account the definition of the Cholesky operator, we deduce that

$$F_{0n} = [\, V_0(0) \quad V_0(1) \quad \ldots \quad V_0(n)\,]. \tag{6.10}$$

Using this relation, we see that the operator Φ_0 defined by (6.7) can be alternatively described as:

$$\Phi_0 : \mathcal{H}_A \longrightarrow \mathcal{H}_0 \oplus \mathcal{D}_{\Gamma_{01}} \oplus \ldots \oplus \mathcal{D}_{\Gamma_{0n}} \tag{6.11}$$

$$\Phi_0[h] = F_{0n}h,$$

where $[h]$ denotes the class in $\mathcal{H}_A$ of the element $h = \oplus_{k=0}^n h_k$ of $\oplus_{k=0}^n \mathcal{H}_k$. Hence, we see that Φ_0 coincides with the unitary operator associated to the Cholesky factorization (5.13) by the formula (2.3). $\qquad\square$

The last issue addressed in this section concerns the connection between the Cholesky factorization and the Gram-Schmidt orthonormalization procedure. In order to simplify our discussion, we suppose that $A = A^{(0n)} = [A_{ij}]_{i,j=0}^n$ is a strictly positive matrix and $A_{ii} = 1$ for $0 \leq i \leq n$. As seen in Example 1.4, the set $\{E_i\}_{i=0}^n$ is an orthonormal basis of $\mathbb{C}^{n+1}$. Since A is invertible, $(\mathbb{C}^{n+1})_A = \mathbb{C}^{n+1}$ and the vectors E_i, $0 \leq i \leq n$, are linearly independent. Using Theorem 5.10, it follows that the Cholesky operator F_{0n} of A is also invertible and we can define the vectors

$$\vartheta_i = \text{the } i\text{-th column of } F_{0n}^{-1}, \qquad i = 0, 1, \ldots n, \tag{6.12}$$

which constitute an orthonormal basis of $(\mathbb{C}^{n+1})_A$. The upper triangular form of F_{0n}^{-1} implies that, for $0 \leq i \leq n$,

$$\vartheta_i = a_{ii}E_i + a_{i-1,i}E_{i-1} + \ldots + a_{0i}E_0, \tag{6.13}$$

where $\{a_{ji}\}_{j=0}^i$ are the non-zero elements of the i-th column of F_{0n}^{-1}. We recognize that the relations given by (6.13) represent the Gram-Schmidt procedure of orthonormalizing the family of vectors $\{E_i\}_{i=0}^n$ in $(\mathbb{C}^{n+1})_A$. This connection with the (inverse of) Cholesky operators justifies why the Gram-Schmidt procedure can be interpreted as a procedure for inverting positive matrices.

6.6 Remark It is useful to know a formula for the computation of the coefficients a_{ii} and a_{0i} in (6.13). The formula for a_{ii} follows directly from the definition (5.6) of F_{0n}. By the same definition of F_{0i}, we see that in order to compute a_{0i} it is convenient to determine the last entry of the row matrix representation of $-L_0^{(i)}(\alpha_{L_0^{(i)}}D_{L_0^{(i)}})^{-1}$. It follows easily that this entry is exactly

$$-L_0^{(i-1)}(\alpha_{L_0^{(i-1)}}D_{L_0^{(i-1)}})^{-1}K_{L_0^{(i-1)}}\Gamma_{0i}D_{\Gamma_{0i}}^{-1} - D_{\Gamma_{01}^*}\ldots D_{\Gamma_{0,i-1}^*}\Gamma_{0i}D_{\Gamma_{0i}}^{-1},$$

where $K_{L_0^{(i-1)}}$ is the matrix defined by (4.6). But now, we remark that the expression $L_0^{(i-1)}(\alpha_{L_0^{(i-1)}} D_{L_0^{(i-1)}})^{-1} K_{L_0^{(i-1)}} + D_{\Gamma_{01}^*} \ldots D_{\Gamma_{0,i-1}^*}$ is a Schur complement of the unitary matrix

$$U = \begin{bmatrix} \alpha_{L_0^{(i-1)}} D_{L_0^{(i-1)}} & -K_{L_0^{(i-1)}} \\ L_0^{(i-1)} & D_{\Gamma_{01}^*} \ldots D_{\Gamma_{0,i-1}^*} \end{bmatrix}.$$

As a consequence of the Frobenius-Schur formula (2.4), we can note the following inversion formula:

$$\begin{bmatrix} a & b \\ c & d \end{bmatrix}^{-1} = \begin{bmatrix} a^{-1} + a^{-1}bs^{-1}ca^{-1} & -a^{-1}bs^{-1} \\ -s^{-1}ca^{-1} & s^{-1} \end{bmatrix}, \tag{6.14}$$

where $s = d - ca^{-1}b$ denotes the Schur complement. Since U is unitary, its inverse is U^* and (6.14) gives

$$(L_0^{(i-1)}(\alpha_{L_0^{(i-1)}} D_{L_0^{(i-1)}})^{-1} K_{L_0^{(i-1)}} + D_{\Gamma_{01}^*} \ldots D_{\Gamma_{0,i-1}^*})^{-1} = D_{\Gamma_{0,i-1}^*} \ldots D_{\Gamma_{01}^*}.$$

Consequently,

$$a_{0i} = -D_{\Gamma_{01}^*}^{-1} \ldots D_{\Gamma_{0,i-1}^*}^{-1} D_{\Gamma_{0i}^*}^{-1} \Gamma_{0i}. \tag{6.15}$$

$\square$

As another illustration of Theorem 6.1, we present a few elements concerning the orthogonal polynomials on the unit circle. Let A be the positive definite Toeplitz kernel such that $A(i,j) = A_{ij} = A_{j-i}$ for $i,j \in \mathbb{Z}$, where $\{A_n\}_{n \in \mathbb{Z}}$ is the set of the Fourier coefficients of the probability measure μ on the interval $[0, 2\pi)$. Suppose that all the matrices $A^{(ij)}$, $i,j \in \mathbb{Z}$, $i \le j$, associated to A by the formula (5.9) are invertible. In this case, $\{\nu_n\}_{k=0}^n$ is a linearly independent family in $L^2(\mu)$ generating a space $\mathcal{P}_n$ of finite dimension. $\mathcal{P}_n$ is the space of polynomials of degree at most n. It follows from the so-called *centro-symmetry* property of $A^{(0n)}$, i.e. from the equality

$$\mathcal{I} A^{(0n)} \mathcal{I} = (A^{(0n)})^\top, \tag{6.16}$$

where $\mathcal{I}$ is defined by (5.14), that the image through Ψ (the unitary operator defined by (3.6)) of the space $\mathcal{P}_n$ is the space $(\mathbb{C}^{n+1})_{(A^{(0n)})^\top}$. Remember that we denoted by $B^{(0n)}$ the matrix $\mathcal{I} A^{(0n)} \mathcal{I}$ and by $\widetilde{F}_{0n}$ we denoted the upper triangular Cholesky factor of $B^{(0n)}$. By the preceding discussion of the Gram-Schmidt procedure, the vectors $\vartheta_k = $ the k-th column of $\widetilde{F}_{0k}^{-1}$, $k = 0, 1, \ldots, n$, constitute an orthonormal basis of $(\mathbb{C}^{n+1})_{B^{(0n)}}$. Then, we define the polynomials

$$\varphi_n = \Psi^{-1} \vartheta_n, \quad n \ge 0, \tag{6.17}$$

usually referred to as the *Szegö polynomials* or the *orthogonal polynomials* associated to the measure μ. It is also useful to define the so-called *monic orthogonal polynomials of first kind* by the formula

$$\psi_n = a_{nn}^{-1} \varphi_n. \tag{6.18}$$

The Schur parameters play a significant role in the developments about Szegö polynomials. Taking into account the formula (6.13), we can write

$$\varphi_n(z) = a_{nn}z^n + a_{n,n-1}z^{n-1} + \ldots + a_{n0}.$$

If $\{\Gamma_n\}_{n \geq 0}$ is the set of the Schur parameters of the kernel A, then the set of the Schur parameters of the positive matrix $B^{(0n)}$ is $\{\Gamma_k^*\}_{k=0}^n$. For a polynomial P of degree n, we introduce the *reverse polynomial* $P^\#$ of P by the formula $P^\#(z) = z^n P^\sim(1/z)$, where $P^\sim(z) = (P(z^*))^*$. In the theory of the orthogonal polynomials a central role is played by the following result.

6.7 Theorem *The following relations hold for all $n \geq 0$ and $z \in \mathbb{D}$: $\psi_0 = 1$,*

$$\psi_{n+1}(z) = z\psi_n(z) - \Gamma_{n+1}^* \psi_n^\#(z) \tag{6.19}$$

and

$$a_{nn}\varphi_{n+1}(z) = za_{n+1,n+1}\varphi_n(z) - a_{n+1,n+1}\Gamma_{n+1}^*\varphi_n^\#(z). \tag{6.20}$$

Proof It is a consequence of (6.12) that $B^{(0n)}\vartheta_n = a_{nn}^{-1}E_n$ and define $\vartheta_n^\# = \Psi\varphi_n^\#$. It follows from (6.16) that $B^{(0n)}\vartheta_n^\# = a_{nn}^{-1}E_0$. From these relations, we deduce that

$$B^{(0,n+1)}a_{n+1,n+1}^{-1}\vartheta_{n+1}^\# - B^{(0,n)}a_{nn}^{-1}\vartheta_n^\# = (a_{n+1,n+1}^{-2} - a_{nn}^{-2})E_0 + \alpha E_{n+1},$$

with a certain constant α. On the other hand, it is obvious that

$$a_{n+1,n+1}\vartheta_{n+1}^\# - a_{nn}^{-1}\vartheta_n^\# = 0 \oplus \oplus_{k=1}^{n+1}\lambda_k,$$

for some complex numbers $\lambda_1, \lambda_2, \ldots, \lambda_{n+1}$. Consequently, for all $0 \leq k \leq n$, we have $\langle h, e_k \rangle_{B^{(0n)}} = 0$, which can be rewritten in the form $a_{nn}\varphi_{n+1}^\#(z) = a_{n+1,n+1}\varphi_n^\#(z) + z\beta\varphi_n(z)$ with a certain constant β. This means that

$$a_{nn}\varphi_{n+1}(z) = a_{n+1,n+1}z\varphi_n(z) + \beta^*\varphi_n^\#(z),$$

where $\beta^* = a_{nn}\varphi_{n+1}(0)(\varphi_n^\#(0))^{-1} = \varphi_{n+1}(0) = a_{n+1,0}$. But, it is a consequence of the equality (6.15) that $a_{n+1,0} = -a_{n+1,n+1}\Gamma_{n+1}^*$, and the equality (6.20) is proven. The equality (6.19) is a direct consequence of (6.20). □

The numbers $\{\psi_n(0)\}_{n \geq 0}$ are known in the theory of orthogonal polynomials as the *Szegö coefficients* of μ. It appears to be an interesting problem to characterize the measure μ in terms of the Szegö coefficients. We present here some remarks in this direction and we will return to this question in Chapter 5. Since the connection between the Schur parameters of A and the Szegö coefficients of μ was clarified by Theorem 6.7, we show now how to obtain the orthogonal polynomials from the Naimark dilation of A. Let W in $\mathcal{L}(\mathcal{K})$ be the Naimark dilation of the positive definite kernel A associated to μ and described in Theorem 6.1. According to (6.2), $\mathcal{K} = \oplus_{j=-\infty}^{-1}\mathcal{D}_* \oplus \mathcal{H} \oplus \oplus_{n=1}^\infty \mathcal{D}_{\Gamma_n}$, where $\mathcal{H} = \mathbb{C}$, $\mathcal{D}_* = \text{cl}\mathcal{R}(H_\infty(L))$ and L is the row contraction associated to the Schur parameters $\{\Gamma_n\}_{n \geq 0}$. Note that $\mathcal{D}_*$ is either $\mathbb{C}$ or 0, and this alternative will be clarified later on in Chapter 5. Since we supposed that all $A^{(ij)}$ are invertible, it follows that $|\Gamma_n| < 1$ for all $n \in \mathbb{N}$ and $\mathcal{D}_{\Gamma_n} = \mathbb{C}$. Now, we can define the operators $w_n = P_{\mathcal{H} \oplus \oplus_{k=1}^n \mathcal{D}_{\Gamma_k}} W / (\mathcal{H} \oplus \oplus_{k=1}^n \mathcal{D}_{\Gamma_k})$. The next result describes the connection between the Naimark dilation and the Szegö polynomials.

6.8 Proposition   *The monic orthogonal polynomials ψ_n are the characteristic polynomials of w_n^*, i.e. $\psi_n(z) = \det(zI - w_n^*)$ for all $n \geq 1$.*

Proof Define the polynomials $q_n(z) = \det(zI - w_n^*)$ and considering the action of the operator Φ_0 defined by (6.7), one obtains:

$$\Phi_0 \Psi q_n = q_n(W^*)[E_0] = [\ldots \quad * \quad * \quad q_n(w_n^*)e_0 \quad * \quad \ldots]^\top,$$

where $q_n(w_n^*)e_0$ is a row vector with $n + 1$ entries which lies on the positions indexed from 0 to n of the vector $\Phi_0 \Psi q_n$, while the entries marked by * play no role here. By Cayley-Hamilton theorem, $q_n(w_n^*) = 0$. Hence, for $0 \leq k < n$,

$$\langle q_n, \nu_k \rangle_{L^2(\mu)} = \langle \Phi_0 \Psi q_n, \Phi_0 \Psi \nu_k \rangle_\mathcal{K} = 0.$$

Since q_n is a monic polynomial of degree n, it follows that $q_n = \psi_n$, for all $n \geq 1$. $\qquad \square$

Finally, we analyze the situation when a certain $A^{(ij)}$ is not invertible, which means that there exists some $n_0 \in \mathbb{N}$ with $|\Gamma_{n_0}| = 1$. Denote by δ_t the Dirac measure with mass 1 at e^{it}, where $t \in [0, 2\pi)$.

6.9 Proposition   *The positive measure μ has finite support, i.e. $\mu = \sum_{k=1}^n a_k \delta_{t_k}$ with $a_k > 0$, if and only if its n-th Schur parameter has modulus 1.*

Proof The proof explores elementary spectral properties of the operators w_n. Suppose that $|\Gamma_n| = 1$ for a certain $n \geq 1$. Then, w_n^* is a unitary operator and remark that it has the cyclic vector $[1 \quad -\Gamma_1^* D_{\Gamma_1}^{-1} \quad \ldots \quad -\Gamma_{n-1}^* D_{\Gamma_1}^{-1} \ldots D_{\Gamma_{n-1}}^{-1}]^\top$. Consequently, it has a simple spectrum, *i.e.* it has n distinct eigenvalues $\lambda_1, \ldots, \lambda_n$ on the unit circle. Let E be the spectral measure of w_n and supp$E$ denotes the support of this measure. Then,

$$\text{supp}E = \sigma(w_n) = \{\lambda_1, \ldots, \lambda_n\},$$

where $\sigma(w_n)$ denotes the spectrum of w_n. The formula $\mu_0(f) = \langle E(f)e_0, e_0 \rangle$, where f is a continuous function on the unit circle, gives rise to a positive measure on the unit circle with supp$\mu_0 \subseteq$ supp$E$. It follows by Theorem 6.1 that this measure μ_0 coincides with μ, hence μ has finite support, *i.e.* there exists $p \leq n$ such that

$$\mu = \sum_{k=1}^p a_k \delta_{t_k},$$

with $e^{it_k} = \lambda_k$ and $a_k > 0$. Consequently,

$$A^{(0p)} = \sum_{k=1}^p a_k \begin{bmatrix} 1 & e^{it_k} & & e^{ipt_k} \\ e^{-it_k} & 1 & & e^{i(p-1)t_k} \\ & & \ddots & \ddots & \\ e^{-ipt_k} & & & 1 \end{bmatrix}$$

and since each of the terms of the sum in the right hand side of the above equality is a matrix of rank 1, it follows that the rank of $A^{(0p)}$ is less or equal than p. By Theorem 5.10, there exists $p_0 \leq p$ such that $|\Gamma_{p_0}| = 1$. Hence, we must have $p_0 = p = n$ and the proof of the proposition is complete.   □

We also introduce the so-called *monic orthogonal polynomials of second kind* of μ as being the monic orthogonal polynomials of first kind associated to the probability measure μ_- determined by the Schur parameters $\{-\Gamma_n\}_{n\geq 0}$ These polynomials will be denoted by $\hat{\psi}_n$, $n \geq 0$, and will be used later.

We conclude this section with a discussion of a so-called *Toeplitz embedding*. This refers to the following construction. Let $\mathbf{H} = \{\mathcal{H}_n\}_{n\in\mathbb{Z}}$ be a family of Hilbert spaces and let A be a positive definite kernel such that $A_{nn} = I_{\mathcal{H}_n}$ for all $n \in \mathbb{Z}$. Let $\{\Gamma_{ij} \mid i,j \in \mathbb{Z}, i \leq j\}$ be the Schur parameters of A and define the following elements: the Hilbert space $\mathcal{H} = \oplus_{n\in\mathbb{Z}}\mathcal{H}_n$, $\bar{\Gamma}_0 = 0_{\mathcal{H}}$ and, for $n \geq 1$,

$$\bar{\Gamma}_n : \mathcal{D}_{\bar{\Gamma}_{n-1}} \longrightarrow \mathcal{D}_{\bar{\Gamma}^*_{n-1}} \tag{6.21}$$

$$(\bar{\Gamma}_n)_{ij} = \begin{cases} \Gamma_{ij} & \text{if} \quad j = i + n, \\ 0 & \text{otherwise.} \end{cases}$$

Let $\bar{A}$ be the Toeplitz kernel associated to the family $\{\bar{\Gamma}_n\}_{n\geq 0}$ of Schur parameters. We can note the following result.

6.10 Proposition   (a) *Let A be a positive definite kernel and let $\bar{A}$ be the positive definite Toeplitz kernel associated to A as above. Then*

$$A_{ij} = P_{\mathcal{H}_j}\bar{A}_{j-i}/\mathcal{H}_i \quad \text{for all} \quad i,j \in \mathbb{Z}.$$

(b) *Let $\bar{W} \in \mathcal{L}(\bar{\mathcal{K}})$ be the Naimark dilation of $\bar{A}$ as described in Example 6.3 and let V_0 be the Kolmogorov decomposition of A, as described in Theorem 6.1. Then,*

$$V_0(n) = P^{\bar{\mathcal{K}}}_{\mathcal{K}_0}\bar{W}^n/\mathcal{H}_n \quad \text{for all} \quad n \in \mathbb{Z}.$$

Proof Part (a) is a transparent application of Theorem 5.3, while part (b) is merely a restatement of the definitions. We can omit the details.   □

This shows that many times, a general result concerning positive definite kernels can be deduced from its Toeplitz counterpart.

1.7 Notes

All the material contained in the first two sections can be found in the books devoted to Hilbert spaces and their linear operators. We mention here, for instance, [Do3], [Fu], [Ha], [Yo]. Theorem 3.1 is a version of a classical result of Kolmogorov (see [EL] for comments on the history of this result), while Theorem 3.2 is another classical result, due to Naimark (see [Na2], [EL], [Sz.-NF2]). The Kolmogorov decomposition can be extended by replacing $\mathbb{Z}$ with an arbitrary set and then it is related to many basic constructions and results concerning positive definite maps, such as the theory of reproducing kernel Hilbert spaces ([Aro], [Sch], [dB], [Dy1]) or the Stinespring theorem ([St], [EL], [Da], [Pau]). The results in Section 4 were systematically used in [ACF], [CeF1-2], and Proposition 4.5 is mentioned in [Co1]. It is difficult to give an exhaustive list of the papers that explore the Schur parameters in connection with the structure of positive matrices. Schur parameters were used for the first time in a classical paper of Schur [Sc]. The Szegö theory of orthogonal polynomials developed in [Sz] also uses Schur parameters, as especially emphasized by Geronimus [Ge]. There are several engineering branches using Schur parameters. For instance, Schur parameters appear in geophysics under the name of reflection coefficients, in connection with waves propagation in layered media—see [Bur], [Cl], [FoF1], [RT]; in speech synthesis, the Schur parameters enter into the structure of the lattice predictor under the name of partial correlation(PARCOR) coefficients—see [MG]; similar lattice structures were used in the domain of adaptive filtering—see [Hay], [HM]. For extensions of the elements of the Szegö theory of orthogonal polynomials to block matrices we mention the papers [DGK1], [DVK]. But the first satisfactory definition of the Schur parameters in matrix or operator case appeared in [CF1] (under the name of *choice sequences*), where the role of the compatibility conditions of type (5.2) was emphasized. Generalizations of the Schur parameters from Toeplitz to arbitrary positive matrices were motivated on the one hand by non-stationary linear least squares estimation problems (see [DM], [Dep], [Ka2], [LAK1], [DGK3], [KL], [DI]), and on the other hand by the study of the multiplicative structure of the Kolmogorov decomposition, as described in Theorem 6.1, which is a dilation theoretic problem—see [Co2]. For the presentation of the material in Section 5 we used the paper [Co2], where the formulae (5.10) are obtained in analogy with some formulae in the contractive intertwining dilations theory (see [CeF1], [ACF] and [FoF1]). Another proof of the formula (5.10), using directly the Kolmogorov decomposition, was noticed in [Ti]. A proof exploiting the transmission-line models was indicated in [KB]. In order to put the parametrization of the positive definite kernels given by Theorem 5.3 in a broader perspective, we mention that this is related to the classical factorization theory of the unitary group (see, for instance, [Mur]). Thus, any real $n \times n$ unitary matrix can be viewed as a generalized rotation and be factorized using $n(n-1)/2$ elementary (or planar) rotations. The dilation theory (Kolmogorov decomposition) allows us to relate the parameters of these planar rotations with the entries of a positive definite kernel as explained in Theorem 5.3. The structure of the Naimark dilation that follows from Theorem 6.1 was previously considered in [Co1]. In this case, the matrix representation of $W_+ = W/\mathcal{H} \oplus \mathcal{D}_{\Gamma_1} \oplus \mathcal{D}_{\Gamma_2} \oplus \dots$ coincides with a so-called *Hessenberg representation* which was already noted, under the name of *adequate isometry* in [CeF2] and under the name of *state-space generator* in [KP]. Essentially the same remarks appear also in [Gr]. The Toeplitz embedding is used or discussed in many places. For instance, we mention here [DGK3], [PP], [Co4].

Chapter 2
Models for Triangular Contractions

In this chapter, two methods which both relate to the structure of triangular contractions are discussed. The first one goes with the main stream of the book and uses Schur parameters. Section 2 is devoted to this material. In Section 3 we solve the realization problem for unitary systems using the associated Schur parameters. Then, we show that the triangular contractions can be viewed as characteristic operators in the Sz.-Nagy-Foias and de Branges-Rovnyak models of families of contractions. Even though the Schur parameters are not directly involved in these models, these considerations offer another perspective on the renorming operation of Hilbert spaces and they will be useful later in connection with the commutant lifting theorem.

2.1 Preliminaries

In this section we review briefly certain facts about some spaces of vector-valued functions and multiplication operators. Let $\mathcal{H}$ be a separable Hilbert space. Then $L^2(\mathcal{H})$ is the Lebesgue space of the square integrable functions on $[0, 2\pi)$ with values in $\mathcal{H}$. $L^2(\mathcal{H})$ is a Hilbert space and each element v of this space has a representation $v = \sum_{n \in \mathbb{Z}} \nu_{-n} v_n$ with $\{v_n\}_{n \in \mathbb{Z}}$ in $l^2(\mathcal{H})$, hence the construction of the unitary operator Ψ in Example 1.3.3 can be extended and provides a unitary operator, denoted by $\Psi_{\mathcal{H}}$, mapping $L^2(\mathcal{H})$ onto $l^2(\mathcal{H})$. The Hardy space $H^2(\mathcal{H})$ of the analytic functions u on the unit disc with values in $\mathcal{H}$, such that

$$\|u\|_{H^2(\mathcal{H})} = \sup_{0 \leq r < 1} \left(\frac{1}{2\pi} \int_0^{2\pi} \|u(re^{it})\|^2 dt \right)^{1/2} < \infty,$$

can be viewed as a closed subspace of $L^2(\mathcal{H})$. This identification is established by the map:

$$u(z) = \sum_{n=0}^{\infty} z^n u_n \longrightarrow v(\zeta) = \sum_{n=0}^{\infty} \zeta^n u_n, \quad |\zeta| = 1,$$

and

$$\|u\|_{H^2(\mathcal{H})}^2 = \|v\|_{L^2(\mathcal{H})}^2 = \sum_{n=0}^{\infty} \|u_n\|^2.$$

The functions u and v related as above are also connected by Poisson's formula: for $0 \le r < 1$,

$$u(re^{it}) = \frac{1}{2\pi} \int_0^{2\pi} P_r(t-s)v(e^{is})ds,$$

where $P_r(t) = (1-r^2)(1-2r\cos t + r^2)^{-1}$ is the Poisson kernel. We also mention that a Fatou type result holds as well in the vector valued case. Thus, for u in $H^2(\mathcal{H})$, $u(z)$ converges strongly, almost everywhere, to $v(e^{it})$ if $|z| < 1$ and z converges non-tangentially to e^{it}.

Let $\mathcal{H}$ and $\mathcal{H}'$ be two separable Hilbert spaces. $L^\infty(\mathcal{L}(\mathcal{H},\mathcal{H}'))$ is the space of the Lebesgue strongly-measurable functions f on the unit circle with values in $\mathcal{L}(\mathcal{H},\mathcal{H}')$, with the property that there exists a constant M such that $\|f(e^{it})\| \le M$ almost everywhere. The norm of this space is defined by

$$\|f\|_{L^\infty(\mathcal{L}(\mathcal{H},\mathcal{H}'))} = \text{ess sup}\{\|f(e^{it})\| \mid t \in [0,2\pi)\},$$

where $\text{ess sup}\{\|f(e^{it})\| \mid t \in [0,2\pi)\}$ denotes the infimum of the real numbers $\lambda \ge 0$ such that the set $\{t \mid \|f(e^{it})\| > \lambda\}$ has Lebesgue measure zero. Another space of interest is $H^\infty(\mathcal{L}(\mathcal{H},\mathcal{H}'))$, the space of bounded analytic functions f on the unit disc with values in $\mathcal{L}(\mathcal{H},\mathcal{H}')$ and with the norm defined by

$$\|f\|_{H^\infty(\mathcal{L}(\mathcal{H},\mathcal{H}'))} = \sup\{\|f(z)\| \mid |z| < 1\}.$$

A Fatou type result holds and leads to an identification of the space $H^\infty(\mathcal{L}(\mathcal{H},\mathcal{H}'))$ as a closed subspace of $L^\infty(\mathcal{L}(\mathcal{H},\mathcal{H}'))$. We denote by $\mathcal{S}(\mathcal{L}(\mathcal{H},\mathcal{H}'))$ the *Schur class* of analytic functions in the unit disc with values contractions in $\mathcal{L}(\mathcal{H},\mathcal{H}')$. An element of $\mathcal{S}(\mathcal{L}(\mathcal{H},\mathcal{H}'))$ will be referred to as a *Schur function*. For two families $\mathbf{H} = \{\mathcal{H}_n\}_{n \in \mathbb{Z}}$ and $\mathbf{H}' = \{\mathcal{H}'_n\}_{n \in \mathbb{Z}}$ of Hilbert spaces, we denote by $\mathcal{S}(\mathbf{H},\mathbf{H}')$ the *Schur class* of the upper triangular contractions mapping $\oplus_{n \in \mathbb{Z}}\mathcal{H}_n$ into $\oplus_{n \in \mathbb{Z}}\mathcal{H}'_n$. For simplicity, we will use the notation $\mathcal{S}$ instead of $\mathcal{S}(\mathcal{L}(\mathbb{C},\mathbb{C}))$.

Also of interest is the space $H^2(\mathcal{L}(\mathcal{H},\mathcal{H}'))$ of the analytic functions F on the unit disc with values in $\mathcal{L}(\mathcal{H},\mathcal{H}')$ and with the property that for any h in $\mathcal{H}$,

$$\sup_{0 \le r < 1} \frac{1}{2\pi} \int_0^{2\pi} \|F(re^{it})h\|^2 dt \le M\|h\|^2.$$

We note that this is a Hilbert space and the previous condition is equivalent to

$$\sum_{n=0}^\infty \|F_n h\|^2 \le M\|h\|^2$$

for all h in $\mathcal{H}$, where F_n are the Taylor coefficients of F about the origin. We may also note that if $\{F_n\}_{n \ge 0}$ is a sequence of operators in $\mathcal{L}(\mathcal{H},\mathcal{H}')$ such that $\sum_{n=0}^\infty \|F_n h\|^2 \le M\|h\|^2$, then $F(z) = \sum_0^\infty z^n F_n$ defines a function in $H^2(\mathcal{H},\mathcal{H}')$. This follows from the inequality

$$\left\| \sum_{k=m}^n z^k F_k h \right\| \le \sum_{k=m}^n |z^k| \|F_k h\| \le (1-|z|^2)^{-1/2} \left(\sum_{k=m}^n \|F_k h\|^2 \right)^{1/2}$$

which is valid for $m \leq n$ and any h in $\mathcal{H}$. However, a basic distinction between $H^2(\mathcal{H})$ and $H^2(\mathcal{L}(\mathcal{H}, \mathcal{H}'))$ is the fact that the Fatou theorem is no longer true for $H^2(\mathcal{L}(\mathcal{H}, \mathcal{H}'))$. This is shown by the following example. Take $\mathcal{H} = H^2$, the scalar Hardy space, $\mathcal{H}' = \mathbb{C}$ and define

$$F(z)f = f(z), \quad f \in H^2, \quad |z| < 1.$$

We remark that $F(z)f = \sum_0^\infty z^n F_n(f)$, where $F_n(f)$ denotes the n-th Taylor coefficient of f. Hence F belongs to $H^2(\mathcal{L}(H^2, \mathbb{C}))$. Now, suppose that there exists $\sigma \subset [0, 2\pi)$, a set of zero Lebesgue measure such that if $t \notin \sigma$, then $\lim_{r \to 1} F(re^{it})f = \lim_{r \to 1} f(re^{it})$ exists for every f in H^2. Obviously, this is a contradiction.

Finally, we introduce the multiplication and Toeplitz operators. For a function f in $L^\infty(\mathcal{L}(\mathcal{H}, \mathcal{H}'))$, the *multiplication operator* M_f is defined from $L^2(\mathcal{H})$ into $L^2(\mathcal{H}')$ by the formula $M_f g = fg$ for g in $L^2(\mathcal{H})$ and one checks that $\|M_f\| = \|f\|_{L^\infty(\mathcal{L}(\mathcal{H}, \mathcal{H}'))}$. Usually, the function f is referred to as the *symbol* of M_f. If one defines $X_f = \Psi_{\mathcal{H}'} M_f \Psi_{\mathcal{H}}^*$ in $\mathcal{L}(l^2(\mathcal{H}), l^2(\mathcal{H}'))$, then one remarks that the matrix representation $X_f = [X_{mn}]_{m,n \in \mathbb{Z}}$, is given by

$$X_{mn} = \frac{1}{2\pi} \int_0^{2\pi} f(e^{it}) e^{-i(n-m)t} dt = X_{n-m},$$

and the diagonals of X_f are constant. Moreover, f belongs to $H^\infty(\mathcal{L}(\mathcal{H}, \mathcal{H}'))$ if and only if X_f has an upper triangular matrix representation with constant diagonals.

For a function f in $L^\infty(\mathcal{L}(\mathcal{H}, \mathcal{H}'))$, the *Toeplitz operator* T_f is defined from $H^2(\mathcal{H})$ into $H^2(\mathcal{H}')$ by the formula $T_f g = P_{H^2(\mathcal{H}')} fg$ for g in $H^2(\mathcal{H})$. A classical result relates the multiplication and the Toeplitz operators with the commutants of the unilateral and bilateral shifts. Thus, define $S_{\mathcal{H}} = M_{\nu_1}$, the multiplication by $\nu_1(e^{it}) = e^{it}$ on $L^2(\mathcal{H})$, and $S_{\mathcal{H}}^+ = T_{\nu_1}$ on $H^2(\mathcal{H})$. An operator B in $\mathcal{L}(L^2(\mathcal{H}), L^2(\mathcal{H}'))$ has the property that $BS_{\mathcal{H}} = S_{\mathcal{H}'} B$ if and only if $B = M_f$ for some f in $L^\infty(\mathcal{L}(\mathcal{H}, \mathcal{H}'))$ and $\|B\| = \|M_f\|$. An operator B in $\mathcal{L}(H^2(\mathcal{H}), L^2(\mathcal{H}'))$ has the property that $BS_{\mathcal{H}}^+ = S_{\mathcal{H}'} B$ if and only if $B = M_f/H^2(\mathcal{H})$ for some f in $L^\infty(\mathcal{L}(\mathcal{H}, \mathcal{H}'))$ and $\|B\| = \|M_f/H^2(\mathcal{H})\| = \|f\|_{L^\infty(\mathcal{L}(\mathcal{H}, \mathcal{H}'))}$. Then, an operator B in $\mathcal{L}(H^2(\mathcal{H}), H^2(\mathcal{H}'))$ has the property that $BS_{\mathcal{H}}^+ = S_{\mathcal{H}'}^+ B$ if and only if $B = T_f$ for some f in $H^\infty(\mathcal{L}(\mathcal{H}, \mathcal{H}'))$ and $\|B\| = \|T_f\| = \|f\|_{H^\infty(\mathcal{L}(\mathcal{H}, \mathcal{H}'))}$.

2.2 The Structure of Triangular Contractions

In this section we use the Schur parameters in order to describe the structure of the triangular contractions. We explore the simple connections between contractions and positive operators mentioned in the previous chapter and Theorem 1.5.3 is the main tool in these developments.

Consider two families $\mathbf{H} = \{\mathcal{H}_n\}_{n \in \mathbb{Z}}$ and $\mathbf{H}' = \{\mathcal{H}_n'\}_{n \in \mathbb{Z}}$ of Hilbert spaces, and the upper triangular contraction $T = [T_{ij}]_{i,j \in \mathbb{Z}}$ in $\mathcal{S}(\mathbf{H}, \mathbf{H}')$. Actually, we suppose without loss of generality and for a better match with the notation used in Section 1.5, that T is *strictly upper triangular*, *i.e.* $T_{ij} = 0$ for $i \geq j$. For $m, n \in \mathbb{Z}$, $m < n$, we define the block matrix contractions

$$T^{(mn)} : \oplus_{k=m+1}^{n} \mathcal{H}_k \longrightarrow \oplus_{k=m}^{n-1} \mathcal{H}'_k \tag{2.1}$$

$$T^{(mn)} = [T_{ij} \mid m \leq i \leq n - 1, \, m + 1 \leq j \leq n].$$

If $\Gamma = \{\Gamma_{ij} \mid i, j \in \mathbb{Z}, i \leq j\}$ belongs to $\Pi(\mathbf{H}, \mathbf{H}')$, then the definitions (1.5.3) and (1.5.4) of the row and column contractions $L_i^{(j)}$ and $C_j^{(i)}$ make sense and, in addition, we introduce the following operators: $Q_{nn} = 0_{\mathcal{H}'_n, \mathcal{H}_n}$ for $n \in \mathbb{Z}$ and for $i < j$,

$$Q_{ij} : \oplus_{k=-j}^{-i} \mathcal{D}_{\Gamma^*_{-k,j}} \longrightarrow \oplus_{k=i}^{j} \mathcal{D}_{\Gamma_{ik}} \tag{2.2}$$

$$Q_{ij} = (0_{\mathcal{H}'_i, \mathcal{H}_i} \oplus I_{\oplus_{k=i+1}^{j} \mathcal{D}_{\Gamma_{ik}}}) R_{ij} (Q_{i+1,j} \oplus I_{\mathcal{D}_{\Gamma^*_{ij}}}),$$

where R_{ij} denotes the unitary operator $R_{L_i^{(j)}}$ defined by (1.4.4). We are now ready to prove the main result of this section.

2.1 Theorem *Let T be a strictly upper triangular contraction in $\mathcal{S}(\mathbf{H}, \mathbf{H}')$. Then there exists a uniquely determined set $\Gamma = \{\Gamma_{ij} \mid i, j \in \mathbb{Z}, i \leq j\}$ of Schur parameters in $\Pi(\mathbf{H}, \mathbf{H}')$ such that $T_{i,i+1} = \Gamma_{i,i+1}$ for $i \in \mathbb{Z}$ and for $i, j \in \mathbb{Z}$, $j > i + 1$,*

$$T_{ij} = L_i^{(j-1)} Q_{i+1,j-1} C_j^{(i+1)} + D_{\Gamma^*_{i,i+1}} \cdots D_{\Gamma^*_{i,j-1}} \Gamma_{ij} D_{\Gamma_{i+1,j}} \cdots D_{\Gamma_{j-1,j}}. \tag{2.3}$$

Proof The idea of the proof is to apply Theorem 1.5.3 to the positive block matrices $C^{(mn)} = \begin{bmatrix} I & T^{(mn)} \\ (T^{(mn)})^* & I \end{bmatrix}$. We proceed by induction on $k = j - i > 0$ and show that for an arbitrary $i \in \mathbb{Z}$, the following statements hold:

$(\alpha)_k$ There exists a uniquely determined contraction $\Gamma_{i,i+k}$ from $\mathcal{D}_{\Gamma_{i+1,i+k}}$ to $\mathcal{D}_{\Gamma^*_{i,i+k-1}}$ satisfying (2.3) for $j = i + k$.

$(\beta)_k$ The Schur parameters $\Gamma_{sr}(C)$, $i \leq s \leq r \leq i + 2k - 1$, of $C^{(i,i+k)}$ are equal to zero if $i \leq s \leq i + 2k - 1$ and $s \leq r \leq \min\{s + k - 1, i + 2k - 1\}$, and otherwise $\Gamma_{sr}(C) = \Gamma_{s,r-k+1}$.

The case $k = 1$ follows easily and suppose that these statements are true for all p, $1 \leq p \leq k$. The positive matrix $C^{(i,i+k+1)}$ can be partitioned in order to exhibit the following matrix representation:

$$C^{(i,i+k+1)} = \begin{bmatrix} I & X & T_{i,i+k+1} \\ X^* & Y & Z \\ T^*_{i,i+k+1} & Z^* & I \end{bmatrix}.$$

It is a consequence of $(\beta)_k$ and Theorem 1.5.3 that the Schur parameters Γ'_{sr}, $i \leq s \leq r \leq i + 2k$, of the positive block matrix $\begin{bmatrix} I & X \\ X^* & Y \end{bmatrix}$ are equal to zero if $i \leq s \leq i + 2k$ and $s \leq r \leq \min\{s + k, i + 2k\}$, and otherwise $\Gamma'_{sr} = \Gamma_{s,r-k}$. Similarly, the Schur parameters Γ_{sr}'', $i+1 \le s \le r \le i+2k+1$, of the positive block matrix $\begin{bmatrix} Y & Z \\ Z^* & I \end{bmatrix}$ are equal to zero if $i+1 \le s \le i+2k+1$ and $s \le r \le \min\{s+k, i+2k+1\}$, and otherwise $\Gamma_{sr}'' = \Gamma_{s,r-k}$. As another consequence of Theorem 1.5.3, one obtains that

$$T_{i,i+k+1} = \begin{bmatrix} 0 & L_i^{(i+k)} \end{bmatrix} \bar{U}_{i+1,i+k} \begin{bmatrix} 0 & C_{i+k+1}^{(i+1)} \end{bmatrix}^{\top}$$
$$+ D_{\Gamma_{i,i+1}^*} \dots D_{\Gamma_{i,i+k}^*} \Gamma_{i,i+k+1} D_{\Gamma_{i+1,i+k+1}} \dots D_{\Gamma_{i+k,i+k+1}},$$

where $\bar{U}_{i+1,i+k}$ is the generalized rotation associated to the family of Schur parameters $\{\bar{\Gamma}_{sr} \mid i+1 \le s \le r \le i+2k\}$ given by

$$\bar{\Gamma}_{sr} = \begin{cases} 0 & \text{if } i+1 \le s \le i+2k \text{ and} \\ & s \le r \le \min\{s+k, i+2k\} \\ \Gamma_{s,r-k} & \text{otherwise.} \end{cases}$$

The proof can be concluded by showing that

$$\bar{U}_{i+1,i+k} = \begin{bmatrix} * & * \\ * & Q_{i+1,i+k} \end{bmatrix}, \tag{2.4}$$

where the entries of $\bar{U}_{i+1,i+k}$ marked by $*$ play no role here. It is convenient to introduce a special notation for the operator $R_n(\Gamma_k)$ defined by (1.4.5) in the case $\Gamma_k = 0$. Thus, this operator is denoted by $t(k, k+1)$, in order to emphasize the fact that it switches the positions k and $k+1$. According to the definition of $\bar{U}_{i+1,i+k}$, it is possible to write $\bar{U}_{i+1,i+k} = u_1 u_2 \dots u_{k-1} u_k$, where

$$u_j = t(1,2)t(2,3) \dots t(k, k+1) R_{2k-1}(\bar{\Gamma}_{i+j,i+k+j+1}) \dots R_{2k-1}(\bar{\Gamma}_{i+j,i+2k})$$

for $1 \le j \le k-1$, and

$$u_k = (t(1,2)t(2,3) \dots t(k, k+1))(t(1,2) \dots t(k-1, k)) \dots t(1,2).$$

Since each factor u_j, $1 \le j \le k$, contains the transposition $t(k, k+1)$, it follows that the operator $\bar{U}_{i+1,i+k}$ has the block matrix representation $\begin{bmatrix} A & B \\ C & D \end{bmatrix}$, where D is a $k \times k$ block matrix such that $D = v_1 \dots v_k$ with $v_j = (0_{\mathcal{H}'_{i+j}, \mathcal{H}_{i+j}} \oplus I) R_{k-1}(\Gamma_{i+j,i+j+1}) \dots R_{k-1}(\Gamma_{i+j,i+k})$ for $1 \le j \le k-1$. It follows from (2.2) that $D = Q_{i+1,i+k}$ and the equality (2.4) is proved. Consequently, the statement $(\alpha)_{k+1}$ holds and $(\beta)_{k+1}$ is also obvious by now. The proof is complete. $\square$

2.2 Remark We may note that taking into account the Cholesky factorization of $C^{(mn)}$ we may produce the Cholesky factorization of $I - (T^{(mn)})^* T^{(mn)}$. Indeed, if F_{mn} is the Cholesky operator of $C^{(mn)}$, then $C^{(mn)} = F_{mn}^* F_{mn}$. Using the structure of the Schur parameters of $C^{(mn)}$ as explained in the proof of Theorem 2.1, we get

$$F_{mn} = \begin{bmatrix} I & X^{(mn)} \\ 0 & Z_{mn} \end{bmatrix}$$

for certain block matrices Z_{mn} and $X^{(mn)}$. On the other hand, we remark that

$$C^{(mn)} = \begin{bmatrix} I & 0 \\ (T^{(mn)})^* & D_{T^{(mn)}} \end{bmatrix} \begin{bmatrix} I & T^{(mn)} \\ 0 & D_{T^{(mn)}} \end{bmatrix}$$

by Lemma 1.2.1, hence $X^{(mn)} = T^{(mn)}$ and $I - (T^{(mn)})^* T^{(mn)} = Z_{mn}^* Z_{mn}$. Since Z_{mn} is upper triangular, it is just the Cholesky factor of $I - (T^{(mn)})^* T^{(mn)}$. We also obtain the formula:

$$Z_{mn} = \begin{bmatrix} Z_{m,n-1} & \begin{bmatrix} 0 & I_{\oplus_{k=m+1}^n D_{\Gamma_{mk}}} \end{bmatrix} Q_{m,n-1} C_n^{(m)} \\ 0 & D_{\Gamma_{mn}} \dots D_{\Gamma_{n-1,n}} \end{bmatrix}. \tag{2.5}$$

In the case that the spaces $\mathcal{H}_n$ and $\mathcal{H}_n'$, $n \in \mathbb{Z}$, are Hilbert spaces of finite dimension we deduce as a consequence of (2.5) (or, directly from Theorem 1.5.10) that

$$\det(I - (T^{(mn)})^* T^{(mn)}) = \prod_{m \le i < j \le n} \det D_{\Gamma_{ij}}^2. \qquad \square$$

It is useful to point out some particular cases of Theorem 2.1. The first one is a consequence of Proposition 1.5.6 and Theorem 2.1, while the second one follows from Proposition 1.5.8 and Theorem 2.1.

2.3 Proposition *Let T be a strictly upper triangular contraction and let Γ be the set of its Schur parameters. The following are equivalent:*

(a) *T is a strictly upper triangular Toeplitz contraction.*

(b) *$\Gamma_{ij} = \Gamma_{i+k,j+k}$ for $i, j \in \mathbb{Z}$, $i \le j$ and $k \in \mathbb{Z}$.* $\qquad \square$

2.4 Proposition *Let T be a strictly upper triangular contraction and let Γ be the set of its Schur parameters. The following are equivalent:*

(a) *$T_{ij} = T_{i,i+1} T_{i+1,i+2} \dots T_{j-1,j}$ for $j > i$.*

(b) *$\Gamma_{i,i+1} = T_{i,i+1}$ for $i \in \mathbb{Z}$ and $\Gamma_{ij} = 0$ for $j > i + 1$.* $\qquad \square$

2.5 Remark We conclude this section with a discussion of the block matrix contractions of the form $T = [T_{ij} \mid 1 \le i \le m, 1 \le j \le n]$. We see that their structure can be easily described using Theorem 2.1. Indeed, we consider the upper triangular operator $T' = [T_{ij}']_{i,j \in \mathbb{Z}}$, such that $T_{ij}' = T_{1-i,j}$ for $1 - m \le i \le 0$, $1 \le j \le n$ and otherwise $T_{ij}' = 0$. T' is a contraction if and only if T is a contraction and, by Theorem 2.1, T' is completely determined by a family $\{\Gamma_{ij}' \mid i, j \in \mathbb{Z}, i \le j\}$ of contractions such that $\Gamma_{ij}' = 0$ for $(i,j) \notin \{1 - m, 2 - m, \dots, 0\} \times \{1, 2, \dots, n\}$. It is useful to rename the parameters Γ_{ij}' by defining $\Gamma_{ij} = \Gamma_{1-i,j}'$ for $1 \le i \le m$, $1 \le j \le n$ and to emphasize the dependence of T on these parameters, by writing $T = T(\{\Gamma_{ij} \mid 1 \le i \le m, 1 \le j \le n\})$.

As an example, consider the block matrix $T = \begin{bmatrix} T_{11} & T_{12} \\ T_{21} & T_{22} \end{bmatrix}$. It follows from the preceding discussion that T is a contraction if and only if $T_{11} = \Gamma_{11}$, $T_{12} = D_{\Gamma_{11}^*} \Gamma_{12}$, $T_{21} = \Gamma_{21} D_{\Gamma_{11}}$ and

$$T_{22} = -\Gamma_{21} \Gamma_{11}^* \Gamma_{12} + D_{\Gamma_{21}^*} \Gamma_{22} D_{\Gamma_{12}}, \tag{2.6}$$

where Γ_{11}, Γ_{12}, Γ_{21} and Γ_{22} are uniquely determined contractions which belong to $\mathcal{L}(\mathcal{H}_1, \mathcal{H}_1')$, $\mathcal{L}(\mathcal{H}_2, \mathcal{D}_{\Gamma_{11}^*})$, $\mathcal{L}(\mathcal{D}_{\Gamma_{11}}, \mathcal{H}_2')$ and, respectively, to $\mathcal{L}(\mathcal{D}_{\Gamma_{12}}, \mathcal{D}_{\Gamma_{21}^*})$. We remark that this result about the structure of the 2×2 block matrices has the following consequence: let T_{11} in $\mathcal{L}(\mathcal{H}_1, \mathcal{H}_1')$, T_{12} in $\mathcal{L}(\mathcal{H}_2, \mathcal{H}_1')$ and T_{21} in $\mathcal{L}(\mathcal{H}_1, \mathcal{H}_2')$ be given operators, then

$$\inf\{\| \begin{bmatrix} T_{11} & T_{12} \\ T_{21} & X \end{bmatrix} \| \mid X \in \mathcal{L}(\mathcal{H}_2, \mathcal{H}_2')\} = \max\{\| \begin{bmatrix} T_{11} \\ T_{21} \end{bmatrix} \|, \| [T_{11} \quad T_{12}] \|\}. \quad (2.7)$$

Indeed, the fact that the left hand side of (2.7) is greater than or equal to the right hand side is obvious, while the reverse inequality is a consequence of (2.6). More precisely, according to (2.6), there exists an operator X such that $\begin{bmatrix} T_{11} & T_{12} \\ T_{21} & X \end{bmatrix}$ is a contraction provided the block-matrices $\begin{bmatrix} T_{11} \\ T_{21} \end{bmatrix}$ and $[T_{11} \quad T_{12}]$ are contractions.

$\square$

2.3 Realization of Triangular Contractions

At this stage we have the opportunity to connect our considerations on the structure of the positive definite kernels and triangular contractions with a general problem in system theory. There are two different ways to describe the evolution of a given process. First, there is an external, input/output description for which only outputs produced by given inputs are available. A second description is internal and it assumes the knowledge of a state-space model. For the class of *discrete time linear systems*, the *state-space model* is given by the following equations:

$$\Omega \begin{cases} x(t) = A(t)x(t+1) + B(t)u(t) \\ y(t) = C(t)x(t+1) + D(t)u(t), \quad t \in \mathbb{Z}, \end{cases} \quad (3.1)$$

where $\mathbf{U} = \{\mathcal{U}(t)\}_{t \in \mathbb{Z}}$ and $\mathbf{V} = \{\mathcal{V}(t)\}_{t \in \mathbb{Z}}$ are given families of Hilbert spaces, referred to as the *input spaces* and, respectively, *output spaces*. Moreover, it is given another family $\mathbf{H} = \{\mathcal{H}(t)\}_{t \in \mathbb{Z}}$ of Hilbert spaces which are referred to as the *state spaces*. The operators $A(t)$ in $\mathcal{L}(\mathcal{H}(t+1), \mathcal{H}(t))$, $B(t)$ in $\mathcal{L}(\mathcal{U}(t), \mathcal{H}(t))$, $C(t)$ in $\mathcal{L}(\mathcal{H}(t+1), \mathcal{Y}(t))$ and $D(t)$ in $\mathcal{L}(\mathcal{U}(t), \mathcal{Y}(t))$ are also given and we will usually refer to the system (3.1) by writing $\Omega = (A(t), B(t), C(t), D(t))_{t \in \mathbb{Z}}$.

The evolution of the system (3.1) can be easily described. Starting with the moment $t_0 \in \mathbb{Z}$, we obtain for $t < t_0$ that

$$y(t) = C(t)A(t+1)A(t+2)\dots A(t_0)x(t_0)$$
$$+ \sum_{k=0}^{t_0-t-2} C(t)A(t+1)\dots A(t_0 - k - 1)B(t_0 - k)u(t_0 - k)$$
$$+ C(t)B(t+1)u(t+1) + D(t)u(t).$$

This equality shows that the external behavior of the system Ω is described by the *transfer map* defined formally as follows:

$$T_\Omega : \oplus_{t \in \mathbb{Z}} \mathcal{U}(t) \longrightarrow \oplus_{t \in \mathbb{Z}} \mathcal{Y}(t) \quad (3.2)$$

$$T_\Omega = [T_{ij}]_{i,j\in\mathbb{Z}},$$

where $T_{ij} = 0$ for $i > j$ and

$$T_{ij} = \begin{cases} D(i) & \text{if } j = i \\ C(i)B(i+1) & \text{if } j = i+1 \\ C(i)A(i+1)\dots A(j-1)B(j) & \text{if } j > i+1. \end{cases}$$

The operators T_{ij} are called the *Markov parameters* of the system Ω. We conclude from the preceding discussion that going from the internal to the external description is straightforward. The inverse problem of deriving the internal description from the Markov parameters is usually referred to as the *realization problem* and play a key role in system theory. In this section we address the realization problem for the class of *unitary systems*, *i.e.* for the systems Ω with the property that all $\begin{bmatrix} A(t) & B(t) \\ C(t) & D(t) \end{bmatrix}$, $t \in \mathbb{Z}$, are unitary block matrices. The transfer maps of these systems belong to a familiar class.

3.1 Lemma   *The transfer map of a unitary system belongs to $\mathcal{S}(\mathbf{U},\mathbf{V})$.*

Proof   Take t_0 arbitrary in $\mathbb{Z}$ and let $\{y(t)\}_{t<t_0}$ be the output generated by the unitary system Ω from the input $\{u(t)\}_{t<t_0}$ supposing $x(t_0) = 0$. By the assumption that Ω is a unitary system, it follows that

$$\|x(t)\|^2 + \|y(t)\|^2 = \|x(t+1)\|^2 + \|u(t)\|^2$$

for $t < t_0$. By induction, we deduce that

$$0 \le \|x(t)\|^2 = \sum_{k=t_0-1}^{t} \|u(k)\|^2 - \sum_{k=t_0-1}^{t} \|y(k)\|^2$$

for $t < t_0$. Since these inequalities hold for arbitrary $t_0 \in \mathbb{Z}$ and arbitrary $t < t_0$, it follows that the transfer map T_Ω of Ω is a contraction, hence T_Ω belongs to $\mathcal{S}(\mathbf{U},\mathbf{V})$. $\square$

We can solve the realization problem for unitary systems using the multiplicative structure of the Kolmogorov decomposition and the structure of the upper triangular contractions established in Theorem 2.1.

3.2 Theorem   *Let T be an upper triangular contraction in $\mathcal{S}(\mathbf{U},\mathbf{V})$. Then there exists a unitary system Ω such that $T_\Omega = T$.*

Proof   In order to be in tune with the notation used in Theorem 2.1 we assume, without loss of generality, that T is strictly upper triangular. We have to prove that in this case there exist families $\{A^0(t)\}_{t\in\mathbb{Z}}$, $\{B^0(t)\}_{t\in\mathbb{Z}}$, $\{C^0(t)\}_{t\in\mathbb{Z}}$ and $\{D^0(t)\}_{t\in\mathbb{Z}}$ of suitably defined operators such that, for $t \in \mathbb{Z}$, $T_{t,t+1} = D^0(t)$, $T_{t,t+2} = C^0(t)B^0(t+1)$, for $i, j \in \mathbb{Z}$, $i+2 < j$,

$$T_{ij} = C^0(i)A^0(i+1)\dots A^0(j-2)B^0(j-1) \tag{3.3}$$

and $\begin{bmatrix} A^0(t) & B^0(t) \\ C^0(t) & D^0(t) \end{bmatrix}$ are unitary block matrices. Let $\Gamma = \{\Gamma_{ij} \mid i,j \in \mathbb{Z}, i \le j\}$ be the set of the Schur parameters associated to T. Define, for $i \in \mathbb{Z}$, the row contractions

$$L_i : \oplus_{k=i+1}^{\infty} \mathcal{D}_{\Gamma_{i+1,k}} \longrightarrow \mathcal{Y}_i$$

$$L_i = \quad \text{the row contraction associated to}$$
$$\text{the parameters} \quad \{\Gamma_{ik} \mid i < k\}.$$

Moreover, define $\mathcal{D}_i = \oplus_{k=i+1}^{\infty} \mathcal{D}_{\Gamma_{ik}}$ and $\mathcal{D}_{i,*} = \mathrm{cl}\mathcal{R}(H_\infty(L_i))$, where $H_\infty(L_i)$ was introduced by (1.4.10). Consider the Hilbert spaces $\mathcal{K}_i = \oplus_{j=-\infty}^{i-1} \mathcal{D}_{j,*} \oplus \mathcal{U}(i) \oplus \mathcal{D}_i$ and $\mathcal{K}_i' = \oplus_{j=-\infty}^{i-1} \mathcal{D}_{j,*} \oplus \mathcal{Y}(i) \oplus \mathcal{D}_i$, and define the unitary operators

$$W_i : \mathcal{K}_{i+1} \longrightarrow \mathcal{K}_i'$$

$$W_i = I \oplus \begin{bmatrix} I & 0 \\ 0 & \alpha_{L_i} \end{bmatrix} R(L_i) \begin{bmatrix} 0 & I \\ \beta_{L_i}^* & 0 \end{bmatrix},$$

with respect to the decompositions $\mathcal{K}_{i+1} = (\oplus_{j=-\infty}^{i-1} \mathcal{D}_{j,*}) \oplus (\mathcal{D}_{i,*} \oplus \mathcal{U}(i+1) \oplus \mathcal{D}_{i+1})$ and, respectively, $\mathcal{K}_i' = (\oplus_{j=-\infty}^{i-1} \mathcal{D}_{j,*}) \oplus (\mathcal{Y}(i) \oplus \mathcal{D}_i)$. The symbol $R(L_i)$ denotes the elementary rotation of L_i, while α_{L_i} and β_{L_i} are the unitary operators defined by (1.4.11) and (1.4.12). Finally, define for $t \in \mathbb{Z}$, the spaces

$$\mathcal{H}(t) = \mathcal{K}_t' \ominus \mathcal{Y}(t) = \mathcal{K}_t \ominus \mathcal{U}(t) \tag{3.4}$$

and the operators

$$\begin{cases} A^0(t) &= P_{\mathcal{H}(t)}^{\mathcal{K}_t'} W_t / \mathcal{H}(t+1) \\ B^0(t) &= P_{\mathcal{H}(t)}^{\mathcal{K}_t'} W_t / \mathcal{U}(t) \\ C^0(t) &= P_{\mathcal{Y}(t)}^{\mathcal{K}_t'} W_t / \mathcal{H}(t+1) \\ D^0(t) &= \Gamma_{t,t+1}. \end{cases} \tag{3.5}$$

Obviously, $T_{t,t+1} = D^0(t)$ and $T_{t,t+2} = C^0(t)B^0(t+1)$ for all $t \in \mathbb{Z}$. By the formula (1.6.4), it follows that $T_{mn} = P_{\mathcal{Y}(m)} \bar{V}_{mn} / \mathcal{U}(n)$, where $\bar{V}_{mn}$ is the generalized rotation associated to the Schur parameters $\bar{\Gamma}_{sr}$, $m \le s \le r \le 2n - m - 1$, given by

$$\bar{\Gamma}_{sr} = \begin{cases} 0 & \text{if} \quad m \le s \le 2n - m - 1 \quad \text{and} \\ & s \le r \le \min\{s + n - m - 1, 2n - m - 1\} \\ \Gamma_{s,r-n+m+1} & \text{otherwise.} \end{cases}$$

By the definition of $\bar{V}_{mn}$, it follows that $\bar{V}_{mn} = u_1 u_2 \ldots u_{n-m-1}, u_{n-m}$, where $u_1, \ldots, u_{n-m}$ are operators as those introduced in the proof of Theorem 1.2, and $\bar{u}_1 = P_{\mathcal{Y}(m)} u_1, \bar{u}_{n-m} = u_{n-m} / \mathcal{U}(n-m)$. It is easy to check that

$$u_1 =$$
$$\begin{bmatrix} 0 & \Gamma_{m,m+1} & D_{\Gamma_{m,m+1}^*} \Gamma_{m,m+2} & \cdots & D_{\Gamma_{m,m+1}^*} & \cdots & D_{\Gamma_{m,n-1}^*} \Gamma_{mn} & D_{\Gamma_{m,m+1}^*} & \cdots & D_{\Gamma_{m,n}^*} \end{bmatrix}$$

and

$$\bar{u}_{n-m} = [\, \Gamma_{n-1,n} \quad 0_{n-1} \quad D_{\Gamma_{n-1,n}} \quad 0_{n-1} \,]^{\top}.$$

Since each factor u_j, $2 \leq j < n - m$, contains the transpositions $t(1, 2)$, $t(2, 3)$, ..., $t(m - 1, m)$, it follows that (3.3) holds for $i, j \in \mathbb{Z}$, $i + 2 < j$. This concludes the proof. $\qquad\square$

It is also of interest to clarify a certain uniqueness property of the construction in the proof of Theorem 3.2. Let $\Omega = (A(t), B(t), C(t), D(t))_{t \in \mathbb{Z}}$ be a unitary system and define for $t \in \mathbb{Z}$, the following *controllability* and, respectively, *observability* maps:

$$\mathcal{C}(t) : \oplus_{k=0}^{\infty} \mathcal{U}(t + k) \longrightarrow \mathcal{H}(t) \tag{3.6}$$

$$\mathcal{C}(t) = [\, B(t) \quad A(t)B(t+1) \quad A(t)A(t+1)B(t+2) \quad \dots \,]$$

and

$$\mathcal{O}(t) : \oplus_{k=-\infty}^{-1} \mathcal{Y}(t + k) \longrightarrow \mathcal{H}(t) \tag{3.7}$$

$$\mathcal{O}(t) = [\, \dots \quad A^{*}(t-1)A^{*}(t-2)C^{*}(t-3) \quad A^{*}(t-1)C^{*}(t-2) \quad C^{*}(t-1) \,].$$

It is easily checked that these maps are well defined contractions and that the unitary system built in the proof of Theorem 3.2 has the property that

$$\mathrm{cl}\mathcal{R}(\mathcal{C}(t)) \vee \mathrm{cl}\mathcal{R}(\mathcal{O}(t)) = \mathcal{H}(t) \tag{3.8}$$

for all $t \in \mathbb{Z}$. A unitary system satisfying (3.8) is called *closely connected*. The significance of this notion comes from the following result.

3.3 Proposition *Let $\Omega_j = (A_j(t), B_j(t), C_j(t), D_j(t))_{t \in \mathbb{Z}}$, $j = 1, 2$, be two closely connected unitary systems. The following are equivalent.*

(a) Ω_1 and Ω_2 have the same transfer map.

(b) $D_1(t) = D_2(t)$ for all $t \in \mathbb{Z}$ and there exists a family $\{\tau(t)\}_{t \in \mathbb{Z}}$ of unitary operators $\tau(t) \in \mathcal{L}(\mathcal{H}_1(t), \mathcal{H}_2(t))$ such that for all $t \in \mathbb{Z}$,

$$\tau(t)A_1(t) = A_2(t)\tau(t+1), \quad \tau(t)B_1(t) = B_2(t), \quad C_1(t) = C_2(t)\tau(t+1).$$

Proof One implication is obvious. Conversely, suppose the two closely connected unitary systems have the same transfer map. It follows that $D_1(t) = D_2(t)$ for all $t \in \mathbb{Z}$. Further, we define the maps:

$$\tau(t) : \mathcal{H}_1(t) \longrightarrow \mathcal{H}_2(t) \tag{3.9}$$

$$\tau(t)(\mathcal{C}_1(t)u + \mathcal{O}_1(t)y) = \mathcal{C}_2(t)u + \mathcal{O}_2(t)y,$$

where u belongs to $\oplus_{k=0}^{\infty} \mathcal{U}(t - k)$ and y belongs to $\oplus_{k=-\infty}^{-1} \mathcal{Y}(t + k)$. We remark that

$$\|\mathcal{C}_2(t)u + \mathcal{O}_2(t)y\|^2 = \|\mathcal{C}_2(t)u\|^2 + 2\mathrm{Re}\langle \mathcal{C}_2(t)u, \mathcal{O}_2(t)y \rangle + \|\mathcal{O}_2(t)y\|^2$$

and it is a matter of same simple computations to verify that $\mathcal{C}_2^*(t)\mathcal{C}_2(t) = \mathcal{C}_1^*(t)\mathcal{C}_1(t)$, $\mathcal{O}_2^*(t)\mathcal{C}_2(t) = \mathcal{O}_1^*(t)\mathcal{C}_1(t)$ and $\mathcal{O}_2^*(t)\mathcal{O}_2(t) = \mathcal{O}_1^*(t)\mathcal{O}_1(t)$. Consequently,

$$\|\mathcal{C}_2(t)u + \mathcal{O}_2(t)y\| = \|\mathcal{C}_1(t)u + \mathcal{O}_1(t)y\|,$$

which implies that the map $\tau(t)$ is well defined and extends to a unitary operator, also denoted by $\tau(t)$, between the state-spaces $\mathcal{H}_1(t)$ and $\mathcal{H}_2(t)$ of the two considered systems. It is easy to check that these operators have the required properties.
□

2.4 Unitary Couplings and Operator Ranges

In the preceding section we have seen that upper triangular contractions appear naturally as transfer functions of unitary systems. This fact may be developed further in order to show connections with two models of non-selfadjoint operators. For this purpose it is convenient to introduce the notion of unitary coupling and show the connection with the construction of renormed Hilbert spaces introduced in Section 1.2 and its relation with the operator ranges. We begin with an analysis of the structure of families of isometries.

Let $\mathbf{L} = \{\mathcal{L}_n\}_{n\in\mathbb{Z}}$ be a family of Hilbert spaces and define *marking operators* as follows: for $n \in \mathbb{Z}$,

$$M_{\mathbf{L}}(n) = M_{\{\mathcal{L}_k\}_{k\in\mathbb{Z}}}(n) : \oplus_{k=n+1}^{\infty}\mathcal{L}_k \longrightarrow \oplus_{k=n}^{\infty}\mathcal{L}_k \qquad (4.1)$$

$$M_{\mathbf{L}}(n)(\oplus_{k=n+1}^{\infty}l_k) = 0 \oplus \oplus_{k=n+1}^{\infty}l_k.$$

It is readily checked that the operators $M_{\mathbf{L}}(n)$ are isometries. Let $\mathbf{E} = \{\mathcal{E}_n\}_{n\in\mathbb{Z}}$ be a family of Hilbert spaces and let $\mathbf{v} = \{v_n\}_{n\in\mathbb{Z}}$ be a family of isometries v_n in $\mathcal{L}(\mathcal{E}_{n+1}, \mathcal{E}_n)$. We say that $\mathbf{v}$ is a *marking family* if there exist unitary operators Φ_n in $\mathcal{L}(\mathcal{E}_n, \oplus_{k=n}^{\infty}\mathcal{L}_k)$ such that $\Phi_n v_n = M_{\mathbf{L}}(n)\Phi_{n+1}$ for all $n \in \mathbb{Z}$. The role of the marking operators is explained by the following *Wold-von Neumann decomposition* of families of isometries.

4.1 Proposition *Let $\mathbf{v} = \{v_n\}_{n\in\mathbb{Z}}$ be an arbitrary family of isometries v_n in $\mathcal{L}(\mathcal{E}_{n+1}, \mathcal{E}_n)$. Then there exists a Wold-von Neumann decomposition of the form $v_n = v_n^{(1)} \oplus v_n^{(2)}$ for all $n \in \mathbb{Z}$, such that $\{v_n^{(1)}\}_{n\in\mathbb{Z}}$ is a marking family and $\{v_n^{(2)}\}_{n\in\mathbb{Z}}$ is a family of unitary operators.*

Proof Define for $n \in \mathbb{Z}$ the spaces $\mathcal{L}_n = \mathcal{E}_n \ominus v_n\mathcal{E}_{n+1}$ and remark that

$$\mathcal{E}_n = (\mathcal{L}_n \oplus v_n\mathcal{L}_{n+1} \oplus \ldots \oplus v_n \ldots v_{n+p-1}\mathcal{L}_{n+p}) \oplus v_n v_{n+1} \ldots v_{n+p}\mathcal{E}_{n+p+1}$$

for every $p \geq 0$. Since $\{v_n v_{n+1} \ldots v_{n+p}\mathcal{E}_{n+p+1}\}_{p\geq 0}$ is a decreasing sequence of subspaces of $\mathcal{E}_n$, we define the *residual subspaces* $\mathcal{R}_n = \cap_{p\geq 0} v_n v_{n+1} \ldots v_{n+p}\mathcal{E}_{n+p+1}$ and obtain the decomposition $\mathcal{E}_n = \oplus_{k=0}^{\infty} v_n v_{n+1} \ldots v_{n+k-1}\mathcal{L}_{n+k} \oplus \mathcal{R}_n$ (with the convention that for $k = 0$, the space $v_n v_{n+1} \ldots v_{n+k-1}\mathcal{L}_{n+k}$ should be interpreted as $\mathcal{L}_n$). Then, let us remark that $v_n\mathcal{R}_{n+1} = \mathcal{R}_n$ and

$$v_n(\oplus_{k=0}^{\infty} v_{n+1}v_{n+2} \ldots v_{n+k}\mathcal{L}_{n+k+1}) \subset \oplus_{k=0}^{\infty} v_n v_{n+1} \ldots v_{n+k-1}\mathcal{L}_{n+k},$$

therefore $v_n^{(2)} = v_n/\mathcal{R}_{n+1}$ belongs to $\mathcal{L}(\mathcal{R}_{n+1}, \mathcal{R}_n)$ and it is a unitary operator. Finally, define the unitary operators Φ_n from $\oplus_{k=0}^{\infty} v_n v_{n+1} \cdots v_{n+k-1} \mathcal{L}_{n+k}$ onto $\oplus_{k=0}^{\infty} \mathcal{L}_{n+k}$ by mapping $v_n v_{n+1} \cdots v_{n+k-1} l_{n+k}$ onto l_{n+k}, where l_{n+k} belongs to $\mathcal{L}_{n+k}$ for every $k \geq 0$. Then, $\{v_n/(\mathcal{E}_{n+1} \ominus \mathcal{R}_{n+1})\}_{n \in \mathbb{Z}}$ is a marking family, *i.e.* $\Phi_n v_n/(\mathcal{E}_{n+1} \ominus \mathcal{R}_{n+1}) = M_{\mathbf{L}}(n)\Phi_{n+1}$ for all $n \in \mathbb{Z}$. $\qquad\square$

4.2 Remark　A similar Wold-von Neumann decomposition holds for families $\mathbf{w} = \{w_n\}_{n \in \mathbb{Z}}$ of isometries w_n in $\mathcal{L}(\mathcal{E}_n, \mathcal{E}_{n+1})$. In this case, we define the spaces $\mathcal{L}_n = \mathcal{E}_n \ominus w_{n-1}\mathcal{E}_{n-1}$ and obtain the decomposition

$$\mathcal{E}_n = \oplus_{k=-\infty}^0 w_{n-1}w_{n-2}\cdots w_{n+k}\mathcal{L}_{n+k} \oplus \mathcal{R}_n,$$

where $\mathcal{R}_n = \cap_{p \geq 1} w_{n-1} \cdots w_{n-p}\mathcal{E}_{n-p}$ (if $k = 0$, then the space $w_{n-1}w_{n-2}\cdots$ $w_{n+k}\mathcal{L}_{n+k}$ should be interpreted as $\mathcal{L}_n$). We introduce another family of *marking operators* by defining, for $n \in \mathbb{Z}$,

$$N_{\mathbf{L}}(n) = N_{\{\mathcal{L}_k\}_{k \in \mathbb{Z}}}(n) : \oplus_{k \leq n-1}\mathcal{L}_k \longrightarrow \oplus_{k \leq n}\mathcal{L}_k \qquad (4.2)$$

$$N_{\mathbf{L}}(n)(\oplus_{k<n} l_k) = \oplus_{k<n} l_k \oplus 0.$$

We see that $N_{\mathbf{L}}(n)$ are isometries and we have the Wold-von Neumann decomposition $w_n = w_n^{(1)} \oplus w_n^{(2)}$ for all $n \in \mathbb{Z}$, where $w_n^{(2)} = w_n/\mathcal{R}_n$ and $w_n^{(1)} = w_n/(\mathcal{E}_n \ominus \mathcal{R}_n)$. We remark that $\{w_n^{(2)}\}_{n \in \mathbb{Z}}$ is a family of unitary operators and there exists a family of unitary operators Φ_n from $\mathcal{E}_n \ominus \mathcal{R}_n$ onto $\oplus_{k \leq n}\mathcal{L}_k$, such that $\Phi_{n+1}w_n/(\mathcal{E}_n \ominus \mathcal{R}_n) = N_{\mathbf{L}}(n+1)\Phi_n$ for all $n \in \mathbb{Z}$. $\qquad\square$

We now discuss the notion of unitary coupling. Suppose there are given two families $\mathbf{w}_+ = \{w_n^+\}_{n \in \mathbb{Z}}$ and $\mathbf{w}_- = \{w_n^-\}_{n \in \mathbb{Z}}$ of isometries w_n^+ in $\mathcal{L}(\mathcal{K}_{n+1}^+, \mathcal{K}_n^+)$ and, respectively, w_n^- in $\mathcal{L}(\mathcal{K}_n^-, \mathcal{K}_{n+1}^-)$. A family $\mathbf{u} = \{u_n\}_{n \in \mathbb{Z}}$ of unitary operators u_n in $\mathcal{L}(\mathcal{K}_{n+1}, \mathcal{K}_n)$ is called a *unitary coupling* of $\mathbf{w}_+$ and $\mathbf{w}_-$ if, for all $n \in \mathbb{Z}$, $\mathcal{K}_n^{\pm} \subset \mathcal{K}_n$ and $u_n/\mathcal{K}_{n+1}^+ = w_n^+$, $u_n^{-1}/\mathcal{K}_n^- = w_n^-$. Taking into account the Wold-von Neumann decompositions of $\mathbf{w}_+$ and $\mathbf{w}_-$, we obtain the equalities $\mathcal{K}_n^+ = \oplus_{k=0}^{\infty} w_n^+ w_{n+1}^+ \cdots w_{n+k-1}^+ \mathcal{L}_{n+k}^+ \oplus \mathcal{R}_n^+$ and $\mathcal{K}_n^- = \oplus_{k=-\infty}^0 w_{n-1}^- w_{n-2}^- \cdots w_{n+k}^- \mathcal{L}_{n+k}^- \oplus \mathcal{R}_n^-$, involving the elements introduced in Proposition 4.1 and Remark 4.2. In addition, it is supposed that:

(Q)　The spaces $\oplus_{k=-\infty}^{-1} u_{n-1}^* u_{n-2}^* \cdots u_{n+k}^* \mathcal{L}_{n+k}^+$ and $\oplus_{k=1}^{-\infty} u_n u_{n+1} \cdots u_{n+k-1}\mathcal{L}_{n+k}^-$ are orthogonal in $\mathcal{K}_n$ for all $n \in \mathbb{Z}$.

We are led to consider the following subspaces of $\mathcal{K}_n$:

$$\mathcal{K}_n^{out} = \oplus_{k=-\infty}^{-1} u_{n-1}^* u_{n-2}^* \cdots u_{n+k}^* \mathcal{L}_{n+k}^+ \oplus (\mathcal{K}_n^+ \ominus \mathcal{R}_n^+) \qquad (4.3)$$

and

$$\mathcal{K}_n^{in} = (\mathcal{K}_n^- \ominus \mathcal{R}_n^-) \oplus \oplus_{k=1}^{\infty} u_n u_{n+1} \cdots u_{n+k-1}\mathcal{L}_{n+k}^-. \qquad (4.4)$$

Moreover, define the subspaces $\mathcal{D}_n^+ = u_n^* \mathcal{L}_n^+$, $\mathcal{D}_{n-1}^- = u_{n-1}\mathcal{L}_n^-$ of $\mathcal{K}_n$, and the operators

$$\pi_n^{\pm} : \mathcal{N}_{\pm} = \oplus_{k \in \mathbb{Z}} \mathcal{D}_k^{\pm} \longrightarrow \mathcal{K}_n \qquad (4.5)$$

$$\pi_n^{\pm}(\oplus_{k\in\mathbb{Z}}d_k^{\pm}) = \oplus_{k=-\infty}^{-1}u_{n-1}^*u_{n-2}^*\cdots u_{n+k}^*d_{n+k}^{\pm} \oplus \oplus_{k=0}^{\infty}u_nu_{n+1}\cdots u_{n+k-1}d_{n+k}^{\pm}.$$

Remark that $\pi_n^{\pm}$ are isometries such that $\pi_n^+\mathcal{N}_+ = \mathcal{K}_n^{out}$ and $\pi_n^-\mathcal{N}_- = \mathcal{K}_n^{in}$. It appears to be of interest to consider the operator angle between $\mathcal{K}_n^{out}$ and $\mathcal{K}_n^{in}$. Thus, one defines the operators $Q_n = c(\mathcal{K}_n^{in},\mathcal{K}_n^{out}) = P_{\mathcal{K}_n^{in}}/\mathcal{K}_n^{out}$ and note the following result which explains the connection between upper triangular contractions and unitary couplings.

4.3 Proposition Let $\mathbf{w}_+$ and $\mathbf{w}_-$ be two families of isometries admitting a unitary coupling with property (Q). Then, for all $n \in \mathbb{Z}$, $(\pi_n^-)^*Q_n\pi_n^+ = \Theta$ for a certain operator Θ which belongs to $\mathcal{S}(\mathcal{N}_+,\mathcal{N}_-)$.

Proof For $i \geq j \geq n$, $d_i^+ = u_i^*l_i^+$ in $\mathcal{D}_i^+$ and $d_j = u_jl_{j+1}^-$ in $\mathcal{D}_j^-$, one obtains that

$$\begin{aligned}
\langle(\pi_n^-)^*Q_n\pi_n^+d_i^+,d_j^-\rangle &= \langle\pi_n^+d_i^+,\pi_n^-d_j^-\rangle\\
&= \langle u_nu_{n+1}\cdots u_id_i^+,u_nu_{n+1}\cdots u_{j-1}d_j^-\rangle\\
&= \langle u_{j+1}u_{j+2}\cdots u_{i-1}l_i^+,l_j^-\rangle,
\end{aligned}$$

and it is seen that the result is independent of n. Similar computations cover all the possible indices $i,j \in \mathbb{Z}$ and it is deduced that, for arbitrary $m,n \in \mathbb{Z}$, $(\pi_n^-)^*Q_n\pi_n^+ = (\pi_m^-)^*Q_m\pi_m^+ = \Theta$. By the definition of Q_n, it follows that Θ is a contraction in $\mathcal{L}(\mathcal{N}_+,\mathcal{N}_-)$. It remains to show that Θ is upper triangular. To that end, note that as a consequence of (Q),

$$\begin{aligned}
\langle(\Theta_{ij}d_j^+,d_i^-\rangle &= \langle\pi_0^+d_j^+,\pi_0^-d_i^-\rangle\\
&= \langle u_0u_1\cdots u_jd_j^+,u_0u_1\cdots u_{i-1}d_i^-\rangle\\
&= \langle l_j^+,u_ju_{j+1}\cdots u_il_{i+1}^-\rangle\\
&= \langle u_j^*l_j^+,u_{j+1}\cdots u_il_{i+1}^-\rangle = 0,
\end{aligned}$$

for $i > j \geq 0$, $d_j^+ = u_j^*l_j^+$ in $\mathcal{D}_j^+$ and $d_i^- = u_il_{i+1}^-$ in $\mathcal{D}_i^-$. Similar computations show that $\Theta_{ij} = 0$ for all $i > j$, hence Θ belongs to $\mathcal{S}(\mathcal{N}_+,\mathcal{N}_-)$. $\qquad\square$

4.4 Example We introduce a method to produce unitary couplings for isometries satisfying a certain commuting property. Thus, let X in $\mathcal{L}(\mathcal{H},\mathcal{H}')$ and Y in $\mathcal{L}(\mathcal{G},\mathcal{G}')$ be two contractions and let w_+, w_- be two isometries in $\mathcal{L}(\mathcal{G},\mathcal{H})$ and $\mathcal{L}(\mathcal{H}',\mathcal{G}')$, respectively. Suppose that

$$Xw_+ = w_-^*Y. \tag{4.6}$$

Define the positive operators

$$A = \begin{bmatrix} I & X \\ X^* & I \end{bmatrix} \quad \text{and} \quad B = \begin{bmatrix} I & Y \\ Y^* & I \end{bmatrix},$$

and remark that the following *coupling property* holds: for g in $\mathcal{G}$ and h' in $\mathcal{H}'$,

$$\langle w_+g,h'\rangle_A = \langle g,w_-h'\rangle_B. \tag{4.7}$$

This equality shows that the map

$$u_0 : w_- \mathcal{H}' \vee \mathcal{G} \longrightarrow \mathcal{H}' \vee w_+ \mathcal{G} \tag{4.8}$$

$$u_0(w_- h' + g) = h' + w_+ g$$

is a well-defined unitary operator. Every unitary extension u of u_0 from $\mathcal{K}_1$ onto $\mathcal{K}_0$, such that $\mathcal{G}' \vee \mathcal{G} = (\mathcal{G}' \oplus \mathcal{G})_B \subset \mathcal{K}_1$ and $\mathcal{H}' \vee \mathcal{H} = (\mathcal{H}' \oplus \mathcal{H})_A \subset \mathcal{K}_0$, gives rise to a unitary coupling of w_+ and w_-.   □

4.5 Remark   The preceding construction is closely related with the elementary rotation of a contraction. Thus, let T in $\mathcal{L}(\mathcal{H}, \mathcal{H}')$ be a contraction. Using the operators Ω and $\tilde{\Omega}$ defined by (1.2.7) and (1.2.8), we introduce the isometries

$$w_+ = \Omega/\mathcal{H} = \begin{bmatrix} T \\ D_T \end{bmatrix} \quad \text{and} \quad w_- = \tilde{\Omega}/\mathcal{H}' = \begin{bmatrix} T^* \\ D_{T^*} \end{bmatrix}.$$

Defining $X = P_{\mathcal{H}'}^{\mathcal{H}' \oplus D_T}$ and $Y =$ the inclusion of $\mathcal{H}$ into $\mathcal{H} \oplus D_T$, it follows that (4.6) holds and some algebra shows that u_0 defined by (4.8) collapses to $R(T)$, the elementary rotation of T. In fact, this connection with the elementary rotation suggests the following continuation of the construction in Example 4.4. Due to (4.6),

$$(I \oplus w_+^*)A(I \oplus w_+) = (w_-^* \oplus I)B(w_- \oplus I) = C,$$

which are factorizations of type (1.2.2). Consequently, unitary operators of type (1.2.3) can be associated to these two factorizations of C. Thus, consider the unitary operators

$$\omega_+ : (\mathcal{H}' \oplus \mathcal{G})_C \longrightarrow \mathcal{F}_+ = (\mathcal{R}(I \oplus w_+))_A = \mathcal{H}' \vee w_+ \mathcal{G}$$

$$\omega_+[k] = [(I \oplus w_+)k]$$

and

$$\omega_- : (\mathcal{H}' \oplus \mathcal{G})_C \longrightarrow \mathcal{F}_- = (\mathcal{R}(w_- \oplus I))_B = w_- \mathcal{H}' \vee \mathcal{G}$$

$$\omega_-[k] = [(w_- \oplus I)k].$$

The unitary operator $R = \omega_+ \omega_-^*$ can be seen as a generalization of the elementary rotation and it is clear that the unitary operator u_0 defined by (4.8) coincides with R.   □

We conclude this section with the introduction of operator ranges. Let A in $\mathcal{L}(\mathcal{H})$ be a positive operator and denote by P the orthogonal projection of $\mathcal{H}$ onto ker A. The formula

$$\langle A^{1/2}h, A^{1/2}g \rangle^A = \langle h, (I - P)g \rangle, \qquad h, g \in \mathcal{H}, \tag{4.9}$$

defines an inner product on $\mathcal{R}(A^{1/2})$. We have the following result.

4.6 Proposition   *Let A be a positive operator on the Hilbert space $\mathcal{H}$. Then:*

(a) $\mathcal{R}(A^{1/2})$ is a Hilbert space, denoted by $\mathcal{H}^A$, with respect to the inner product defined by (4.9).

(b) $\mathcal{R}(A)$ is a dense subspace of $\mathcal{H}^A$.

Proof It remains to show that $\mathcal{R}(A^{1/2})$ is complete with respect to the norm induced by the inner product (4.9). Suppose that $\{A^{1/2}h_n\}_{n\geq 1}$ is a Cauchy sequence with respect to this norm. It follows that $\{(I-P)h_n\}_{n\geq 1}$ is a Cauchy sequence in $\mathcal{H}$, hence there exists $\lim_{n\to\infty}(I-P)h_n = h$ in $\mathcal{H}$. Moreover, $(I-P)h = h$, therefore $\|A^{1/2}h_n - A^{1/2}h\|^A = \|(I-P)(h_n-h)\| \to 0$ as $n \to \infty$ i.e. the sequence $\{A^{1/2}h_n\}_{n\geq 1}$ has a limit in $\mathcal{R}(A^{1/2})$. In order to prove (b), we first remark that $\mathcal{R}(A) \subset \mathcal{R}(A^{1/2})$. Then, pick g in $\mathcal{H}$ such that $\langle Ah, A^{1/2}g\rangle^A = 0$ for all h in $\mathcal{H}$, i.e. $\langle A^{1/2}h, (I-P)g\rangle = 0$ for all h in $\mathcal{H}$. Since $\ker A = \ker A^{1/2}$, it follows that $A^{1/2}(I-P) = A^{1/2}$, therefore $\langle h, A^{1/2}g\rangle = 0$ for all h in $\mathcal{H}$. This means that $A^{1/2}g = 0$, consequently $\mathcal{R}(A)$ is dense in $\mathcal{H}^A$. $\square$

The Hilbert space $\mathcal{H}^A$ is called the *operator range* of A. Next we point out the connection between the space $\mathcal{H}^A$ and the space $\mathcal{H}_A$ introduced in Section 1.2. Since $\mathcal{R}(A)$ is a dense subspace of $\mathcal{H}^A$, we can define the map δ' from $\mathcal{R}(A)$ to $\mathrm{cl}\mathcal{R}(A^{1/2})$ by the formula $\delta'Ah = A^{1/2}h$ for h in $\mathcal{H}$, and remark that δ' is a well-defined operator. Moreover, for h and g in $\mathcal{H}$, $\langle \delta'Ah, \delta'Ag\rangle = \langle A^{1/2}h, A^{1/2}g\rangle = \langle Ah, Ag\rangle^A$, hence δ' can be extended by continuity to a unitary operator, also denoted by δ', from $\mathcal{H}^A$ into $\mathrm{cl}\mathcal{R}(A^{1/2})$. But, $\mathrm{cl}\mathcal{R}(A^{1/2})$ is an identification of $\mathcal{H}_A$, i.e. there exists the unitary operator w_A mapping $\mathcal{H}_A$ onto $\mathrm{cl}\mathcal{R}(A^{1/2})$. Hence, $\delta = \delta'w_A^*$ is a unitary operator relating the operator range $\mathcal{H}^A$ with the Hilbert space $\mathcal{H}_A$. The construction of the operator range can be extended to operators T in $\mathcal{L}(\mathcal{H},\mathcal{K})$ in the following way. The formula

$$\langle Th, Tg\rangle^T = \langle h, (I - P_{\ker T})g\rangle, \qquad h,g \in \mathcal{H}, \tag{4.10}$$

defines an inner product on $\mathcal{R}(T)$. The same proof as of Lemma 4.6 (b) shows that $\mathcal{R}(T)$ is a Hilbert space with respect to the norm induced by (4.10). Denote by $\mathcal{K}^T$ this Hilbert space and let us notice that it can be identified with the operator range space produced by the positive operator $A = T^*T$. To see this, define the map η' from $\mathcal{K}^T$ to $\mathcal{H}^A$ by the formula $\eta'Th = A^{1/2}h$ for h in $\mathcal{H}$, and remark that, since $\ker T = \ker A = \ker A^{1/2}$, then η' is well-defined. Besides, the equalities $\langle \eta'Th, \eta'Tg\rangle^A = \langle A^{1/2}h, A^{1/2}g\rangle^A = \langle h, (I - P_{\ker A})g\rangle = \langle Th, Tg\rangle^T$ hold for h and g in $\mathcal{H}$, therefore η' is a unitary operator.

Of special interest is the case when T is injective. We obtain the following result.

4.7 Proposition Let T be an injective operator in $\mathcal{L}(\mathcal{H},\mathcal{K})$. Then there exists a unitary operator η from $\mathcal{K}_B$ onto $\mathcal{K}^T$, where $B = TT^*$.

Proof Since $\ker T = 0$, $\langle Th, Tg\rangle^T = \langle h, g\rangle$ for all h and g in $\mathcal{H}$, which shows that T is a unitary operator from $\mathcal{H}$ onto $\mathcal{K}^T$. Then, define the map

$$\eta : \mathcal{R}(T) \longrightarrow \mathcal{R}(T) \tag{4.11}$$

$$\eta Th = TAh, \qquad h \in \mathcal{H},$$

where $A = T^*T$. We see that this map is well-defined and note that

$$\langle \eta Th, \eta Tg \rangle^T = \langle TAh, TAg \rangle^T = \langle Ah, Ag \rangle = \langle T^*Th, T^*Tg \rangle = \langle Th, Tg \rangle_B.$$

If the element g in $\mathcal{H}$ is such that $\langle TAh, Tg \rangle^T = 0$ for all h in $\mathcal{H}$, then $Ag = 0$. Hence $Tg = 0$ and $\mathcal{R}(TA)$ is a dense subspace of $\mathcal{K}^T$. Similarly, $\mathcal{R}(T)$ is dense in $\mathcal{K}_B$. We conclude that η can be extended by continuity to a unitary operator, also denoted by η, from $\mathcal{K}_B$ onto $\mathcal{K}^T$. $\qquad\square$

2.5 Modeling Families of Contractions

In this section we describe two models for families of contractions. The first one, the Sz.-Nagy-Foias model, employs the unitary coupling method, while the second one, the de Branges-Rovnyak model, employs operator ranges. Moreover, connections with the realization problem are emphasized. It is useful to begin by noting the following result.

5.1 Proposition Let $\mathbf{H} = \{\mathcal{H}_n\}_{n \in \mathbb{Z}}$ be a family of Hilbert spaces and let $\{T_n\}_{n \in \mathbb{Z}}$ be a family of contractions T_n in $\mathcal{L}(\mathcal{H}_{n+1}, \mathcal{H}_n)$. Then, for all $n \in \mathbb{Z}$, there exists a reducing decomposition of the form $T_n = T_n^{(1)} \oplus T_n^{(2)}$ on $\mathcal{H}_n = \mathcal{H}_n^{(1)} \oplus \mathcal{H}_n^{(2)}$, such that $T_n^{(2)}/\mathcal{H}_n^{(2)}$ is unitary and $\mathcal{H}_n^{(2)}$ is maximal with these properties.

Proof Define

$$\mathcal{H}_n^{(2)} = \{h_n \in \mathcal{H}_n \mid \ldots = \|T_{n-1}h_n\| = \|h_n\| = \|T_n^*h_n\| = \|T_{n+1}^*T_n^*h_n\| = \ldots\}, \tag{5.1}$$

and remark that for a contraction T, the equality $\|Th\| = \|h\|$ is equivalent to $T^*Th = h$. Consequently, $\mathcal{H}_n^{(2)} = \cap_{k=0}^\infty \ker D_{T_{n+k}^* \cdots T_n^*} \cap \cap_{k=1}^\infty \ker D_{T_{n-k} \cdots T_{n-1}}$ and $\mathcal{H}_n^{(2)}$ is a subspace of $\mathcal{H}_n$ for all $n \in \mathbb{Z}$. If h_n belongs to $\mathcal{H}_n^{(2)}$, then for $k > 1$, $\|T_{n-k} \cdots T_{n-2}T_{n-1}h_n\| = \|h_n\| = \|T_{n-1}h_n\|$ and for $p \geq -1$,

$$\|T_{n+p}^* \cdots T_{n-1}^* T_{n-1}h_n\| = \|T_{n+p}^* \cdots T_n^* h_n\| = \|h_n\| = \|T_{n-1}h_n\|$$

since $T_{n-1}^*T_{n-1}h_n = h_n$. Therefore, $T_{n-1}\mathcal{H}_n^{(2)} \subset \mathcal{H}_{n-1}^{(2)}$ and, in fact, $T_{n-1}\mathcal{H}_n^{(2)} = \mathcal{H}_{n-1}^{(2)}$. Obviously, $T_{n-1}/\mathcal{H}_n^{(2)}$ is unitary. Finally, it is readily checked that if $\{\mathcal{H}_n'\}_{n \in \mathbb{Z}}$ is another family of subspaces such that for all $n \in \mathbb{Z}$, $\mathcal{H}_n' \subseteq \mathcal{H}_n$ and $T_{n-1}/\mathcal{H}_n'^{(2)}$ is unitary, then $\mathcal{H}_n' \subseteq \mathcal{H}_n^{(2)}$. $\qquad\square$

A family $\{T_n\}_{n \in \mathbb{Z}}$ of contractions will be called *completely non-unitary* if the spaces defined by (5.1) are all trivial.

5.2 Remark (a) We can notice several particular cases of Proposition 5.1. First, if $\{T_n\}_{n \in \mathbb{Z}}$ is a family of isometries, then the subspaces defined by (5.1) reduce to the residual subspaces of the family. Then, another case of interest is when $\mathcal{H}_n = \mathcal{H}$ and $T_n = T$ for all $n \in \mathbb{Z}$, with T a contraction on $\mathcal{H}$. Using Proposition 5.1, it follows that T admits a (unique) decomposition of the form $T = T^{(1)} \oplus T^{(2)}$ on $\mathcal{H} = \mathcal{H}^{(1)} \oplus \mathcal{H}^{(2)}$, such that $T^{(2)}$ is unitary and $T^{(1)}$ is *completely non-unitary*, *i.e.*

there is no nonzero reducing subspace $\mathcal{H}'$ of $\mathcal{H}^{(1)}$ such that $T/\mathcal{H}'$ is unitary. Also of interest there is the case when the given family of Hilbert spaces is such that $\mathcal{H}_n = 0$ for all $n \in \mathbb{Z} - \{1, 2\}$. In this case, there is only one nontrivial operator in the family $\{T_n\}_{n\in\mathbb{Z}}$, namely $T = T_1$. By Proposition 5.1, we obtain the reducing decomposition $T : \mathcal{D}_T \oplus \ker D_T \longrightarrow \mathcal{D}_{T^*} \oplus \ker D_{T^*}$, such that $T/\ker D_T$ is unitary in $\mathcal{L}(\ker D_T, \ker D_{T^*})$ and $T/\mathcal{D}_T$ is a *pure contraction*, i.e. $\|Th\| < \|h\|$ for all h in $\mathcal{H}_2 - \{0\}$.

(b)   Given a family $\{T_n\}_{n\in\mathbb{Z}}$ of contractions, we consider the following unitary system:

$$\Omega \begin{cases} x(n) = T_n x(n+1) + D_{T_n^*} u(n) \\ y(n) = D_{T_n} x(n+1) - T_n^* u(n), \quad t \in \mathbb{Z}. \end{cases} \tag{5.2}$$

Its controllability and observability maps are given by:

$$\mathcal{C}_n : \oplus_{k \geq n} \mathcal{D}_{T_k^*} \longrightarrow \mathcal{H}_n \tag{5.3}$$

$$\mathcal{C}_n = \begin{bmatrix} D_{T_n^*} & T_n D_{T_{n+1}^*} & T_n T_{n+1} D_{T_{n+2}^*} & \cdots \end{bmatrix}$$

and, respectively,

$$\mathcal{O}_n : \oplus_{k \leq n-1} \mathcal{D}_{T_k} \longrightarrow \mathcal{H}_n \tag{5.4}$$

$$\mathcal{O}_n = \begin{bmatrix} \cdots & T_{n-1}^* T_{n-2}^* D_{T_{n-3}} & T_{n-1}^* D_{T_{n-2}} & D_{T_{n-1}} \end{bmatrix}$$

and we see that the system Ω is closely connected if and only if the family $\{T_n\}_{n\in\mathbb{Z}}$ is completely non-unitary. $\qquad\square$

Motivated by Remark 4.5, we introduce the following elements. Let $\{W_n\}_{n\in\mathbb{Z}}$ be the Kolmogorov decomposition described by Theorem 1.6.1 of the positive definite kernel A associated to the family $\{T_n\}_{n\in\mathbb{Z}}$ of contractions as in Corollary 1.5.9. The operators W_n are given by the formulae:

$$W_n : \mathcal{K}_{n+1} \longrightarrow \mathcal{K}_n \tag{5.5}$$

$$W_n = \begin{bmatrix} \ddots & \vdots & \vdots & & \\ & I & 0 & 0 & 0 \\ \cdots & 0 & D_{T_n^*} & T_n & 0 & \cdots \\ \cdots & 0 & -T_n^* & D_{T_n} & 0 & \cdots \\ & 0 & 0 & 0 & I \\ & \vdots & \vdots & & \ddots \end{bmatrix},$$

where $\mathcal{K}_n = \oplus_{j<n} \mathcal{D}_{T_j^*} \oplus \mathcal{H}_n \oplus \oplus_{k \geq n} \mathcal{D}_{T_k}$. We also define the spaces $\mathcal{K}_n^+ = \mathcal{H}_n \oplus \oplus_{k \geq n} \mathcal{D}_{T_k}$ and $\mathcal{K}_n^- = \oplus_{k > -n} \mathcal{D}_{T_k^*} \oplus \mathcal{H}_n$, and remark that $W_n \mathcal{K}_{n+1}^+ \subset \mathcal{K}_n^+$, $W_n^* \mathcal{K}_n^- \subset \mathcal{K}_{n+1}^-$. Therefore, the family $\mathbf{u} = \{W_n\}_{n\in\mathbb{Z}}$ is a unitary coupling of the two families of isometries, $\mathbf{w}_+ = \{W_n^+ = W_n/\mathcal{K}_{n+1}^+\}_{n\in\mathbb{Z}}$ and $\mathbf{w}_- = \{W_n^- = W_n^*/\mathcal{K}_n^-\}_{n\in\mathbb{Z}}$. Due to the special form of the operators W_n, we can simply obtain the description of the Wold-von Neumann decompositions of $\mathbf{w}_+$ and $\mathbf{w}_-$. Thus, $\mathcal{L}_n^+ = \mathcal{K}_n^+ \ominus W_n^+ \mathcal{K}_{n+1}^+ = W_n \mathcal{D}_{T_n^*}$ (the space $\mathcal{D}_{T_n^*}$ is viewed as embedded in $\mathcal{K}_n$) and $\mathcal{L}_n^- = \mathcal{K}_n^- \ominus W_{n-1}^- \mathcal{K}_{n-1}^- = W_{n-1}^* \mathcal{D}_{T_{n-1}}$ (the space $\mathcal{D}_{T_{n-1}}$ is viewed as embedded in $\mathcal{K}_{n-1}$).

Moreover, $\mathcal{D}_n^+ = \mathcal{D}_{T_n^*}$ and $\mathcal{D}_n^- = \mathcal{D}_{T_{n-1}}$. The pair $(\mathbf{w}_+, \mathbf{w}_-)$ has property (Q) and the upper triangular contraction Θ in $\mathcal{L}(\oplus_{n \in \mathbb{Z}} \mathcal{D}_{T_n^*}, \oplus_{n \in \mathbb{Z}} \mathcal{D}_{T_n})$ introduced in Proposition 4.3 can be explicitly described by the formula:

$$
\Theta_{ij} = \begin{cases}
-T_i^* & \text{if } j = i \\
D_{T_i} D_{T_{i+1}^*} & \text{if } j = i+1 \\
D_{T_i} T_{i+1} \ldots T_{j-1} D_{T_j^*} & \text{if } j > i+1 \\
0 & \text{otherwise.}
\end{cases}
$$

Comparing with formula (3.2), it follows that Θ is the transfer map of the unitary system defined by (5.2). The map Θ is also referred to as the *characteristic operator* of the family $\{T_n\}_{n \in \mathbb{Z}}$. We note now the following result.

5.3 Lemma *Let $\{T_n\}_{n \in \mathbb{Z}}$ be a completely non-unitary family of contractions. Then, for all $n \in \mathbb{Z}$, $\mathcal{K}_n = \mathcal{K}_n^{in} \vee \mathcal{K}_n^{out}$.*

Proof Let f be a vector in $\mathcal{K}_n$ which is orthogonal to $\mathcal{K}_n^{in} \vee \mathcal{K}_n^{out}$. In particular, f is orthogonal to $W_n \ldots W_{n+k-1} \mathcal{D}_{T_{n+k}}$ for all $k \geq 0$ and f is orthogonal to $W_{n-1}^* \ldots W_{n-p+1}^* \mathcal{D}_{T_{n-p}^*}$ for all $p \geq 1$, which implies that f belongs to $\mathcal{H}_n$. Then, use the fact that f is orthogonal to $W_{n-1}^* \mathcal{D}_{T_{n-1}}$ and take into account the definition of W_{n-1} to get $D_{T_{n-1}} f = 0$. Continuing to explore the fact that f is orthogonal to $W_{n-1}^* \ldots W_{n-k}^* \mathcal{D}_{T_{n-k}^*}$ for all $k \geq 2$, deduce that f belongs to $\cap_{k=1}^{\infty} \ker D_{T_{n-k} \ldots T_{n-1}}$. Similarly, since f is orthogonal to $W_n \ldots W_{n+k} \mathcal{D}_{T_{n+k}^*}$ for all $k \geq 0$, it follows that f belongs to $\cap_{k=0}^{\infty} \ker D_{T_{n+k}^* \ldots T_n^*}$. In conclusion, f belongs to $\mathcal{H}_n^{(2)}$, the space defined by (5.1). Since $\{T_n\}_{n \in \mathbb{Z}}$ is a completely non-unitary family, this implies that $f = 0$. $\qquad\square$

As a consequence of Lemma 5.3, it follows that the space $\mathcal{R}_n^-$ appearing in the Wold-von Neumann decomposition of $\mathbf{w}_-$ is $\mathrm{cl}(I - Q_n) \mathcal{K}_n^{out}$, therefore we can define the unitary operator

$$
\Phi_{\mathcal{R}_n^-} : \mathcal{R}_n^- \longrightarrow \mathcal{D}_\Theta \tag{5.6}
$$

$$
\Phi_{\mathcal{R}_n^-} (I - Q_n) k = D_\Theta (\pi_n^+)^* k, \qquad k \in \mathcal{K}_n^{out}.
$$

It follows that the unitary operator $\Psi_n = (\pi_n^-)^* / \mathcal{K}_n^{in} \oplus \Phi_{\mathcal{R}_n^-}$ from $\mathcal{K}_n$ onto $\mathcal{N}_- \oplus \mathcal{D}_\Theta$ gives rise to an identification of the space $\mathcal{K}_n$ in terms of the characteristic operator Θ. It turns out that the position of $\mathcal{H}_n$ inside $\mathcal{K}_n$ can be also described in terms of Θ.

5.4 Lemma *Let $\{T_n\}_{n \in \mathbb{Z}}$ be a completely non-unitary family of contractions with the characteristic operator Θ. Then, for all $n \in \mathbb{Z}$,*

$$
\Psi_n \mathcal{H}_n = (\oplus_{k \leq n-1} \mathcal{D}_{T_k} \oplus \mathcal{D}_\Theta) \ominus \{\Theta v \oplus D_\Theta v \mid v \in \oplus_{k \leq n-1} \mathcal{D}_{T_k^*}\}.
$$

Proof One remarks that $\Psi_n \mathcal{K}_n^- = \oplus_{k \leq n-1} \mathcal{D}_{T_k} \oplus \mathcal{D}_\Theta$ and

$$
\mathcal{H}_n = \mathcal{K}_n^- \ominus \{Q_n u \oplus (I - Q_n) u \mid u \in \oplus_{p=-\infty}^0 W_{n-1}^* \ldots W_{n-p}^* \mathcal{D}_{T_{n-p-1}^*}\},
$$

therefore the formula for $\Psi_n \mathcal{H}_n$ is a consequence of the definition of Ψ_n. $\qquad\square$

Now we are ready to describe the *Sz.-Nagy-Foias model* of a family of contractions. We say that the family $\{T_n^0\}_{n \in \mathbb{Z}}$, T_n^0 in $\mathcal{L}(\mathcal{H}_{n+1}^0, \mathcal{H}_n^0)$, is a *model* of $\{T_n\}_{n \in \mathbb{Z}}$ if there exist unitary operators ψ_n in $\mathcal{L}(\mathcal{H}_n, \mathcal{H}_n^0)$ satisfying $T_n^0 = \psi_n T_n \psi_{n+1}^*$ for all $n \in \mathbb{Z}$ (we also say that $\{T_n\}_{n \in \mathbb{Z}}$ and $\{T_n^0\}_{n \in \mathbb{Z}}$ are *unitarily equivalent families*).

5.5 Theorem *Let $\{T_n\}_{n \in \mathbb{Z}}$ be a completely non-unitary family of contractions with the characteristic operator Θ. Then, $\{T_n\}_{n \in \mathbb{Z}}$ has the following Sz.-Nagy-Foias model:*

$$\mathcal{H}_n^0 = (\oplus_{k \leq n-1} \mathcal{D}_{T_k} \oplus \mathcal{D}_\Theta) \ominus \{\Theta v \oplus D_\Theta v \mid v \in \oplus_{k \leq n-1} \mathcal{D}_{T_k^*}\}$$

and

$$T_n^0 : \mathcal{H}_{n+1}^0 \longrightarrow \mathcal{H}_n^0 \qquad (5.7)$$

$$T_n^0 = P_{\mathcal{H}_n^0}(N_{\{\mathcal{D}_{T_k}\}_{k \in \mathbb{Z}}}^*(n) \oplus I_{\mathcal{D}_\Theta})/\mathcal{H}_{n+1}^0.$$

Proof Remark that, for u in $\oplus_{k \leq n-1} \mathcal{D}_{T_k}$ and v in $\mathcal{N}_+$,

$$\Psi_{n+1} W_n^- \Psi_n^*(u \oplus D_\Theta v) = \Psi_{n+1} W_n^-(\pi_n^- u \oplus \Phi_{\mathcal{R}_n^-}^* D_\Theta v)$$

$$= \Psi_{n+1}(W_n^- \pi_n^- u \oplus W_n^- \Phi_{\mathcal{R}_n^-}^* D_\Theta v)$$

$$= \Psi_{n+1} W_n^- \pi_n^- u \oplus \Psi_{n+1} \Phi_{\mathcal{R}_{n+1}^-}^* D_\Theta v$$

$$= N_{\{\mathcal{D}_{T_k}\}_{k \in \mathbb{Z}}}(n) u \oplus D_\Theta v,$$

therefore $\Psi_{n+1} W_n^- \Psi_n^* / \Psi_n \mathcal{K}_n^- = N_{\{\mathcal{D}_{T_k}\}_{k \in \mathbb{Z}}}(n) \oplus I_{\mathcal{D}_\Theta}$. Since $\{W_n\}_{n \in \mathbb{Z}}$ is the Kolmogorov decomposition associated to the family $\{T_n\}_{n \in \mathbb{Z}}$, it follows that $T_n = (W_n^-)^*/\mathcal{H}_{n+1}$ for all $n \in \mathbb{Z}$. By Lemma 5.4, $\mathcal{H}_n^0 = \Psi_n \mathcal{H}_n$ and for h in $\mathcal{H}_{n+1}$,

$$T_n h = (W_n^-)^* h$$

$$= \Psi_n^* \Psi_n (W_n^-)^* \Psi_{n+1}^* \Psi_{n+1} h$$

$$= \Psi_n^*(N_{\{\mathcal{D}_{T_k}\}_{n \in \mathbb{Z}}}^*(n) \oplus I_{\mathcal{D}_\Theta}) \Psi_{n+1} h$$

$$= \Psi_n^* T_n^0 \Psi_{n+1} h.$$

This shows that $\{T_n^0\}_{n \in \mathbb{Z}}$ is a model of the family $\{T_n\}_{n \in \mathbb{Z}}$. $\qquad\square$

It follows from Theorem 5.5 that the characteristic operator is an invariant for the models of a completely non-unitary family of contractions. More precisely, we say that two lower triangular contractions T in $\mathcal{L}(\oplus_{n \in \mathbb{Z}} \mathcal{E}_n, \oplus_{n \in \mathbb{Z}} \mathcal{F}_n)$ and T' in $\mathcal{L}(\oplus_{n \in \mathbb{Z}} \mathcal{E}_n', \oplus_{n \in \mathbb{Z}} \mathcal{F}_n')$ *coincide* if there exist unitary operators τ_n in $\mathcal{L}(\mathcal{E}_n, \mathcal{E}_n')$ and τ_n' in $\mathcal{L}(\mathcal{F}_n, \mathcal{F}_n')$, such that $T = (\oplus_{n \in \mathbb{Z}} \tau_n')^* T'(\oplus_{n \in \mathbb{Z}} \tau_n)$ for all $n \in \mathbb{Z}$, and we have the following consequence of Theorem 5.5.

5.6 Corollary *Two completely non-unitary families of contractions are unitarily equivalent if and only if their characteristic operators coincide.*   □

In order to describe the *de Branges-Rovnyak* model of a completely-nonunitary family $\{T_n\}_{n\in\mathbb{Z}}$ of contractions, we introduce the following operators: for $n \in \mathbb{Z}$,

$$Z_n : \mathcal{H}_n \longrightarrow \oplus_{k\leq n-1}\mathcal{D}_{T_k} \oplus \oplus_{k\geq n}\mathcal{D}_{T_k^*} \tag{5.8}$$

$$Z_n = \begin{bmatrix} \mathcal{O}_n^* \\ \mathcal{C}_n^* \end{bmatrix},$$

where $\mathcal{C}_n$ and $\mathcal{O}_n$ are the controllability and, respectively, the observability maps of the unitary system Ω defined by (5.2). By Remark 5.2, Z_n is injective since the family $\{T_n\}_{n\in\mathbb{Z}}$ is completely non-unitary. If the operator Θ belongs to $\mathcal{L}(\mathcal{N}_+,\mathcal{N}_-)$, then define

$$\Theta(n) = P_{\oplus_{k\geq n}\mathcal{D}_{T_k}}\Theta / \oplus_{k\geq n}\mathcal{D}_{T_k^*} \quad \text{and} \quad \widetilde{\Theta}(n) = P_{\oplus_{k\leq n-1}\mathcal{D}_{T_k}}\Theta / \oplus_{k\leq n-1}\mathcal{D}_{T_k^*}.$$

5.7 Theorem *Let $\{T_n\}_{n\in\mathbb{Z}}$ be a completely non-isometric family of contractions with the characteristic operator Θ. Then, $\{T_n\}_{n\in\mathbb{Z}}$ has the following de Branges-Rovnyak model:*

$$\mathcal{F}_n^0 = (\oplus_{k\leq n-1}\mathcal{D}_{T_k} \oplus \oplus_{k\geq n}\mathcal{D}_{T_k^*})_{\Psi(n)}$$

and

$$R_n^0 : \mathcal{F}_{n+1}^0 \longrightarrow \mathcal{F}_n^0 \tag{5.9}$$

$$R_n^0 = \begin{bmatrix} N_{\{\mathcal{D}_{T_k}\}_{k\in\mathbb{Z}}}^*(n) & 0 \\ -\Theta^* P_{\mathcal{D}_{T_n}} & M_{\{\mathcal{D}_{T_k}\}_{k\in\mathbb{Z}}}(n) \end{bmatrix},$$

where

$$\Psi(n) : \oplus_{k\leq n-1}\mathcal{D}_{T_k} \oplus \oplus_{k\geq n}\mathcal{D}_{T_k^*} \longrightarrow \oplus_{k\leq n-1}\mathcal{D}_{T_k} \oplus \oplus_{k\geq n}\mathcal{D}_{T_k^*}$$

$$(\Psi(n))_{ij} = \begin{cases} (I - \widetilde{\Theta}(n)\widetilde{\Theta}^*(n))_{ij} & \text{if } i,j \leq n-1 \\ (I - \Theta^*(n)\Theta(n))_{ij} & \text{if } i,j \geq n \\ \Theta_{ij} & \text{if } i < j, i \leq n-1, j \geq n \\ \Theta_{ij}^* & \text{if } i > j, i \geq n, j \leq n-1. \end{cases}$$

Proof First, we find a model of the family $\{T_n\}_{n\in\mathbb{Z}}$ in terms of the operators ranges of Z_n, $n \in \mathbb{Z}$. Thus, it is easy to check that for all $n \in \mathbb{Z}$,

$$Z_n T_n = R_n' Z_{n+1},$$

where

$$R_n' : \oplus_{k\leq n-1}\mathcal{D}_{T_k} \oplus \oplus_{k\geq n}\mathcal{D}_{T_k^*} \longrightarrow \oplus_{k\leq n-1}\mathcal{D}_{T_k} \oplus \oplus_{k\geq n}\mathcal{D}_{T_k^*}$$

$$R_n' = \begin{bmatrix} (N_{\{\mathcal{D}_{T_k}\}_{k\in\mathbb{Z}}}^*(n)) & 0 \\ -\Theta^* P_{\mathcal{D}_{T_n}} & M_{\{\mathcal{D}_{T_k}\}_{k\in\mathbb{Z}}}(n) \end{bmatrix}.$$

It follows that $R_n' \mathcal{R}(Z_{n+1}) \subset \mathcal{R}(Z_n)$ and since Z_n are injective operators, it follows that they are unitary from $\mathcal{H}_n$ onto $(\oplus_{k \leq n-1} \mathcal{D}_{T_k} \oplus \oplus_{k \geq n} \mathcal{D}_{T_k^*})^{Z_n}$. Consequently, $\{R_n'\}_{n \in \mathbb{Z}}$ is a model of the family $\{T_n\}_{n \in \mathbb{Z}}$. By Proposition 4.7, there exist unitary operators η_n from $\mathcal{F}_n^0 = (\oplus_{k \leq n-1} \mathcal{D}_{T_k} \oplus \oplus_{k \geq n} \mathcal{D}_{T_k^*})_{B_n}$ onto $(\oplus_{k \leq n-1} \mathcal{D}_{T_k} \oplus \oplus_{k \geq n} \mathcal{D}_{T_k^*})^{Z_n}$, where $B_n = Z_n Z_n^*$. Moreover, if $A_n = Z_n^* Z_n$, then for h in $\mathcal{H}_{n+1}$ and g in $\mathcal{H}_n$,

$$
\begin{aligned}
\langle T_n A_{n+1} h, A_n g \rangle_{\mathcal{H}_n} &= \langle T_n Z_{n+1}^* Z_{n+1} h, Z_n^* Z_n g \rangle_{\mathcal{H}_n} \\
&= \langle Z_{n+1}^* Z_{n+1} h, Z_{n+1}^* R_n'^* Z_n g \rangle_{\mathcal{H}_{n+1}} \\
&= \langle Z_{n+1} h, R_n'^* Z_n g \rangle^{Z_{n+1}} \\
&= \langle Z_n T_n h, Z_n g \rangle^{Z_n} \\
&= \langle Z_n^* Z_n T_n h, Z_n^* Z_n g \rangle_{\mathcal{H}_n} \\
&= \langle A_n T_n h, A_n g \rangle_{\mathcal{H}_n}.
\end{aligned}
$$

This equality shows that

$$
\begin{aligned}
R_n' \eta_{n+1}^* Z_{n+1} A_{n+1} &= R_n' Z_{n+1} = Z_n T_n = \eta_n^* Z_n A_n T_n \\
&= \eta_n^* Z_n T_n A_{n+1} = \eta_n^* R_n' Z_{n+1} A_{n+1},
\end{aligned}
$$

therefore $\{R_n^0\}_{n \in \mathbb{Z}}$ is a model of the family $\{T_n\}_{n \in \mathbb{Z}}$. As a consequence of the definitions of $\mathcal{C}_n$, $\mathcal{O}_n$ and Θ, it follows that $B_n = \Psi(n)$, thereby concluding the proof. □

The two models presented in this section can be used in order to obtain another proof of Theorem 3.2. Since Remark 5.2 makes this approach clear, the details can be omitted.

2.6 Notes

A detailed presentation of the material contained in Section 1 can be found in the books [Do3], [FoF1], [Fu], [Ho], [RR1], [RR2], [Sz.-NF2]. As stated here, Theorem 2.1 was noticed in [ACC] and [Co2]. However, many particular cases appeared before in connection with various applications. The selfadjoint case of (2.6) goes back to the work of M.G.Krein [Kr3] and it was used to a systematic study of the positive selfadjoint extensions of the symmetric positive operators. Details will be presented in Chapter 3. The general case of (2.6) was proved in [AGh], [DKW] and [SY]. Formula (2.7) appeared in [Ph] in connection with extension problems for dual pairs in Krein spaces, and later in [Pa]. It can be used to obtain several results such as the Arveson distance formula for nest algebras (see [Arv], [Da]), or the lifting theorem of Sarason-Sz.-Nagy-Foias (see [BC2], [Da], [FoF1], [Pa]). Moreover, it can be used to develop a quick approach to several interpolation problems, including Nehari problem (see [Pow], [Yo]). The structure of $m \times n$ block contractions was studied in [Dav] and then in [BG1], [BG2], [ACC].

The realization theorem for unitary, time invariant systems, was proved in
[Br2]. In the case of time-varying, unitary systems, Theorem 3.2 was noticed in
[ACC] and the proof follows an idea in [KB]. The classical realization theory refers
more to the possibility of modeling by finite-dimensional systems – see [Ka1],
[KFA] for time-invariant systems and [GKL], [Ve] for recent results concerning
the time-varying case. Unitary couplings were introduced in [AA1-2]. A recent
discussion of operator ranges and of Proposition 4.6 can be found in [ADD2]. The
Sz.-Nagy-Foias model is presented in [Sz.-NF2] and the de Branges-Rovnyak is
presented in [dBR1], [dBR2].

In connection with this chapter see [AG1], [AG2], [BaC], [Be], [BGK], [dB],
[Br1], [BL], [Do4], [DP], [EL], [GoK1], [GoK2], [Ha], [He], [Kam], [KKP], [LP],
[Ni], [NV], [Rot].

Chapter 3
Moment Problems and Interpolation

In this chapter we see how the use of the Kolmogorov decomposition leads to a general method to solve interpolation and moment problems. We solve all the problems mentioned in Preface and we show the connection with the commutant lifting method. Two parametrizations of the solutions are indicated. One uses Schur parameters and the other upper triangular contractions. The connection between these two methods is explained, as well as the connection with the realization problem for unitary systems.

3.1 A Survey on Completion Problems

We introduce several problems known in the literature as interpolation, moment or extension problems (we will use the term of *completion problem* to denote any of them), and we suggest a general approach to their solution exploiting the Kolmogorov decomposition of positive definite kernels.

We begin our presentation with a classical problem concerning the existence of selfadjoint extensions of unbounded symmetric operators. Let $T : D(T) \longrightarrow \mathcal{H}$ be a linear unbounded operator whose *domain* $D(T)$ a dense subspace of the Hilbert space $\mathcal{H}$. The *adjoint* of T, denote by T^*, is defined according the following rules: $D(T^*)$ is the space of the elements φ in $\mathcal{H}$ for which there exists a ψ in $\mathcal{H}$ with $\langle Th, \varphi \rangle = \langle h, \psi \rangle$ for all h in $D(T)$, and for each such φ in $D(T^*)$ we define $T^*\varphi = \psi$. An unbounded operator T is called *closed* if the graph $G(T) = \{\varphi \oplus T\varphi \mid \varphi \in D(T)\}$ is a closed subset of $\mathcal{H} \oplus \mathcal{H}$. An unbounded operator T is called an *extension* of S (and we write $S \subset T$) if $D(S) \subset D(T)$ and $Th = Sh$ for h in $D(S)$. A densely defined operator T is called *symmetric* if $T \subset T^*$ and an operator T is called *selfadjoint* if $T = T^*$.

1.1 Problem *Given a closed symmetric operator S, it is required to find conditions for the existence of selfadjoint extensions of S.*

This problem was motivated by H. Weyl's work on second-order differential operators and its complete solution is provided by the so-called *von Neumann theory*. For a closed symmetric operator S, we define $\mathcal{K}_{\pm}(S) = \ker(S^* \mp i)$. We have the following result.

1.2 Theorem   *Let S be a closed symmetric operator. Then*

(a)   *S has selfadjoint extensions if and only if $\dim \mathcal{K}_+(S) = \dim \mathcal{K}_-(S)$.*

(b)   *Suppose $\dim \mathcal{K}_+(S) = \dim \mathcal{K}_-(S)$, then there exists a one-to-one correspondence between the selfadjoint extensions of S and the unitary operators in $\mathcal{L}(\mathcal{K}_+(S), \mathcal{K}_-(S))$.*

Proof The proof is based on the construction of the *Cayley transform*. For this purpose, note that for φ in $D(S)$, $\|(S \pm i)\varphi\|^2 = \|S\varphi\|^2 + \|\varphi\|^2$. Thus, $\mathcal{R}(S \pm i)$ are closed subspaces, $S + i$ is injective and the map

$$V(S) : \mathcal{R}(S + i) \longrightarrow \mathcal{R}(S - i) \tag{1.1}$$

$$V(S) = (S - i)(S + i)^{-1}$$

is well defined. Actually, $V(S)$ is a unitary operator with the additional property that $\mathrm{cl}((I - V(S))\mathcal{R}(S+i)) = \mathcal{H}$. Now suppose that S has a selfadjoint extension A. By a well-known result concerning unbounded selfadjoint operators, $\mathcal{K}_\pm(A) = 0$, hence the Cayley transform of A is a unitary operator in $\mathcal{L}(\mathcal{H})$ with the property that $V(A)\varphi = V(S)\varphi$ for φ in $\mathcal{R}(S + i)$. Therefore, $\dim \mathcal{K}_+(S) = \dim \mathcal{K}_-(S)$. Conversely, suppose $\dim \mathcal{K}_+(S) = \dim \mathcal{K}_-(S)$ and let U_0 be a unitary operator in $\mathcal{L}(\mathcal{K}_+(S), \mathcal{K}_-(S))$. Then $U = V(S) \oplus U_0$ is a unitary operator in $\mathcal{L}(\mathcal{H})$ such that $\mathrm{cl}(I - U)\mathcal{H} = \mathcal{H}$, hence the inverse of the Cayley transform is well defined for U and produces a selfadjoint extension of S. $\square$

Also connected with the theory of differential operators there is the problem of selfadjoint extension of positive operators. An unbounded operator S is *positive* if S is symmetric and $\langle S\varphi, \varphi \rangle \geq 0$ for all φ in $D(S)$. There is a well-known result that $\dim \mathcal{K}_+(S) = \dim \mathcal{K}_-(S)$ for every positive operator S, therefore every positive operator has selfadjoint extensions. The interesting question in this direction regards the existence of positive selfadjoint extensions.

1.3 Problem   *Given a closed positive S, it is required to find conditions for the existence of positive selfadjoint extensions of S.*

As it turns out, this problem is always solvable and a complete solution is given by the so-called *Krein theory*. Let us mention that for two unbounded positive selfadjoint operators A and B, we write $A \leq B$ if $(A + I)^{-1} \geq (B + I)^{-1}$ in the ordinary sense of bounded operators. We have the following result.

1.4 Theorem   *Let S be a closed positive operator. Then*

(a)   *S always has positive selfadjoint extensions.*

(b)   *There exist two distinguished positive selfadjoint extensions A_0 and A_∞ of S such that the set of all the positive selfadjoint extensions of S is precisely the set of the positive selfadjoint operators A with $A_0 \leq A \leq A_\infty$.*

Proof The construction of the Cayley transform may be adapted to the present setting. Note that for φ in $D(S)$, $\|(S+I)\varphi\|^2 = \|S\varphi\|^2 + 2\langle S\varphi, \varphi \rangle + \|\varphi\|^2 \geq \|\varphi\|^2$. Thus, $\mathcal{R}(S + I)$ is a closed subspace. Moreover,

$$\|(S + I)\varphi\|^2 \geq \|S\varphi\|^2 - 2\langle S\varphi, \varphi \rangle + \|\varphi\|^2 \geq \|\varphi\|^2 = \|(S - I)\varphi\|^2,$$

therefore the map (also called the Cayley transform of S),

$$C(S) : \mathcal{R}(S + I) \longrightarrow \mathcal{H} \tag{1.2}$$

$$C(S) = (S - I)(S + I)^{-1}$$

is a well defined contraction in $\mathcal{L}(\mathcal{R}(S+I), \mathcal{H})$. For φ and ψ in $\mathcal{R}(S+I)$, we have $\langle C(S)\varphi, \psi \rangle = \langle \varphi, C(S)\psi \rangle$, hence $C(S)$ has the block matrix representation

$$C(S) = \begin{bmatrix} T \\ R \end{bmatrix} : \mathcal{R}(S + I) \longrightarrow \mathcal{R}(S + I) \oplus \mathcal{R}(S + I)^{\perp},$$

where T is a selfadjoint contraction. According to Remark 2.2.5, there exists a selfadjoint contraction $K = \begin{bmatrix} T & R^* \\ R & X \end{bmatrix}$ in $\mathcal{L}(\mathcal{H})$, and since it is easy to verify that $\mathrm{cl}(I - C(S))\mathcal{R}(S+I) = \mathcal{H}$, it follows that $\mathrm{cl}(I - K)\mathcal{H} = \mathcal{H}$. Therefore, the inverse of the Cayley transform of type (1.2) is well defined for K and produces a positive selfadjoint operator which is an extension of S. According to the formula (2.2.6), a parametrization of all the selfadjoint contractions of the form $\begin{bmatrix} T & R^* \\ R & X \end{bmatrix}$ is given by $K(\Gamma) = \begin{bmatrix} T & R^* \\ R & X(\Gamma) \end{bmatrix}$, where $X(\Gamma) = -\Gamma_R T^* \Gamma_R^* + D_{\Gamma_R^*} \Gamma D_{\Gamma_R^*}$. In this formula of $X(\Gamma)$, Γ_R is the unique contraction in $\mathcal{L}(\mathcal{D}_T, \mathcal{R}(S+I)^{\perp})$ with $R = \Gamma_R D_T$, and Γ is an arbitrary selfadjoint contraction in $\mathcal{L}(\mathcal{D}_{\Gamma_R^*})$. This means that the set of all the selfadjoint positive extensions of S is the set of all the operators of the form $A(\Gamma) = (I + K(\Gamma))(I - K(\Gamma))^{-1}$, with Γ a selfadjoint contraction in $\mathcal{L}(\mathcal{D}_{\Gamma_R^*})$. Defining $A_0 = (I + K(-I))(I - K(-I))^{-1}$ and $A_\infty = (I + K(I))(I - K(I))^{-1}$, it follows that

$$A_0 \leq A(\Gamma) \leq A_\infty.$$

The extension A_∞ is known in the literature as *Friedrichs' extension* of S, while A_0 is usually referred to as the *von Neumann-Krein extension* of S. $\qquad\square$

While Problem 1.1 was clearly reduced to an extension of a partial isometry to a unitary operator, apparently this is not the case with Problem 1.3. However, it turns out that the problem of deciding whether there exists an operator X such that $\begin{bmatrix} C & X \\ A & B \end{bmatrix}$ is a contraction, provided that the operators $[\,A \;\; B\,]$ and $\begin{bmatrix} C \\ A \end{bmatrix}$ are contractions, can be also reformulated as a problem of extending a partial isometry to a unitary operator. In order to see this, suppose that the contraction $[\,A \;\; B\,]$ belongs to $\mathcal{L}(\mathcal{H}_1 \oplus \mathcal{H}_2, \mathcal{H}_1')$ and the contraction $\begin{bmatrix} C \\ A \end{bmatrix}$ belongs to $\mathcal{L}(\mathcal{H}_1, \mathcal{H}_2' \oplus \mathcal{H}_1')$, and define the positive operators

$$H_0 = \begin{bmatrix} I & 0 & C \\ 0 & I & A \\ C^* & A^* & I \end{bmatrix} \quad \text{and} \quad H_1 = \begin{bmatrix} I & A & B \\ A^* & I & 0 \\ B^* & 0 & I \end{bmatrix}.$$

The spaces $\mathcal{H}_2' \oplus \mathcal{H}_1' \oplus \mathcal{H}_1$ and $\mathcal{H}_1' \oplus \mathcal{H}_1 \oplus \mathcal{H}_2$ may be renormed by mean of these two positive operators as in Section 1.2 and two Hilbert spaces $\mathcal{G}_1$ and, respectively,

$\mathcal{G}_2$ are obtained after this operation. In order to avoid working with classes, we may actually suppose that $H_0 > 0$ and $H_1 > 0$. Then, denote by $\mathcal{E}_1$ the subspace of $\mathcal{G}_1$ generated by the set $\{h_1' \oplus h_1 \oplus 0 \mid h_1' \in \mathcal{H}_1', h_1 \in \mathcal{H}_1\}$ and denote by $\mathcal{F}_0$ the subspace of $\mathcal{G}_0$ generated by the set $\{0 \oplus h_1' \oplus h_1 \mid h_1' \in \mathcal{H}_1', h_1 \in \mathcal{H}_1\}$. Define the operator

$$v_0 : \mathcal{E}_1 \longrightarrow \mathcal{F}_0$$

$$v_0(h_1' \oplus h_1 \oplus 0) = 0 \oplus h_1' \oplus h_1, \qquad h_1' \oplus h_1 \oplus 0 \in \mathcal{E}_1,$$

and it is easy to see that v_0 is a unitary operator in $\mathcal{L}(\mathcal{E}_1, \mathcal{F}_0)$. Moreover, it is a simple remark to see that there exists a unitary operator w_0 in $\mathcal{L}(\mathcal{K}_1, \mathcal{K}_0)$ extending v_0. Hence, we can define the contraction

$$T = P_{\mathcal{H}_2' \oplus \mathcal{H}_1'}^{\mathcal{K}_0} w_0 / \mathcal{H}_1 \oplus \mathcal{H}_2.$$

For h_1 in $\mathcal{H}_1$ and $h_2' \oplus h_1'$ in $\mathcal{H}_2' \oplus \mathcal{H}_1'$, we obtain that

$$
\begin{aligned}
\langle Th_1, h_2' \oplus h_1' \rangle_{\mathcal{K}_0} &= \langle w_0(0 \oplus h_1 \oplus 0), h_2' \oplus h_1' \oplus 0 \rangle_{\mathcal{K}_0} \\
&= \langle v_0(0 \oplus h_1 \oplus 0), h_2' \oplus h_1' \oplus 0 \rangle_{\mathcal{G}_0} \\
&= \langle H_0(0 \oplus 0 \oplus h_1), h_2' \oplus h_1' \oplus 0 \rangle = \langle \begin{bmatrix} C \\ A \end{bmatrix} h_1, h_2' \oplus h_1' \rangle,
\end{aligned}
$$

hence $Th_1 = \begin{bmatrix} C \\ A \end{bmatrix} h_1$. Similarly, we obtain that $T^* h_1' = \begin{bmatrix} A^* \\ B^* \end{bmatrix} h_1'$, therefore $T = \begin{bmatrix} C & X \\ A & B \end{bmatrix}$.

Apparently unrelated with the problems considered so far there is the classical *trigonometric moment problem*.

1.5 Problem *Given the complex numbers $\{A_k\}_{k=0}^N$, where N is a positive integer, it is required to find conditions for the existence of a positive finite measure μ on $[0, 2\pi]$, such that the given numbers coincide with the first $N+1$ Fourier coefficients of μ.*

It follows from Proposition 1.5.6 that the trigonometric moment problem for $\{A_k\}_{k=0}^N$ is solvable if and only if the Toeplitz matrix

$$
A^{(0N)} = \begin{bmatrix}
A_0 & A_1 & & A_N \\
A_1^* & A_0 & \ddots & \\
& \ddots & \ddots & \\
A_N^* & & & A_0
\end{bmatrix}
\tag{1.3}
$$

is positive. What is important for our purpose is to remark that this problem can be also reduced to an extension of a partial isometry to a unitary operator, on a possibly larger space. For a certain simplicity, suppose that $A^{(0N)}$ is invertible and consider the Hilbert space $(\mathbb{C}^{N+1})_{A^{(0N)}}$ obtained by renorming $\mathbb{C}^{N+1}$ with $A^{(0N)}$. Since $A^{(0N)}$ is invertible, $(\mathbb{C}^{N+1})_{A^{(0N)}}$ and $\mathbb{C}^{N+1}$ coincide as sets. Furthermore, denote by $\mathcal{E}$ the subspace of $(\mathbb{C}^{N+1})_{A^{(0N)}}$ generated by the set $\{\oplus_{k=0}^{n-1} x_k \oplus 0 \mid x_k \in \mathbb{C},\ k = 0, 1, \ldots, n-1\}$ and denote by $\mathcal{F}$ the subspace of $(\mathbb{C}^{N+1})_{A^{(0N)}}$ generated by the set $\{0 \oplus \oplus_{k=1}^{n} x_k \mid x_k \in \mathbb{C},\ k = 1, \ldots, n\}$. Then, define the operator

$$v : \mathcal{E} \longrightarrow \mathcal{F} \tag{1.4}$$

$$v(\oplus_{k=0}^{n-1} x_k \oplus 0) = 0 \oplus \oplus_{k=1}^{n} x_k, \qquad \oplus_{k=0}^{n-1} x_k \oplus 0 \in \mathcal{E}.$$

Since $A^{(0N)}$ is a Toeplitz matrix, it follows that v is a partial isometry on $(\mathbb{C}^{N+1})_{A^{(0N)}}$. Let w be a unitary extension of v to the Hilbert space $\mathcal{K}$. Using this unitary operator, we can define the probability measure μ by the formula $\mu(f) = \langle f(w^*) E_0, E_0 \rangle_{\mathcal{K}}$, for every continuous function f on $C(\mathbb{T})$. Finally, remark that, for $0 < k \leq N$,

$$
\begin{aligned}
\mu(e^{-ikt}) &= \langle w^k E_0, E_0 \rangle_{\mathcal{K}} = \langle w E_0, (w^*)^{k-1} E_0 \rangle_{\mathcal{K}} \\
&= \langle v E_0, (w^*)^{k-1} E_0 \rangle_{\mathcal{K}} = \langle E_1, (w^*)^{k-1} E_0 \rangle_{\mathcal{K}} \\
&= \langle E_k, E_0 \rangle_{A^{(0N)}} = A_k.
\end{aligned}
$$

Obviously, $\mu(1) = A_0$, hence μ is a solution of the considered trigonometric moment problem. We will show in the next sections that a fair number of apparently unrelated completion problems can be solved by reducing them to a problem of extending a family of partial isometries to a family of unitary operators. This method will settle not only the question of solvability of the problems, but it will also allow us to decide whether there is a unique solution or to give a convenient parametrization of all the solutions. The next two problems, of Hamburger and of M.G. Krein, are related to the theory of moments.

1.6 Problem   *Given the set $\{s_n\}_{n=0}^{\infty}$ of real numbers, it is required to find conditions for the existence of a positive Borel measure μ on $\mathbb{R}$, such that $s_n = \int_{-\infty}^{\infty} t^n d\mu(t)$, $n = 0, 1, 2, \ldots$.*

1.7 Problem   *Given a continuous function $Q : (-a, a) \longrightarrow \mathbb{R}$, $a > 0$, it is required to find conditions for the existence of a positive Borel measure μ on $\mathbb{R}$ such that $Q(x) = \int_{-\infty}^{\infty} e^{-ixt} d\mu(t)$ for $x \in (-a, a)$.*

Also related with the trigonometric moment problem, in fact a generalization of it, there is the so-called *band completion problem*. For an increasing sequence $r = \{r_n\}_{n \in \mathbb{Z}}$ of integers with $r_n \geq n$ for $n \in \mathbb{Z}$, we define the set $\alpha(r) = \{(i,j) \in \mathbb{Z} \times \mathbb{Z} \mid i \leq j \leq r_i\}$.

1.8 Problem   *Given an increasing sequence $r = \{r_n\}_{n \in \mathbb{Z}}$ of integers with $r_n \geq n$ for $n \in \mathbb{Z}$ and a family $\{\widetilde{A}_{ij}\}_{(i,j) \in \alpha(r)}$ of operators, it is required to find conditions for the existence of a positive definite kernel $A = [A_{ij}]_{i,j \in \mathbb{Z}}$ such that $A_{ij} = \widetilde{A}_{ij}$ for $(i,j) \in \alpha(r)$.*

We now introduce some bounded interpolation problems. The first one, referred to as the *Schur problem*, can be easily solved using Theorem 2.2.1 or relating it to the trigonometric moment problem.

1.9 Problem   *Given the complex numbers $\{T_k\}_{k=0}^n$, it is required to find conditions for the existence of a function f in S, such that the given numbers are the first $n + 1$ Taylor coefficients of f.*

This problem has an operator version.

1.10 Problem   *Given an increasing sequence $r = \{r_n\}_{n \in \mathbb{Z}}$ of integers with $r_n \geq n$ for $n \in \mathbb{Z}$ and a family $\{\widetilde{X}_{ij}\}_{(i,j) \in \alpha(r)}$ of operators, it is required to find conditions for the existence of an upper triangular contraction $T = [X_{ij}]_{i,j \in \mathbb{Z}}$ such that $X_{ij} = \widetilde{X}_{ij}$ for $(i, j) \in \alpha(r)$.*

For the next *Nevanlinna-Pick problem*, the connection with the trigonometric moment problem is not so obvious.

1.11 Problem   *Given a set $\{z_k\}_{k=0}^n$ of distinct complex numbers inside the unit disc and a set $\{w_k\}_{k=0}^n$ of complex numbers, it is required to find conditions for the existence of a function f in S such that $f(z_k) = w_k$ for $k = 0, 1, \ldots, n$.*

However, the proof of the following result shows that the Nevanlinna-Pick problem can be also related to the trigonometric moment problem.

1.12 Theorem   *The Nevanlinna-Pick problem for $\{z_k\}_{k=0}^n$ and $\{w_k\}_{k=0}^n$ is solvable if and only if the Pick matrix*

$$
P^{(0n)} = \begin{bmatrix} \dfrac{1 - |w_0|^2}{1 - |z_0|^2} & \dfrac{1 - w_0 w_1^*}{1 - z_0 z_1^*} & \cdots & \dfrac{1 - w_0 w_n^*}{1 - z_0 z_n^*} \\[2mm] \dfrac{1 - w_0^* w_1}{1 - z_0^* z_1} & \dfrac{1 - |w_1|^2}{1 - |z_1|^2} & & \dfrac{1 - w_1 w_n^*}{1 - z_1 z_n^*} \\[2mm] \vdots & & \ddots & \\[2mm] \dfrac{1 - w_0^* w_n}{1 - z_0^* z_n} & & & \dfrac{1 - |w_n|^2}{1 - |z_n|^2} \end{bmatrix}
$$

is positive.

Proof   First, remark that it is more convenient to reformulate the Nevanlinna-Pick problem for functions in the Carathéodory class. This is the class C of analytic functions on the unit disc with positive real part. The connection between the class C and the class S is obtained by using the conformal mapping of the unit disc into the half plane $\mathrm{Re}\, z > 0$ given by $z \longrightarrow (1 - z)(1 + z)^{-1}$. Then, the reformulation of the Nevanlinna-Pick problem for the class C is the following: given a set $\{z_k\}_{k=0}^n$ of distinct complex numbers inside the unit disc and a set $\{w_k'\}_{k=0}^n$ of complex numbers, it is required to find conditions for the existence of a function g in C such that $g(z_k) = w_k'$, $k = 0, 1, \ldots, n$. Suppose this problem has a solution g in

$\mathcal{C}$. According to an integral representation result of G. Herglotz, there exists a positive measure μ on $[0, 2\pi)$ such that

$$g(z) = i\mathrm{Im}g(0) + \frac{1}{4\pi}\int_0^{2\pi}\frac{e^{it}+z}{e^{it}-z}d\mu(t),$$

hence

$$w_k' = i\mathrm{Im}g(0) + \frac{1}{4\pi}\int_0^{2\pi}\frac{e^{it}+z_k}{e^{it}-z_k}d\mu(t), \quad k = 0,1,\dots n.$$

It follows that

$$\frac{w_k' + w_j'^*}{1 - z_k z_j^*} = \frac{1}{2\pi}\int_0^{2\pi}\frac{1}{(e^{it}-z_k)(e^{-it}-z_j^*)}d\mu(t). \tag{1.5}$$

Now the Lagrange interpolation can be used as follows. Define $q(z) = \prod_{k=0}^n(z - z_k)$, then the Lagrange formula gives that

$$u(z) = \sum_{k=0}^n\frac{u(z_k)}{q'(z_k)}\frac{q(z)}{z - z_k} \tag{1.6}$$

for every polynomial u of degree $\le n$. Therefore, the equality (1.5) becomes:

$$\frac{w_k' + w_j'^*}{1 - z_k z_j^*} = \frac{1}{2\pi}\int_0^{2\pi}\frac{q(e^{it})}{e^{it}-z_k}\times\frac{q(e^{it})^*}{e^{-it}-z_j^*}\frac{d\mu(t)}{|q(e^{it})|^2},$$

which is equivalent to the equality $Q^{(0n)} = L^{-1}A^{(0n)}(L^{-1})^*$, where $Q^{(0n)} = \left[\frac{w_k'+w_j'^*}{1-z_k z_j^*}\right]_{j,k=0}^n$, $L = \left[\frac{z_k^p}{q'(z_k)}\right]_{p,k=0}^n$ and $A^{(0n)}$ is the Toeplitz matrix of the first $n + 1$ Fourier coefficients of the measure $|q|^{-2}\mu$. In other words, if the considered Nevanlinna-Pick problem has a solution, then the trigonometric moment problem with the matrix data $A^{(0n)} = LQ^{(0n)}L^*$ has a solution. It is easy to fill in the details concerning the converse and they can be omitted. The connection between the Nevanlinna-Pick problem for the class $\mathcal{C}$ and the Nevanlinna-Pick problem for the class $\mathcal{S}$ is obtained using the afore mentioned connection between the Schur and the Carathéodory classes. $\qquad\square$

A generalization of both the Schur problem and the Nevanlinna-Pick problem is the following *Hermite-Fejér problem*, mentioned here in a so-called *tangential* formulation. For a function f in $\mathcal{S}(\mathcal{H}, \mathcal{H}')$, a positive integer n and a complex number z inside the unit disc, we define

$$\mathcal{H}_f(z) = \begin{bmatrix} f(z) & \frac{1}{1!}f^{(1)}(z) & & & \frac{1}{(n-1)!}f^{(n-1)}(z) \\ & f(z) & \frac{1}{1!}f^{(1)}(z) & \ddots & \\ & & \ddots & \ddots & \\ & 0 & & & \frac{1}{1!}f^{(1)}(z) \\ & & & & f(z) \end{bmatrix}. \tag{1.7}$$

1.13 Problem   *Given two sets $\{u_j^{(i)} \mid 0 \le i \le m - 1, 0 \le j < r_i\}$ and $\{v_j^{(i)} \mid 0 \le i \le m - 1, 0 \le j < r_i\}$ of operators in $\mathcal{L}(\mathcal{H}', \mathcal{G})$ and $\mathcal{L}(\mathcal{H}, \mathcal{G})$ respectively, it is required to find conditions for the existence of a function f in $S(\mathcal{H}, \mathcal{H}')$ such that*

$$\begin{bmatrix} v_1^{(i)} & \cdots & v_{r_i}^{(i)} \end{bmatrix} = \begin{bmatrix} u_1^{(i)} & \cdots & u_{r_i}^{(i)} \end{bmatrix} \mathcal{H}_f(z_i), \qquad 0 \le i \le m - 1.$$

Another problem which generalizes in other direction both Schur and Nevanlinna-Pick problems, is the following *Nehari problem*.

1.14 Problem   *Given a sequence $\{c_n\}_{n=-\infty}^{-1}$ of complex numbers, it is required to find conditions for the existence of a function F in L^∞ such that $\|F\|_\infty \le 1$ and $c_n = \frac{1}{2\pi} \int_0^{2\pi} e^{-int} F(e^{it}) dt$ for $n = -1, -2, \ldots$.*

The Nehari problem can be reformulated as an approximation problem. More precisely, it will be easy to see that a solution of Problem 1.14 will lead to the computation of the distance from a function in L^∞ to the Hardy space H^∞. It turns out that this computation is quite useful in answering some questions in control theory. For instance, let there be given two Schur functions T_1 and T_2. The *model-matching problem* requires to find a Schur function Q so as to minimize $\|T_1 - T_2 Q\|_\infty$. In other words, if T_1 is a model and T_2 is a plant, a controller Q is required so as $T_2 Q$ approximates T_1 in L^∞. After some manipulations, the model matching problem turns into a particular case of the following *four block problem*.

1.15 Problem   *Given a function $L = [L_{ij}]_{i,j=1}^2$ in $L^\infty(\mathcal{L}(\mathcal{H}_1 \oplus \mathcal{H}_2, \mathcal{H}_1' \oplus \mathcal{H}_2'))$, it is required to find conditions for the existence of a function Q in $H^\infty(\mathcal{L}(\mathcal{H}_1, \mathcal{H}_1'))$ such that*

$$\left\| \begin{bmatrix} L_{11} - Q & L_{12} \\ L_{21} & L_{22} \end{bmatrix} \right\|_{L^\infty(\mathcal{L}(\mathcal{H}_1 \oplus \mathcal{H}_2, \mathcal{H}_1' \oplus \mathcal{H}_2'))} \le 1.$$

In conclusion, several completion problems were introduced and a general framework for their solution was suggested. We will develop this framework in the following sections.

3.2   Extensions of Partial Isometries

In this section we formulate and solve a problem of extending families of partial isometries to unitary operators and we explain the role played by the Schur parameters in the description of the solutions of this problem.

2.1 Problem   *Given families $\mathbf{E} = \{\mathcal{E}_n\}_{n \in \mathbb{Z}}$, $\mathbf{F} = \{\mathcal{F}_n\}_{n \in \mathbb{Z}}$ and $\mathbf{G} = \{\mathcal{G}_n\}_{n \in \mathbb{Z}}$ of Hilbert spaces such that $\mathcal{E}_n$ and $\mathcal{F}_n$ are subspaces of $\mathcal{G}_n$ for every $n \in \mathbb{Z}$, and a family $\mathbf{v} = \{v_n\}_{n \in \mathbb{Z}}$ of unitary operators v_n in $\mathcal{L}(\mathcal{E}_{n+1}, \mathcal{F}_n)$, it is required to find families $\mathbf{K} = \{\mathcal{K}_n\}_{n \in \mathbb{Z}}$ of Hilbert spaces such that*

(α) $\mathcal{G}_n \subseteq \mathcal{K}_n$ for all $n \in \mathbb{Z}$,

and families $\mathbf{w} = \{w_n\}_{n \in \mathbb{Z}}$ of unitary operators w_n in $\mathcal{L}(\mathcal{K}_{n+1}, \mathcal{K}_n)$ such that

(β) $w_n / \mathcal{E}_{n+1} = v_n$ for all $n \in \mathbb{Z}$

(γ) $\mathcal{K}_0 = \vee_{n=-\infty}^{-1} w_{-1}^* \dots w_n^* \mathcal{G}_n \vee \mathcal{G}_0 \vee \vee_{n=1}^{\infty} w_0 w_1 \dots w_{n-1} \mathcal{G}_n.$

We introduce an equivalence relation on the set of solutions of this problem in the following way: two families $\mathbf{w} = \{w_n\}_{n \in \mathbb{Z}}$, $\mathbf{w}' = \{w_n'\}_{n \in \mathbb{Z}}$ of unitary operators w_n in $\mathcal{L}(\mathcal{K}_{n+1}, \mathcal{K}_n)$ and, respectively, w_n' in $\mathcal{L}(\mathcal{K}_{n+1}', \mathcal{K}_n')$ satisfying (α)-(γ) are equivalent if there exists another family $\{\varphi_n\}_{n \in \mathbb{Z}}$ of unitary operators φ_n in $\mathcal{L}(\mathcal{K}_n, \mathcal{K}_n')$ such that for all $n \in \mathbb{Z}$, $\varphi_n / \mathcal{G}_n = I_{\mathcal{G}_n}$ and $w_n' \varphi_{n+1} = \varphi_n w_n$. We denote by $\mathcal{E}(\mathbf{E}, \mathbf{F}; \mathbf{v})$ the set of equivalence classes determined by this relation. For $n \in \mathbb{Z}$, we define the spaces $\mathcal{Q}_n = \mathcal{G}_n \ominus \mathcal{E}_n$ and $\mathcal{R}_n = \mathcal{G}_n \ominus \mathcal{F}_n$. The main result concerning Problem 2.1 is the following.

2.2 Theorem   *There exists a one-to-one correspondence between the set $\mathcal{E}(\mathbf{E}, \mathbf{F}; \mathbf{v})$ and $\Pi(\{\mathcal{Q}_n\}_{n \in \mathbb{Z}}, \{\mathcal{R}_n\}_{n \in \mathbb{Z}})$.*

Proof   The proof uses the construction of the Kolmogorov decomposition presented in Section 1.6. Thus, consider a family $\Gamma = \{\Gamma_{ij} \mid i, j \in \mathbb{Z}, i \leq j\}$ of Schur parameters in $\Pi(\{\mathcal{Q}_n\}_{n \in \mathbb{Z}}, \{\mathcal{R}_n\}_{n \in \mathbb{Z}})$ and introduce, for $n \in \mathbb{Z}$, the row contractions

$$L_n : \oplus_{k=n+1}^{\infty} \mathcal{D}_{\Gamma_{n+1,k}} \longrightarrow \mathcal{R}_n \tag{2.1}$$

$$L_n = \quad \text{the row contraction associated to} \atop \text{the parameters} \quad \{\Gamma_{nk} \mid n < k\}.$$

Define $\mathcal{D}_n = \oplus_{k=n}^{\infty} \mathcal{D}_{\Gamma_{nk}}$ and $\mathcal{D}_{n,*} = \mathrm{cl}\mathcal{R}(H_\infty(L_n))$, where $H_\infty(L_n)$ was introduced by (1.4.10). Consider the Hilbert spaces $\mathcal{K}_n^1 = \oplus_{k=-\infty}^{n-1} \mathcal{D}_{k,*} \oplus \mathcal{Q}_n \oplus \mathcal{D}_n$ and $\mathcal{K}_n^2 = \oplus_{k=-\infty}^{n-1} \mathcal{D}_{k,*} \oplus \mathcal{R}_n \oplus \mathcal{D}_n$, and the unitary operators

$$w_n^1 : \mathcal{K}_{n+1}^1 \longrightarrow \mathcal{K}_n^2 \tag{2.2}$$

$$w_n^1 = I \oplus \begin{bmatrix} I & 0 \\ 0 & \alpha_{L_n} \end{bmatrix} R(L_n) \begin{bmatrix} 0 & I \\ \beta_{L_n}^* & 0 \end{bmatrix},$$

with respect to the decompositions $\mathcal{K}_{n+1}^1 = (\oplus_{k=-\infty}^{n-1} \mathcal{D}_{k,*}) \oplus (\mathcal{D}_{n,*} \oplus \mathcal{Q}_{n+1} \oplus \mathcal{D}_{n+1})$ and, respectively, $\mathcal{K}_n^2 = (\oplus_{k=-\infty}^{n-1} \mathcal{D}_{k,*}) \oplus (\mathcal{R}_n \oplus \mathcal{D}_n)$. Remember that $R(L_n)$ denotes the elementary rotation of L_n, while α_{L_n} and β_{L_n} are the unitary operators defined by (1.4.11) and (1.4.12). Finally, define for $n \in \mathbb{Z}$, the Hilbert spaces

$$\mathcal{K}_n = \mathcal{K}_n^1 \oplus \mathcal{E}_n = \mathcal{K}_n^2 \oplus \mathcal{F}_n \tag{2.3}$$

and the unitary operators

$$w_n(\Gamma) : \mathcal{K}_{n+1} \longrightarrow \mathcal{K}_n \tag{2.4}$$

$$w_n(\Gamma) = w_n^1 \oplus v_n.$$

It is easy to see that $\mathbf{w}(\Gamma) = \{w_n(\Gamma)\}_{n \in \mathbb{Z}}$ is a solution of Problem 2.1 and define the map

$$\Phi : \Pi(\{\mathcal{Q}_n\}_{n \in \mathbb{Z}}, \{\mathcal{R}_n\}_{n \in \mathbb{Z}}) \longrightarrow \mathcal{E}(\mathbf{E}, \mathbf{F}; \mathbf{v}) \tag{2.5}$$

$$\Phi(\Gamma) = [\mathbf{w}(\Gamma)],$$

where $[\mathbf{w}(\Gamma)]$ denotes the class of $\mathbf{w}(\Gamma)$ in $\mathcal{E}(\mathbf{E}, \mathbf{F}; \mathbf{v})$. It will be shown that the map Φ establishes a one-to-one correspondence between the sets $\Pi(\{\mathcal{Q}_n\}_{n \in \mathbb{Z}}, \{\mathcal{R}_n\}_{n \in \mathbb{Z}})$ and $\mathcal{E}(\mathbf{E}, \mathbf{F}; \mathbf{v})$.

First, suppose that $\Phi(\Gamma) = \Phi(\Gamma')$ for two sets Γ, Γ' in $\Pi(\{\mathcal{Q}_n\}_{n \in \mathbb{Z}}, \{\mathcal{R}_n\}_{n \in \mathbb{Z}})$. Therefore, there exists a family $\{\varphi_n\}_{n \in \mathbb{Z}}$ of unitary operators φ_n in $\mathcal{L}(\mathcal{K}_n, \mathcal{K}'_n)$ such that $\varphi_n / \mathcal{G}_n = I_{\mathcal{G}_n}$ and $w_n(\Gamma')\varphi_{n+1} = \varphi_n w_n(\Gamma)$ for all $n \in \mathbb{Z}$. Consequently, for $i, j \in \mathbb{Z}$ and $i < j$,

$$
\begin{aligned}
P_{\mathcal{G}_i}^{\mathcal{K}_i} w_i(\Gamma) \ldots w_{j-1}(\Gamma) / \mathcal{G}_j &= P_{\mathcal{G}_i}^{\mathcal{K}_i} w_i(\Gamma) \ldots w_{j-1}(\Gamma) \varphi_j^* / \mathcal{G}_j \\
&= P_{\mathcal{G}_i}^{\mathcal{K}_i} \varphi_i^* w_i(\Gamma') \ldots w_{j-1}(\Gamma') / \mathcal{G}_j \\
&= P_{\mathcal{G}_i}^{\mathcal{K}_i} w_i(\Gamma') \ldots w_{j-1}(\Gamma') / \mathcal{G}_j.
\end{aligned}
$$

By Theorem 1.5.3 and Theorem 1.6.1, it follows that $\Gamma = \Gamma'$. This shows that Φ is injective. Then, consider $\mathbf{w} = \{w_n\}_{n \in \mathbb{Z}}$ a representative of a class in $\mathcal{E}(\mathbf{E}, \mathbf{F}; \mathbf{v})$. Remark that for all $n \in \mathbb{Z}$, the operator $A_{n,n+1} = P_{\mathcal{G}_n}^{\mathcal{K}_n} w_n / \mathcal{G}_{n+1}$ does not depend on the chosen representative, and $A_{n,n+1} = v_n \oplus \Gamma_{n,n+1}$, where $\Gamma_{n,n+1}$ is a contraction in $\mathcal{L}(\mathcal{Q}_{n+1}, \mathcal{R}_n)$. Further on, define for $i, j \in \mathbb{Z}$ and $i < j$, the operators $A_{ij} = P_{\mathcal{G}_i}^{\mathcal{K}_i} w_i \ldots w_{j-1} / \mathcal{G}_j$ which, in their turn, do not depend on the chosen representative $\mathbf{w}$. Now, define the kernel $A = [A_{ij}]_{i,j \in \mathbb{Z}}$ by

$$
A_{ij} = \begin{cases}
A_{ii} = I_{\mathcal{G}_i} & \text{if } i = j \\
P_{\mathcal{G}_i}^{\mathcal{K}_i} w_i \ldots w_{j-1} / \mathcal{G}_j & \text{if } i < j \\
A_{ji}^* & \text{if } i > j.
\end{cases}
$$

This is obviously a positive definite kernel and let $\Gamma' = \{\Gamma'_{ij} \mid i, j \in \mathbb{Z}, i \leq j\}$ be the set of the Schur parameters of A. Then, for all $n \in \mathbb{Z}$, the following equalities hold: $\Gamma'_{n,n+1} = v_n \oplus \Gamma_{n,n+1}$ and $\mathcal{D}_{\Gamma'_{n,n+1}} = 0 \oplus \mathcal{D}_{\Gamma_{n,n+1}}$, $\mathcal{D}_{\Gamma'^*_{n,n+1}} = 0 \oplus \mathcal{D}_{\Gamma^*_{n,n+1}}$. Consequently, the set $\Gamma = \{\Gamma_{n,n+1}\}_{n \in \mathbb{Z}} \cup \{\Gamma'_{ij} \mid i, j \in \mathbb{Z}, i < j + 1\}$ belongs to $\Pi(\{\mathcal{Q}_n\}_{n \in \mathbb{Z}}, \{\mathcal{R}_n\}_{n \in \mathbb{Z}})$ and $\Phi(\Gamma) = [\mathbf{w}]$, i.e. Φ is surjective. $\qquad \square$

We can note as an immediate consequence of Theorem 2.2.

2.3 Corollary *The set $\mathcal{E}(\mathbf{E}, \mathbf{F}; \mathbf{v})$ contains exactly one element if and only if for each $n \in \mathbb{Z}$, either $\mathcal{E}_n = \mathcal{G}_n$ or $\mathcal{F}_n = \mathcal{G}_n$.* $\qquad \square$

We will always distinguish a special element in $\mathcal{E}(\mathbf{E}, \mathbf{F}; \mathbf{v})$, namely the class of $\mathbf{w}^0 = \{w_n^0\}_{n \in \mathbb{Z}}$, where $\mathbf{w}^0$ is the solution of Problem 2.1 associated to the set Γ^0 of the Schur parameters $\Gamma_{ij}^0 = 0$ for all $i \leq j$. This will be referred to as the *central solution* of Problem 2.1.

Finally, we mention the following particular case of Theorem 2.2. Consider a partial isometry v on the Hilbert space $\mathcal{G}$, then a unitary operator w on $\mathcal{K}$ is called *minimal unitary extension* of v if $\mathcal{G} \subset \mathcal{K}$, $w / \mathcal{G} = v$ and $\vee_{n \in \mathbb{Z}} w^n \mathcal{G} = \mathcal{K}$. Two minimal unitary extensions w on $\mathcal{K}$ and w' on $\mathcal{K}'$ are equivalent if there exists a unitary operator φ in $\mathcal{L}(\mathcal{K}, \mathcal{K}')$ such that $\varphi / \mathcal{G} = I_{\mathcal{G}}$ and $w'\varphi = \varphi w$. We denote by

$\mathcal{E}(v)$ the set of the minimal unitary extensions of v, factorized by this equivalence relation. Define $\mathcal{E} = \mathcal{G} \ominus \ker v$, $\mathcal{F} = \mathcal{R}(v)$. We have the following result.

2.4 Corollary   *There exists a one-to-one correspondence between the set $\mathcal{E}(v)$ and the set of Schur parameters $\Pi^0(\mathcal{G} \ominus \mathcal{E}, \mathcal{G} \ominus \mathcal{F})$.*

Proof We remark that the map Φ defined by (2.5) has the property that $\Phi / \Pi^0(\mathcal{G} \ominus \mathcal{E}, \mathcal{G} \ominus \mathcal{F})$ establishes a one-to-one correspondence between $\Pi^0(\mathcal{G} \ominus \mathcal{E}, \mathcal{G} \ominus \mathcal{F})$ and $\mathcal{E}(v)$. $\qquad\square$

3.3 Krein's Formula

In this section we obtain another parametrization of the set $\mathcal{E}(\mathbf{E}, \mathbf{F}; \mathbf{v})$, based on the notion of generalized coresolvent. For this purpose we need to introduce some additional notation. Suppose $\mathbf{H} = \{\mathcal{H}_n\}_{n \in \mathbb{Z}}$ is a family of Hilbert spaces and for $i, j \in \mathbb{Z}$, $i \leq j$, we define the Hilbert space $\mathcal{H}^{(ij)} = \oplus_{k=i}^j \mathcal{H}_k$. Moreover, if a family $\{T_n\}_{n \in \mathbb{Z}}$ of operators T_n in $\mathcal{L}(\mathcal{H}_{n+1}, \mathcal{H}'_n)$ is given, then one defines $T^{(ij)} = \oplus_{k=i}^j T_k$, which is a bounded operator from $\mathcal{H}^{(i+1, j+1)}$ into $\mathcal{H}'^{(ij)}$. The *truncated marking operators* are defined by the formula:

$$N^{(ij)} = N^{(ij)}(\mathbf{H}) : \mathcal{H}^{(ij)} \longrightarrow \mathcal{H}^{(i+1, j+1)} \qquad (3.1)$$

$$N^{(ij)}(\oplus_{k=i}^j h_k) = \oplus_{k=j+1}^j h_k \oplus 0.$$

If $\mathbf{w} = \{w_n\}_{n \in \mathbb{Z}}$ is a representative of a class in $\mathcal{E}(\mathbf{E}, \mathbf{F}; \mathbf{v})$, then the *generalized coresolvent* of this class is defined by

$$\mathcal{C}^{(ij)}([\mathbf{w}]) = P_{\mathcal{G}^{(ij)}}^{\mathcal{K}^{(ij)}} (I + w^{(ij)} N^{(ij)})(I - w^{(ij)} N^{(ij)})^{-1} / \mathcal{G}^{(ij)}. \qquad (3.2)$$

It is readily seen that this definition does not depend on the chosen representative $\mathbf{w}$. Notice now that each unitary operator of a family $\mathbf{w}$ which is a solution of Problem 2.1 has the matrix representation

$$w_n = \begin{bmatrix} \hat{w}_n & \hat{C}_n \\ \hat{B}_n & \hat{A}_n \end{bmatrix} \qquad (3.3)$$

with respect to the decompositions $\mathcal{G}_{n+1} \oplus (\mathcal{K}_{n+1} \ominus \mathcal{G}_{n+1})$ and $\mathcal{G}_n \oplus (\mathcal{K}_n \ominus \mathcal{G}_n)$. The next result contains the basic computations regarding the generalized coresolvent. It exhibits a connection between the generalized coresolvents of an arbitrary solution of Problem 2.1 and of the central solution, in terms of the matrix representation (3.3).

3.1 Lemma   *If $\mathbf{w}$ is a solution of Problem 2.1, then for $i, j \in \mathbb{Z}$, $i < j$, the following formula holds:*

$$\mathcal{C}^{(ij)}([\mathbf{w}]) = \mathcal{C}^{(ij)}([\mathbf{w}^0]) + 2(I - (\hat{w}^0)^{(ij)} N^{(ij)})^{-1} \hat{\Theta}^{(ij)} N^{(ij)}$$
$$\times [I - (I - (\hat{w}^0)^{(ij)} N^{(ij)})^{-1} \hat{\Theta}^{(ij)} N^{(ij)}]^{-1} (I - (\hat{w}^0)^{(ij)} N^{(ij)})^{-1},$$

where $\hat{\Theta}^{(ij)} = \hat{D}^{(ij)} + \hat{C}^{(ij)} N^{(ij)} (I - \hat{A}^{(ij)} N^{(ij)})^{-1} \hat{B}^{(ij)}$ and $\hat{D}_n = \hat{w}_n - \hat{w}_n^0$ for $n \in \mathbb{Z}$.

Proof Fix $i, j \in \mathbb{Z}$, $i < j$. It is convenient to introduce the notation $S_{ij} = (I - (\hat{w}^0)^{(ij)} N^{(ij)})^{-1}$ and $T_{ij} = (I - \hat{w}^{(ij)} N^{(ij)})^{-1}$. Using the matrix representation (3.3) of w_n and the inversion formula (1.6.14), it follows that

$$
\begin{aligned}
\mathcal{C}^{(ij)}([\mathbf{w}]) &= -I_{\mathcal{G}^{(ij)}} + 2 P_{\mathcal{G}^{(ij)}}^{\mathcal{K}^{(ij)}} (I - w^{(ij)} N^{(ij)})^{-1} / \mathcal{G}^{(ij)} \\
&= -I + 2[T_{ij} + T_{ij} \hat{C}^{(ij)} N^{(ij)} (I - \hat{A}^{(ij)} N^{(ij)}) \\
&\quad - \hat{B}^{(ij)} N^{(ij)} T_{ij} \hat{C}^{(ij)} N^{(ij)})^{-1} \hat{B}^{(ij)} N^{(ij)} T_{ij}] \\
&= -I + 2\{T_{ij} + T_{ij} \hat{C}^{(ij)} N^{(ij)} (I - \hat{A}^{(ij)} N^{(ij)})^{-1} \hat{B}^{(ij)} N^{(ij)} \\
&\quad \times [I - T_{ij} \hat{C}^{(ij)} N^{(ij)} (I - \hat{A}^{(ij)} N^{(ij)})^{-1} \hat{B}^{(ij)} N^{(ij)}]^{-1} T_{ij}\}.
\end{aligned}
$$

The computation of $\mathcal{C}^{(ij)}([\mathbf{w}^0])$ is simpler and one obtains $\mathcal{C}^{(ij)}([\mathbf{w}^0]) = -I_{\mathcal{G}^{(ij)}} + 2S_{ij}$, hence the generalized coresolvent of the class $[\mathbf{w}]$ can be expressed in terms of the generalized coresolvent of the class $[\mathbf{w}^0]$ and the matrix representation (3.3) according to the following formula:

$$
\begin{aligned}
\mathcal{C}^{(ij)}([\mathbf{w}]) - \mathcal{C}^{(ij)}([\mathbf{w}^0]) &= 2(T_{ij} - S_{ij}) + 2T_{ij} \hat{C}^{(ij)} N^{(ij)} \\
&\quad \times (I - \hat{A}^{(ij)} N^{(ij)})^{-1} \hat{B}^{(ij)} N^{(ij)} \\
&\quad \times [I - T_{ij} \hat{C}^{(ij)} N^{(ij)} (I - \hat{A}^{(ij)} N^{(ij)})^{-1} \hat{B}^{(ij)} N^{(ij)}]^{-1} T_{ij}.
\end{aligned}
$$

Using now the operators $\hat{D}_n = \hat{w}_n - \hat{w}_n^0$ and the identity $T_{ij} - S_{ij} = S_{ij} \hat{D}^{(ij)} N^{(ij)} T_{ij}$, it follows that

$$
T_{ij} = (I - S_{ij} \hat{D}^{(ij)} N^{(ij)})^{-1} S_{ij} = S_{ij} (I - \hat{D}^{(ij)} N^{(ij)} S_{ij})^{-1}.
$$

The computation of the generalized coresolvent of the class $[\mathbf{w}]$ can be continued and one obtains:

$$
\begin{aligned}
\mathcal{C}^{(ij)}([\mathbf{w}]) - \mathcal{C}^{(ij)}([\mathbf{w}^0]) &= 2S^{(ij)} [\hat{D}^{(ij)} N^{(ij)} (I - S^{(ij)} \hat{D}^{(ij)} N^{(ij)})^{-1} \\
&\quad + (I - \hat{D}^{(ij)} N^{(ij)} S^{(ij)})^{-1} \hat{C}^{(ij)} N^{(ij)} (I - \hat{A}^{(ij)} N^{(ij)})^{-1} \\
&\quad \times \hat{B}^{(ij)} N^{(ij)} Q_{ij} (I - S^{(ij)} \hat{D}^{(ij)} N^{(ij)})^{-1}] S^{(ij)},
\end{aligned}
$$

where we defined

$$
Q_{ij} = (I - (I - S_{ij} \hat{D}^{(ij)} N^{(ij)})^{-1} S_{ij} \hat{C}^{(ij)} N^{(ij)} (I - \hat{A}^{(ij)} N^{(ij)})^{-1} \hat{B}^{(ij)} N^{(ij)})^{-1}.
$$

Finally, the term $\hat{\Theta}^{(ij)}$ can be isolated in the previous expression of $\mathcal{C}^{(ij)}([\mathbf{w}]) - \mathcal{C}^{(ij)}([\mathbf{w}^0])$. Thus, one obtains:

$$\mathcal{C}^{(ij)}([\mathbf{w}]) - \mathcal{C}^{(ij)}([\mathbf{w}^0]) = 2S_{ij}(I - \hat{D}^{(ij)}N^{(ij)}S_{ij})^{-1}[\hat{D}^{(ij)}N^{(ij)}$$
$$+\hat{C}^{(ij)}N^{(ij)}(I - \hat{A}^{(ij)}N^{(ij)})^{-1}\hat{B}^{(ij)}N^{(ij)}(I - S_{ij}\hat{\Theta}^{(ij)}N^{(ij)})^{-1}]S_{ij}$$
$$= 2S_{ij}(I - \hat{D}^{(ij)}N^{(ij)}S_{ij})^{-1}[\hat{D}^{(ij)}N^{(ij)}(I - S_{ij}\hat{\Theta}^{(ij)}N^{(ij)})$$
$$+\hat{C}^{(ij)}N^{(ij)}(I - \hat{A}^{(ij)}N^{(ij)})^{-1}\hat{B}^{(ij)}N^{(ij)}](I - S_{ij}\hat{\Theta}^{(ij)}N^{(ij)})^{-1}S_{ij}$$
$$= 2S_{ij}(I - \hat{D}^{(ij)}N^{(ij)}S_{ij})^{-1}[\hat{\Theta}^{(ij)}N^{(ij)} - \hat{D}^{(ij)}N^{(ij)}(I - S_{ij}\hat{\Theta}^{(ij)}N^{(ij)})]$$
$$\times(I - S_{ij}\hat{\Theta}^{(ij)}N^{(ij)})^{-1}S_{ij} = 2S_{ij}\hat{\Theta}^{(ij)}N^{(ij)}(I - S_{ij}\hat{\Theta}^{(ij)}N^{(ij)})^{-1}S_{ij},$$

which is exactly the required formula. □

Now, we can obtain the main result of this section.

3.2 Theorem   *There exists a one-to-one correspondence between the set* $\mathcal{E}(\mathbf{E}, \mathbf{F}; \mathbf{v})$ *and the set* $\mathcal{S}(\{\mathcal{Q}_n\}_{n \in \mathbb{Z}}, \{\mathcal{R}_n\}_{n \in \mathbb{Z}})$.

Proof Let $\mathbf{w} = \{w_n\}_{n \in \mathbb{Z}}$ be a solution of Problem 2.1 and it follows from (3.3) that each w_n has the matrix representation

$$w_n = \begin{bmatrix} v_n & 0 & 0 \\ 0 & \Gamma_{n,n+1} & C_n \\ 0 & B_n & \hat{A}_n \end{bmatrix} \tag{3.4}$$

with respect to the decompositions $\mathcal{E}_{n+1} \oplus \mathcal{Q}_{n+1} \oplus (\mathcal{K}_{n+1} \ominus \mathcal{G}_{n+1})$ and $\mathcal{F}_n \oplus \mathcal{R}_n \oplus (\mathcal{K}_n \ominus \mathcal{G}_n)$, where Γ is the set of Schur parameters associated to the class $[\mathbf{w}]$ in $\mathcal{E}(\mathbf{E}, \mathbf{F}; \mathbf{v})$ by Theorem 2.2. Consequently, the formula obtained in Lemma 3.1 can be rewritten in the following form, which will be referred to as the *generalized coresolvent formula* (or *Krein's formula*):

$$\mathcal{C}^{(ij)}([\mathbf{w}]) = \mathcal{C}^{(ij)}([\mathbf{w}^0)]) + 2(I - (\hat{w}^0)^{(ij)}N^{(ij)})^{-1}$$
$$\times P_{\mathcal{R}^{(ij)}}\Theta^{(ij)}N^{(ij)}(I - X^{(ij)}\Theta^{(ij)}N^{(ij)})^{-1}P_{\mathcal{Q}^{(ij)}}(I - (\hat{w}^0)^{(ij)}N^{(ij)})^{-1},$$

where

$$\Theta^{(ij)} : \mathcal{Q}^{(i+1,j+1)} \longrightarrow \mathcal{R}^{(ij)} \tag{3.5}$$
$$\Theta^{(ij)} = \Gamma^{(ij)} + C^{(ij)}N^{(ij)}(I - \hat{A}^{(ij)}N^{(ij)})^{-1}B^{(ij)}$$

and

$$X^{(ij)} : \mathcal{R}^{(ij)} \longrightarrow \mathcal{Q}^{(ij)} \tag{3.6}$$
$$X^{(ij)} = P_{\mathcal{Q}^{(ij)}}(I - (\hat{w}^0)^{(ij)}N^{(ij)})^{-1}/\mathcal{R}^{(ij)}.$$

Since $\begin{bmatrix} \Gamma_{n,n+1} & C_n \\ B_n & \hat{A}_n \end{bmatrix}$ is a unitary matrix for every $n \in \mathbb{Z}$ and the operator Θ introduced by (3.5) is the transfer map of the unitary system

$$\Omega \begin{cases} x_n = \hat{A}_n x_{n+1} + B_n u_n \\ y_n = C_n x_{n+1} + \Gamma_{n,n+1}, \quad n \in \mathbb{Z}, \end{cases}$$

it follows by Lemma 2.3.1 that Θ belongs to $\mathcal{S}(\{Q_n\}_{n\in\mathbb{Z}}, \{\mathcal{R}_n\}_{n\in\mathbb{Z}})$. Moreover, it is readily checked that the operator X defined by (3.6) belongs to $\mathcal{S}(\{\mathcal{R}_n\}_{n\in\mathbb{Z}}, \{Q_n\}_{n\in\mathbb{Z}})$. Consequently, we can define the map

$$\Psi : \mathcal{E}(\mathbf{E}, \mathbf{F}; \mathbf{v}) \longrightarrow \mathcal{S}(\{Q_n\}_{n\in\mathbb{Z}}, \{\mathcal{R}_n\}_{n\in\mathbb{Z}}) \tag{3.7}$$

$$\Psi([\mathbf{w}]) = \Theta,$$

where Θ is associated to the class $[\mathbf{w}]$ by the generalized coresolvent formula. This map is well defined and it remains to show that it establishes a one-to-one correspondence between the sets $\mathcal{E}(\mathbf{E}, \mathbf{F}; \mathbf{v})$ and $\mathcal{S}(\{Q_n\}_{n\in\mathbb{Z}}, \{\mathcal{R}_n\}_{n\in\mathbb{Z}})$.

First, suppose $\Psi([\mathbf{w}]) = \Psi([\mathbf{w}'])$, then $\Theta = \Theta'$ and $\mathcal{C}([\mathbf{w}]) = \mathcal{C}([\mathbf{w}'])$ for all $i, j \in \mathbb{Z}$, $i < j$. By the definition of $\mathcal{C}([\mathbf{w}])$, it follows that for all $i, j \in \mathbb{Z}$, $i < j$,

$$P^{\mathcal{K}_i}_{\mathcal{G}_i} w_i \dots w_{j-1}/\mathcal{G}_i = P^{\mathcal{K}_i}_{\mathcal{G}_i} w'_i \dots w'_{j-1}/\mathcal{G}_i. \tag{3.8}$$

This equality suggests to consider the operators $V(n)$ in $\mathcal{L}(\mathcal{G}_n, \mathcal{K}_0)$ defined by the formula

$$V(n) = \begin{cases} w^*_{-1} \dots w^*_n/\mathcal{G}_n & \text{if } n < 0 \\ P^{\mathcal{K}_0}_{\mathcal{G}_0}/\mathcal{G}_0 & \text{if } n = 0 \\ w_0 w_1 \dots w_{n-1}/\mathcal{G}_n & \text{if } n > 0. \end{cases}$$

A similar family $\{V'(n)\}_{n\in\mathbb{Z}}$ of operators $V'(n)$ in $\mathcal{L}(\mathcal{G}_n, \mathcal{K}_0)$ is defined for $\mathbf{w}'$. Then, the minimality condition (γ) in the statement of Problem 2.1 can be restated as $\mathcal{K}_0 = \vee_{k\in\mathbb{Z}} V(k)\mathcal{G}_k$ and, respectively, $\mathcal{K}'_0 = \vee_{k\in\mathbb{Z}} V'(k)\mathcal{G}_k$. Consequently, we can define the map

$$\varphi_0 : \mathcal{K}_0 \longrightarrow \mathcal{K}'_0 \tag{3.9}$$

$$\varphi_0\left(\sum_{k\in\mathbb{Z}} V(k)g_k\right) = \sum_{k\in\mathbb{Z}} V'(k)g_k,$$

where only a finite number of vectors g_k in $\mathcal{G}_k$ are different from zero. By (3.8), it follows that φ_0 is a unitary operator. On the other hand, since $\mathcal{K}_n = w^*_{n-1} \dots w^*_0 \mathcal{K}_0$ for $n > 0$ and $\mathcal{K}_n = w_n \dots w_{-1}\mathcal{K}_0$ for $n < 0$, it follows that it is possible to define the operators $\varphi_n = w'^*_{n-1} \dots w'^*_0 \varphi_0 w_0 \dots w_{n-1}$ for $n > 0$, and $\varphi_n = w'_n \dots w'_{-1}\varphi_0 w^*_{-1} \dots w^*_n$ for $n < 0$. It is clear that $\{\varphi_n\}_{n\in\mathbb{Z}}$ is a family of unitary operators φ_n in $\mathcal{L}(\mathcal{K}_n, \mathcal{K}'_n)$ with the properties $\varphi_n/\mathcal{G}_n = I_{\mathcal{G}_n}$ and $w'_n\varphi_{n+1} = \varphi_n w_n$ for all $n \in \mathbb{Z}$, i.e. $\mathbf{w}$ and $\mathbf{w}'$ are equivalent solutions of Problem 2.1. This means that Ψ is an injective map.

In order to prove that Ψ is surjective, choose Θ in $\mathcal{S}(\{Q_n\}_{n\in\mathbb{Z}}, \{\mathcal{R}_n\}_{n\in\mathbb{Z}})$. Let Γ be the set of Schur parameters associated by Theorem 2.2.1 to Θ. Then, using the construction in the proof of Theorem 2.3.2, it follows that $\Psi(\Phi(\Gamma)) = \Theta$, where Φ is the map defined by (2.5). This concludes the proof. $\qquad\square$

Finally, we consider again the set $\mathcal{E}(v)$ introduced in Section 2, where v is a partial isometry on $\mathcal{G}$. This time, we can define the generalized coresolvent of a representative w on $\mathcal{K}$ of a class in $\mathcal{E}(v)$, by the formula

$$\mathcal{C}([w], z) = P^{\mathcal{K}}_{\mathcal{G}}(I + zw)(I - zw)^{-1}/\mathcal{G}, \quad |z| < 1. \tag{3.10}$$

Since $|z| < 1$, the operator $(I - zw)$ is invertible, hence $\mathcal{C}([w], z)$ is well defined and it is not necessary to use truncated marking operators in this case. Using the notation $\mathcal{E} = \mathcal{G} \ominus \ker v$ and $\mathcal{F} = \mathcal{R}(v)$, we can complement Corollary 2.4 with the following result which is a consequence of Theorem 3.2.

3.3 Corollary   *There exists a one-to-one correspondence between the set $\mathcal{E}(v)$ and the set $\mathcal{S}(\mathcal{L}(\mathcal{G} \ominus \mathcal{E}, \mathcal{G} \ominus \mathcal{F}))$, given by the generalized coresolvent formula.*   □

We must mention that in this case, the generalized coresolvent formula is written in the following form:

$$\mathcal{C}([w], z) = \mathcal{C}([w^0], z) + 2(I - z(\hat{w}^0))^{-1}$$
$$\times P_{\mathcal{G} \ominus \mathcal{F}} \Theta(z)(I - zX(z)\Theta(z))^{-1} P_{\mathcal{G} \ominus \mathcal{E}}(I - z(\hat{w}^0))^{-1}.$$

The elements involved in this formula are defined as follows:

$$\Theta(z) = D + zC(I - zA)^{-1}B,$$

where the matrix representation

$$w = \begin{bmatrix} v & 0 & 0 \\ 0 & D & C \\ 0 & B & A \end{bmatrix}$$

with respect to the decompositions $\mathcal{K} = \mathcal{E} \oplus (\mathcal{G} \ominus \mathcal{E}) \oplus (\mathcal{K} \ominus \mathcal{G})$ and, respectively, $\mathcal{K} = \mathcal{F} \oplus (\mathcal{G} \ominus \mathcal{F}) \oplus (\mathcal{K} \ominus \mathcal{G})$ is taken into account;

$$X(z) = P_{\mathcal{G} \ominus \mathcal{E}}(I - z(\hat{w}^0))^{-1}/(\mathcal{G} \ominus \mathcal{F})$$

and w^0 is the central solution, *i.e.* the minimal unitary extension of v associated to the Schur parameters $\Gamma_n^0 = 0$ for all $n \geq 1$. In the next two sections we will show how to use Theorem 2.2 and Theorem 3.2 in order to solve the completion problems introduced in Section 1.

3.4   Moment Problems

We begin with a discussion of the classical trigonometric moment problem. We have the following result.

4.1 Theorem   *Let $\{A_k\}_{k=0}^N$ be a set of complex numbers and suppose that $A_0 = 1$. Then*

(i)   *The trigonometric moment problem for $\{A_k\}_{k=0}^N$ is solvable if and only if the matrix $A^{(0N)}$ defined by (1.3) is positive.*

(ii)   *If solvable, then the trigonometric moment problem for $\{A_k\}_{k=0}^N$ either has a unique solution or it has an infinity of solutions.*

(iii) *Suppose $A^{(0N)}$ is a positive matrix and let $\{\Gamma_k\}_{k=0}^N$ be the Schur parameters of the matrix $A^{(0N)}$. Then, the following are equivalent:*

(a)   *The trigonometric moment problem for $\{A_k\}_{k=0}^N$ has a unique solution.*

(b)   *There exists a positive integer $r \leq N$, such that $\Gamma_r = 1$.*

(c) $\text{rank} A^{(0N)} = r \leq n.$

(d) $\det A^{(0N)} = 0.$

(e)   *There exists a positive integer $r \leq N$ and two sets $\{a_k\}_{k=1}^r$, $\{\lambda_k\}_{k=1}^r$ of strictly positive numbers and, respectively, of distinct complex numbers of modulus one, such that*

$$A_m = \sum_{k=0}^r a_k \lambda_k^m, \quad m = 0, \ldots, N.$$

(iv) *Suppose that the trigonometric moment problem for $\{A_k\}_{k=0}^N$ has more then one solution. Let $\{\psi_k\}_{k=0}^N$ and $\{\hat\psi_k\}_{k=0}^N$ be the monic orthogonal polynomials of first and second kind associated to the Schur parameters $\{\Gamma_k\}_{k=0}^N$. Then there exists a one-to-one correspondence between the solutions μ of the considered problem and $\mathcal{S}$, given by the formula*

$$\frac{1}{4\pi} \int_0^{2\pi} \frac{e^{it} + z}{e^{it} - z} d\mu(t) = \frac{a(z) + c(z)g(z)}{b(z) + d(z)g(z)}, \quad z \in \mathbb{D}, g \in \mathcal{S}, \tag{4.1}$$

where a, b, c and d are certain analytic functions on the unit disc. Another one-to-one correspondence between the solutions μ and $\widetilde{\Pi}^0$ is given by

$$\mu = \text{the probability measure associated to the Schur} \atop \text{parameters} \quad \{\Gamma_0, \Gamma_1, \Gamma_2, \ldots, \Gamma_n, \gamma_0, \gamma_1, \ldots\}, \quad \{\gamma_n\}_{n \geq 0} \in \widetilde{\Pi}^0. \tag{4.2}$$

Proof It was already explained in Section 1 that the positivity of the matrix $A^{(0N)}$ is a necessary and sufficient condition for the solvability of the trigonometric moment problem for $\{A_k\}_{k=0}^N$. Proposition 1.5.6 and Theorem 1.5.10 show immediately that the trigonometric moment problem for $\{A_k\}_{k=0}^N$ has a unique solution if and only if the matrix $A^{(0N)}$ is singular. Otherwise, the problem has an infinity of solutions. So, the equivalence of (a), (b), (c) and (d) is proven. The equivalence of these statements with (e) is a consequence of Proposition 1.6.8 and Proposition 1.6.9.

   If $A^{(0N)}$ is strictly positive and $v : \mathcal{E} \longrightarrow \mathcal{F}$ is the unitary operator defined by (1.4), then it follows that $\dim[(\mathbb{C}^{N+1})_{A^{(0N)}} \ominus \mathcal{E}] = \dim[(\mathbb{C}^{N+1})_{A^{(0N)}} \ominus \mathcal{F}] = 1$, hence the parametrization (4.2) is a consequence of Corollary 2.4 (or Proposition 1.5.6). Taking w in $\mathcal{L}(\mathcal{K})$ a minimal unitary extension of v, we have seen that $\mu(f) = \langle f(w^*)e_0, e_0 \rangle_{\mathcal{K}}$, $f \in C(\mathbb{T})$, is a solution of the trigonometric moment problem for $\{A_k\}_{k=0}^N$. Moreover, the generalized coresolvent of the class of w is

$$\mathcal{C}([w], z) = \langle (I + zw)(I - zw)^{-1} E_0, E_0 \rangle_{\mathcal{K}} = \langle \varphi_z(w^*) E_0, E_0 \rangle_{\mathcal{K}},$$

where $\varphi_z(e^{it}) = (e^{it} + z)(e^{it} - z)^{-1}$, hence

$$C([w], z) = \mu(\varphi_z) = \frac{1}{2\pi} \int_0^{2\pi} \frac{e^{it} + z}{e^{it} - z} d\mu(t).$$

Therefore, formula (4.1) is a consequence of Krein's formula. In Chapter 5 we will obtain the exact form of the functions a, b, c and d. $\qquad\square$

We continue this section with an analysis of the Hamburger moment problem. We note first another result concerning selfadjoint extensions of symmetric operators, which may be viewed as a version of the result that every partial isometry has a unitary extension, possibly on a larger space.

4.2 Proposition *If S is a closed symmetric operator on a Hilbert space $\mathcal{H}$, then there always exist selfadjoint extensions of S acting on a Hilbert space $\mathcal{K}$ containing $\mathcal{H}$.*

Proof The Cayley transform $V(S)$ of S, defined by (1.1), is a unitary operator in $\mathcal{L}(\mathcal{R}(S + i), \mathcal{R}(S - i))$. If $\dim \mathcal{K}_+(S) = \dim \mathcal{K}_-(S)$, then the proof can be concluded using Theorem 1.2. Otherwise, let w in $\mathcal{L}(\mathcal{K})$ be a minimal unitary extension of $V(S)$ and we claim that $\mathrm{cl}(I - w)\mathcal{K} = \mathcal{K}$. Pick k in $\mathcal{K}$ such that k is orthogonal to the space $(I - w)\mathcal{K}$. On the one hand, it follows that k is orthogonal to the space $(I - V(S))\mathcal{R}(S + i)$ and it was remarked in the proof of Theorem 1.2 that $\mathrm{cl}(I - V(S))\mathcal{R}(S + i) = \mathcal{H}$. Therefore h is orthogonal to $\mathcal{H}$. On the other hand, k being orthogonal to the space $(I - w)\mathcal{K}$ implies $k = w^n k = w^{*n} k$ for all $n \in \mathbb{Z}$. Hence, k is orthogonal to the space $\vee_{n \in \mathbb{Z}} w^n \mathcal{H} = \mathcal{K}$, which implies that $k = 0$ and the claim is proved. According to this claim, the inverse of the Cayley transform is well defined for w and an operator A can be defined by the formula $A = i(I + w)(I - w)^{-1}$ on the domain $D(A) = \mathcal{R}(I - w)$. It is readily seen that A is selfadjoint, since $\mathcal{K}_\pm(A) = 0$. Finally, it is remarked that if A is restricted to $D(S)$, then it coincides with S, which means that A is an extension of S. $\qquad\square$

We can address now the Hamburger problem. A necessary condition for the solvability of this problem can be easily obtained. Thus, suppose the Hamburger moment problem for $\{s_n\}_{n\geq0}$ has a solution μ and let $\lambda_1, \lambda_2, \ldots \lambda_n$ be complex numbers. Then

$$\sum_{j,k=0}^n \lambda_j \lambda_k^* s_{j+k} = \sum_{j,k=0}^n \lambda_j \lambda_k^* \int_{-\infty}^\infty t^{j+k} d\mu(t)$$

$$= \int_{-\infty}^\infty \left(\sum_{j,k=0}^n \lambda_j \lambda_k^* t^{j+k} \right) d\mu(t)$$

$$= \int_{-\infty}^\infty \left| \sum_{j=0}^n \lambda_j t^j \right|^2 d\mu(t) \geq 0,$$

hence the Hankel kernel $\{s_{j+k}\}_{j,k\geq0}$ is positive definite. Conversely, suppose the Hankel kernel $\{s_{j+k}\}_{j,k\geq0}$ is positive definite. Let $\mathcal{P}$ denote the set of polynomials with complex coefficients and define an inner product on $\mathcal{P}$ by the formula:

$$\langle \sum_{j=0}^{m} a_j t^j, \sum_{k=0}^{n} b_k t^k \rangle = \sum_{j=0}^{m} \sum_{k=0}^{n} a_j b_k^* s_{j+k}. \tag{4.3}$$

Renorming $\mathcal{P}$ with respect to this inner product we obtain a Hilbert space $\mathcal{H}$. Define the operator

$$S : \mathcal{P} \longrightarrow \mathcal{P} \tag{4.4}$$

$$S(\sum_{j=0}^{m} a_j t^j) = \sum_{j=0}^{m} a_j t^{j+1},$$

which is symmetric with respect to the inner product (4.3) since $\{s_{j+k}\}_{j,k\geq 0}$ is a Hankel kernel. S extends to a densely defined symmetric operator on $\mathcal{H}$, hence it is closable, *i.e.* the *closure* $\bar{S}$ of S can be defined by the following rules: $D(\bar{S})$ is the space of the elements φ in $\mathcal{H}$ with the property that there exists an element ψ in $\mathcal{H}$ such that $(\varphi, \psi) \in \mathrm{cl}(G(S))$, and $\bar{S}\varphi = \psi$ for φ in $D(\bar{S})$. By this definition, $\bar{S}$ is the least closed extension of S and the set of selfadjoint extensions of $\bar{S}$ coincides with the set of the selfadjoint extensions of S. Before stating the main result concerning the Hamburger moment problem we need to introduce another class of analytic functions. Thus, let $\mathcal{N}$ denote the *Nevanlinna class* of the analytic functions f on the half plane $\mathrm{Im} z < 0$, with values in the closed half plane $\mathrm{Im} z \geq 0$. By convention, $\mathcal{N}$ contains the constant ∞.

4.3 Theorem   *Let $\{s_n\}_{n\geq 0}$ be a given set of real numbers. Then*

(a)   *The Hamburger problem for $\{s_n\}_{n\geq 0}$ is solvable if and only if the Hankel kernel $\{s_{j+k}\}_{j,k\geq 0}$ is positive definite.*

(b)   *If solvable, then the Hamburger problem for $\{s_n\}_{n\geq 0}$ either has a unique solution or it has an infinity of solutions.*

(c)   *Suppose the Hankel kernel $\{s_{j+k}\}_{j,k\geq 0}$ is positive definite. Then, the Hamburger problem for $\{s_n\}_{n\geq 0}$ has a unique solution if and only if $\mathcal{K}_{\pm}(\bar{S}) = 0$.*

(d)   *Suppose that the Hamburger problem for $\{s_n\}_{n\geq 0}$ has more then one solution. Then, there exists a one-to-one correspondence between the solutions μ of the considered problem and $\mathcal{N}$, given by the formula*

$$\int_{-\infty}^{\infty} \frac{d\mu(t)}{t - \lambda} = \frac{a(\lambda)f(\lambda) + c(\lambda)}{b(\lambda)f(\lambda) + d(\lambda)}, \qquad \mathrm{Im}\lambda < 0,\ f \in \mathcal{N}, \tag{4.5}$$

where a, b, c, d, are certain analytic functions in the domain $\mathrm{Im}\lambda < 0$.

Proof   By Proposition 4.2, $\bar{S}$ has selfadjoint extensions and let A be one of them. Let E be the spectral measure of A and define $\mu(f) = \langle E(f)[1], [1] \rangle$ for the continuous functions f on $\mathbb{R}$, where $[1]$ denotes the class of 1 in $\mathcal{H}$. Then, for $n \geq 0$,

$$\int_{-\infty}^{\infty} t^n d\mu(t) = \langle A^n[1], [1] \rangle = \langle t^n, 1 \rangle = s_n,$$

hence μ is a solution of the considered Hamburger moment problem. The statements (b) and (c) are obvious by now, hence we can focus on the proof of (d).

Consider for $k = 0, 1, 2, \ldots$, the determinants

$$D_k = \det \begin{bmatrix} s_0 & s_1 & \cdots & s_k \\ s_1 & s_2 & \cdots & s_{k+1} \\ \vdots & & \ddots & \\ s_k & s_{k+1} & \cdots & s_{2k} \end{bmatrix}$$

and suppose that $D_k > 0$ for all $k \geq 0$. Otherwise, the space $\mathcal{H}$ constructed by renorming $\mathcal{P}$ with respect to the inner product (4.3) is finite dimensional and S is selfadjoint, hence the Hamburger moment problem has a unique solution. A sequence $\{P_n\}_{n \geq 0}$ of polynomials on the real line is introduced such that each P_n is a polynomial of degree n and

$$\langle P_m, P_n \rangle = \delta_{mn} = \begin{cases} 0 & \text{if } m \neq n \\ 1 & \text{if } m = n. \end{cases} \tag{4.6}$$

It follows that these polynomials obey a second order finite difference equation of the form:

$$t P_k(t) = b_{k-1} P_{k-1}(t) + a_k P_k(t) + b_k P_{k+1}(t) \tag{4.7}$$

for $k \geq 0$, where $b_{-1} = 0$. Indeed, remark that

$$t P_k(t) = a_{k,k+1} P_{k+1}(t) + a_{k,k} P_k(t) + \ldots + a_{k,0} P_0(t).$$

Hence, $a_{k,k+1} = (D_{k-1} D_{k+1})^{\frac{1}{2}} D_k^{-1}$ and $a_{k,i} = 0$ for $i = 0, 1, \ldots, k-2$. Since $\{s_{j+k}\}_{j,k \geq 0}$ is a Hankel kernel, it also follows that $a_{k,k-1} = a_{k-1,k}$, and the relation (4.7) is obtained using the notation $a_k = a_{k,k}$, $b_k = a_{k,k+1}$. Due to (4.7), the operator $\bar{S}$ has the following *Jacobi matrix representation* with respect to the orthonormal basis $\{P_n\}_{n \geq 0}$ in $\mathcal{H}$:

$$\begin{bmatrix} a_0 & b_0 & 0 & 0 & \cdots \\ b_0 & a_1 & b_1 & 0 & \cdots \\ 0 & b_1 & a_2 & b_2 & \cdots \\ \vdots & \vdots & \vdots & \vdots & \ddots \end{bmatrix}.$$

Moreover, it is easy to deduce from (4.7) that $\dim \mathcal{K}_+(\bar{S}) = \dim \mathcal{K}_-(\bar{S}) = 1$. Let $V(\bar{S})$ be the Cayley transform of $\bar{S}$. This is a partial isometry on $\mathcal{H}$ and there exists a one-to-one correspondence between the set of the solutions of the considered Hamburger problem and $\mathcal{E}(V(\bar{S}))$. Then, let A be a selfadjoint extension of $\bar{S}$, let E be its spectral measure and define $\mu(f) = \langle E(f)[1], [1] \rangle$ for f in $C(\mathbb{R})$. Consequently,

$$\int_{-\infty}^{\infty} \frac{d\mu(t)}{t - \lambda} = \langle (A - \lambda)^{-1}[1], [1] \rangle, \qquad \text{Im} \lambda < 0.$$

Based on this remark, we show that the generalized coresolvent formula for the Cayley transform of A reduces, essentially, to the formula (4.5). Denote by w the

Cayley transform of A and the map $z = z(\lambda) = (-\lambda - i)(-\lambda + i)^{-1}$ is a conformal map of the lower half plane onto the unit disc. By direct computations, it follows that

$$(I + zw)(I - zw)^{-1} = i(1 + \lambda^2)(A - \lambda)^{-1} + \lambda I.$$

As a consequence of the generalized coresolvent formula, one obtains:

$$i(1 + \lambda^2)\langle (A - \lambda)^{-1}[1], [1]\rangle - i(1 + \lambda^2)\langle (A_0 - \lambda)^{-1}[1], [1]\rangle =$$

$$= 2\Theta(z)(1 - zX(z)\Theta(z))^{-1}\|(I - zw^0)^{-1}[1]\|^2,$$

where A_0 is the inverse Cayley transform of the central solution w^0. Finally, one uses the formula

$$f(z) = i\frac{1 + g(\lambda)}{1 - g(\lambda)}, \qquad z = \frac{-\lambda - i}{-\lambda + i},$$

which establishes a one-to-one correspondence between the class $\mathcal{N}$ and the class $\mathcal{S}$, and a formula of type (4.5) easily follows. $\qquad\square$

4.4 Remark   A more detailed analysis leads to some other properties of the functions a, b, c, d in the formula (4.5). Another important issue is the translation of the criterion (c) into a more tractable form. All of these are classical results in the theory of the Hamburger moment problem and some references will be indicated in the notes at the end of this chapter. $\qquad\square$

The next problem we deal with is the Krein problem. We approach it within the framework of extending partial isometries and a few preliminaries about generalized semigroups of contractions are necessary. Using this approach makes it clear that the Krein problem may be viewed as a "continuous" analogue of the trigonometric moment problem.

4.5 Remark   We have seen the connection between the trigonometric moment problem and the Naimark dilation. The same construction can be used to characterize the Fourier transform of positive measures on the real line. Namely, let μ be a finite positive measure on the real line and define

$$Q(x) = \int_{-\infty}^{\infty} e^{-itx} d\mu(t).$$

Then Q is a continuous function on $\mathbb{R}$ which is also positive definite, since

$$\sum_{j,k=0}^{n} \lambda_j \lambda_k^* Q(x_j - x_k) = \sum_{j,k=0}^{n} \lambda_j \lambda_k^* \int_{-\infty}^{\infty} e^{-it(x_k - x_j)} d\mu(t)$$

$$= \int_{-\infty}^{\infty} |\sum_{j=0}^{n} \lambda_j e^{-itx_j}|^2 d\mu(t) \geq 0$$

for all $n \in \mathbb{N}$, arbitrary complex numbers λ_0, ..., λ_n and arbitrary real numbers $x_0, \ldots, x_n$. Consider now a positive definite continuous function Q on $\mathbb{R}$ and denote by $\mathcal{F}$ the vector space of the complex functions on $\mathbb{R}$ with finite support. Define the inner product $\langle \cdot, \cdot \rangle_Q$ on $\mathcal{F}$ by the formula

$$\langle \varphi, \psi \rangle_Q = \sum_{x,y \in \mathbb{R}} Q(x-y)\varphi(x)\psi^*(y), \quad \varphi, \psi \in \mathcal{F}.$$

Factoring out the subspace $\mathcal{N} = \{\varphi \in \mathcal{F} \mid \langle \varphi, \varphi \rangle_Q = 0\}$, we get a space $\mathcal{F}/\mathcal{N}$ endowed with an inner product, also denoted by $\langle \cdot, \cdot \rangle_Q$, and let $\mathcal{H}_Q$ be the completion of $\mathcal{F}/\mathcal{N}$ with respect to the norm induced by this inner product. This is a Hilbert space and we also define the operators

$$(U_t\varphi)(x) = \varphi(x-t), \quad \varphi \in \mathcal{F}, t \in \mathbb{R}.$$

For fixed $t \in \mathbb{R}$ and φ, ψ in $\mathcal{F}$, we have

$$\langle U_t\varphi, U_t\psi \rangle_Q = \sum_{x,y \in \mathbb{R}} Q(x-y)(U_t\varphi)(x)(U_t\psi)^*(y)$$

$$= \sum_{x,y \in \mathbb{R}} Q(x-y)\varphi(x-t)\psi^*(y-t) = \langle \varphi, \psi \rangle_Q,$$

hence U_t extends to a unitary operator on $\mathcal{H}_Q$. Actually, $\{U_t\}_{t \in \mathbb{R}}$ is a strongly-continuous group of unitary operators, i.e. $U_0 = I$, $U_{t+s} = U_t U_s$ for $t, s \in \mathbb{R}$ and $s\text{-}\lim_{t \to 0} U_t = I$. By Stone's theorem, there is a (possible unbounded) selfadjoint operator A such that

$$U_t = e^{-itA} = E(e_t),$$

where $e_t(\lambda) = e^{-it\lambda}$ and E is the spectral measure of A. We define

$$\delta_0(x) = \begin{cases} 1 & \text{if} \quad x = 0 \\ 0 & \text{if} \quad x \neq 0, \end{cases} \tag{4.8}$$

hence δ_0 belongs to $\mathcal{F}$ and then the finite positive measure μ on $\mathbb{R}$ is defined by the formula:

$$\mu(f) = \langle E(f)[\delta_0], [\delta_0] \rangle_Q, \quad f \in C(\mathbb{R}).$$

It follows that

$$Q(t) = \langle U_t[\delta_0], [\delta_0] \rangle_Q = \langle E(e_t)[\delta_0], [\delta_0] \rangle_Q$$

$$= \mu(e_t) = \int_{-\infty}^{\infty} e^{-itx} d\mu(x).$$

In conclusion, every positive definite continuous function on $\mathbb{R}$ is the Fourier transform of a finite positive measure on $\mathbb{R}$. This is a well known result of S. Bochner. As it was the case with the trigonometric moment problem, we can expect that this construction can be properly adapted to solve Problem 1.7. $\qquad \square$

When Q is defined only on the interval $(-a, a)$, $0 < a < \infty$, we can define only a "truncation" of the unitary operator U_t in Remark 4.5. The formal definition is the following. A family $\{T_t\}_{t\geq 0}$ of contractions on a Hilbert space $\mathcal{H}$ is called a *generalized semigroup of contractions* if

(1)   The domains $D(T_t) = \mathcal{E}_t$ are closed subspaces of $\mathcal{H}$ and $\mathcal{E}_t \subset \mathcal{E}_s$ for $0 \leq s \leq t$, $\mathrm{cl}(\cup_{t>0}\mathcal{E}_t) = \mathcal{H}$ and $T_s\mathcal{E}_{s+t} \subset \mathcal{E}_t$.

(2) $T_0 = I_{\mathcal{H}}$ and $T_{s+t} = T_tT_s$ for $t, s \geq 0$.

(3)   If $t_0 > 0$, h belongs to $\mathcal{E}_{t_0}$ and $0 \leq t, s \leq t_0$, then $\lim_{s \to t} T_s h = T_t h$.

An infinitesimal generator of $\{T_t\}_{t\geq 0}$ can be defined by the formula:

$$D(G) = \{h \in \cup_{t>0}\mathcal{E}_t \mid \lim_{t\downarrow 0} \frac{1}{t}(T_t h - h) \quad \text{exists}\}$$

$$Gh = \lim_{t\downarrow 0} \frac{1}{t}(T_t h - h), \qquad h \in D(G).$$

Several results from the theory of semigroups of contractions remain true in this more general situation.

4.6 Proposition   Let $\{T_t\}_{t\geq 0}$ be a generalized semigroup of contractions. Then

(a)   If h belongs to $D(G) \cap \mathcal{E}_t$ for some $t > 0$, then $T_t h$ belongs to $D(G)$, Gh belongs to $\mathcal{E}_t$ and

$$\frac{d}{dt}T_t h = GT_t h = T_t Gh.$$

(b)   If h belongs to $\mathcal{E}_{t_0}$ and $0 < t < t_0$, then $\int_0^t T_s h ds$ belongs to $D(G)$ and

$$G(\int_0^t T_s h ds) = T_t h - h.$$

(c) $D(G)$ is dense in $\mathcal{H}$.

(d)   Supposing in addition that T_t are isometries for all $t > 0$, then $S = -iG$ is a symmetric operator.

Proof   (a) Consider h in $D(G) \cap \mathcal{E}_t$ and $s > 0$. Then h belongs to $\mathcal{E}_{t+s}$, $T_s\mathcal{E}_{t+s} \subset \mathcal{E}_t$, $T_t\mathcal{E}_{t+s} \subset \mathcal{E}_s$ and

$$\frac{1}{s}(T_{t+s}h - T_t h) = \frac{1}{s}T_t(T_s h - h) = \frac{1}{s}(T_s - I)T_t h.$$

Therefore, $T_t h$ is an element of $D(G)$, Gh belongs to $\mathcal{E}_t$ and letting $s \to 0$, it follows that $\frac{d}{dt}T_t h = GT_t h = T_t Gh$. In order to prove (b) we remark that if h belongs to $\mathcal{E}_{t_0}$, then h also belongs to $\mathcal{E}_t$ for $0 < t < t_0$. Define $h_t = \int_0^t T_s h ds$ and pick $r > 0$ such that $t + r < t_0$. Then

$$\frac{1}{r}(T_r h_t - h_t) = \frac{1}{r}\int_t^{t+r} T_s h ds - \frac{1}{r}\int_0^t T_s h ds \to T_t h - h,$$

when $r \downarrow 0$. Consequently, $\int_0^t T_s h ds$ belongs to $D(G)$ and $G(\int_0^t T_s h ds) = T_t h - h$. The density of $D(G)$ in $\mathcal{H}$ follows immediately from (b) and the assumption that

$\mathrm{cl}(\cup_{t>0}\mathcal{E}_t) = \mathcal{H}$. In order to prove (d), one defines $S = -iG$. Pick $t_0 > 0$ and h, g in $D(G) \cap \mathcal{E}_{t_0}$. Then, for $0 < t < t_0$,

$$\langle \frac{1}{t}(T_t - i)h, g \rangle = \langle \frac{1}{t}T_t(T_t - i)h, T_tg \rangle = \langle T_t h, \frac{1}{t}(g - T_tg) \rangle$$

and letting $t \downarrow 0$, it follows that $\langle Sh, g \rangle = \langle h, Sg \rangle$, i.e. S is a symmetric operator.
□

After these preliminaries we can obtain a solution of Problem 1.7 by properly adapting the construction used in the proof of Bochner's theorem indicated in Remark 4.5. Thus, consider $Q : (-a, a) \rightarrow \mathbb{R}$ a positive definite continuous function. Let $\mathcal{F}$ be the vector space of all complex functions on $(-a, a)$ which are different from zero only for a finite subset of points in $(-a, a)$. Define the inner product $\langle \cdot \cdot \rangle_Q$ on $\mathcal{F}$ by the formula

$$\langle \varphi, \psi \rangle_Q = \sum_{x,y \in \mathbb{R}} Q(x - y)\varphi(x)\psi(y)^*, \quad \varphi, \psi \in \mathcal{F}, \quad x, y \in (-a, a). \qquad (4.9)$$

Renorming $\mathcal{F}$ with respect to this inner product, we get a Hilbert space $\mathcal{H}_Q$ and define for $t > 0$, the subspaces $\mathcal{E}_t$ of $\mathcal{H}_Q$ generated by the sets $\mathcal{E}_t^0 = \{\varphi \in \mathcal{F} \mid t + \mathrm{supp}\varphi \subset (-a, a)\}$, where $\mathrm{supp}\varphi$ denotes the support of the function φ. Moreover, for f in $\mathcal{E}_t^0$, define

$$(V_t f)(s) = \begin{cases} f(s - t) & \text{if} \quad -a + t \leq s < a \\ 0 & \text{if} \quad -a < s \leq -a + t. \end{cases} \qquad (4.10)$$

We can easily verify that every V_t extends to an isometry on $\mathcal{H}_Q$ such that $\{V_t\}_{t \geq 0}$ is a generalized semigroup of isometries. Let G be its infinitesimal generator and $S = -iG$. The operator S is not necessarily closed and let $\bar{S}$ be its closure. We have the following result concerning Problem 1.7

4.7 Theorem   *Let $Q : (-a, a) \rightarrow \mathbb{R}$ be a continuous function. Then*

(a)   *The Krein problem for Q is solvable if and only if Q is positive definite.*

(b)   *If solvable, then the Krein problem for Q either has a unique solution or it has an infinity of solutions.*

(c)   *Suppose the function Q is positive definite. Then, the Krein problem for Q has a unique solution if and only if $\mathcal{K}_{\pm}(\bar{S}) = 0$.*

(d)   *Suppose that the Krein problem for Q has more then one solution. Then, there exists a one-to-one correspondence between the solutions μ of the considered problem and $\mathcal{N}$, given by the formula*

$$-i \int_0^\infty e^{-i\lambda t}\widetilde{Q}(-t)dt = \frac{a(\lambda)f(\lambda) + c(\lambda)}{b(\lambda)f(\lambda) + d(\lambda)}, \quad \mathrm{Im}\lambda < -\gamma, \, f \in \mathcal{N}, \qquad (4.11)$$

for some $\gamma \geq 0$, where a, b, c, d are certain functions analytic in the domain $\mathrm{Im}\lambda < 0$ and

$$\widetilde{Q}(x) = \int_{-\infty}^\infty e^{-xt}d\mu(t). \qquad (4.12)$$

Proof (a) By Proposition 4.2, the operator $\bar{S}$ has selfadjoint extensions and let A be one of them. Let E be the spectral measure of A and define the positive measure μ on $\mathbb{R}$ by the formula $\mu(f) = \langle E(f)[\delta_0], [\delta_0]\rangle_Q$, where δ_0 is given by (4.8), the inner product $\langle \cdot, \cdot \rangle_Q$ is defined by (4.9) and f belongs to $C(\mathbb{R})$. We obtain that for $t \in (-a, a)$,

$$Q(t) = \langle V_t[\delta_0], [\delta_0]\rangle_Q = \langle E(e_t)[\delta_0], [\delta_0]\rangle_Q$$

$$= \mu(e_t) = \int_{-\infty}^{\infty} e^{-itx} d\mu(x).$$

The statements (b) and (c) are obvious by now, hence we can focus on the proof of (d). Let $V(\bar{S})$ be the Cayley transform of $\bar{S}$, which is a partial isometry on $\mathcal{H}_Q$. Therefore, there exists a one-to-one correspondence between the set of the solutions of the considered Krein problem and $\mathcal{E}(V(\bar{S}))$. Denote by $AC[-a, a]$ the set of all absolutely continuous functions φ in $L^2(-a, a)$ such that φ' belongs to $L^2(-a, a)$ and define the operator:

$$D(T) = \{\varphi \in AC[-a, a] \mid \varphi(-a) = \varphi(a) = 0\} \tag{4.13}$$

$$T\varphi = -i\varphi',$$

which is a symmetric operator such that $T \subset \bar{S}$. To determine $\mathcal{K}_{\pm}(T)$, we remark that $D(T^*) = AC[-a, a]$ and then, the equality $-\varphi' = \pm\varphi$ shows that φ is infinitely differentiable and $\mathcal{K}_{\pm}(T) = \{ce^{\pm x} \mid c \in \mathbb{C}\}$. Therefore, $\dim \mathcal{K}_+(\bar{S}) = \dim \mathcal{K}_-(\bar{S}) = 1$. Then, let A be a selfadjoint extension of $\bar{S}$, let E be the spectral measure of A and $\mu(f) = \langle E(f)[\delta_0], [\delta_0]\rangle_Q$, for f in $C(\mathbb{R})$. Define

$$\widetilde{Q}(x) = \int_{-\infty}^{\infty} e^{-ixt} d\mu(t).$$

Then, for $\mathrm{Im}\lambda < -\gamma$, where $\gamma \geq 0$, it follows that

$$-i\int_{-\infty}^{\infty} e^{-i\lambda t} \widetilde{Q}(-t) dt = \int_{-\infty}^{\infty} \frac{d\mu(t)}{t - \lambda} = \langle (A - \lambda)^{-1}[\delta_0], [\delta_0]\rangle_Q.$$

From now on, we use the generalized coresolvent formula as in the proof of Theorem 4.3. The details can be omitted.   □

4.8 Remark   As it was the case with the Hamburger problem, a more detailed analysis of the functions a, b, c and d can be done. This analysis, as well as many other ramifications of the theory can be found in papers mentioned at the end of this chapter.   □

We conclude this section with a brief discussion of Problem 1.8. Let $r = \{r_n\}_{n\in\mathbb{Z}}$ be an increasing sequence of integers with $r_n \geq n$ and define $\alpha(r) = \{(i, j) \in \mathbb{Z} \times \mathbb{Z} \mid i \leq j \leq r_i\}$, $\beta(r) = \{(i, j) \in \mathbb{Z} \times \mathbb{Z} \mid (i, j) \in \alpha(r) \text{ or } (j, i) \in \alpha(r)\}$. Let $\{\widetilde{A}_{ij}\}_{(i,j)\in\alpha(r)}$ be a given family of operators and suppose, without loss of generality, that $\tilde{A}_{ii} = I$ for all $i \in \mathbb{Z}$. An obvious necessary condition for the solvability of Problem 1.8 for this family of operators is that $[\tilde{A}_{ij}]_{(i,j)\in\gamma\times\gamma} \geq 0$ for all $\gamma \subset \mathbb{Z}$ with $\gamma \times \gamma \subset \beta(r)$, where $\tilde{A}_{ij} = \tilde{A}_{ji}^*$ for $i > j$. By Theorem 1.5.3, there is a uniquely determined family $\{\tilde{\Gamma}_{ij}\}_{(i,j)\in\alpha(r)}$ of Schur parameters associated to $\{\tilde{A}_{ij}\}_{(i,j)\in\alpha(r)}$. As a consequence of the same Theorem 1.5.3, we obtain the following result.

4.9 Theorem   Let $\{\tilde{A}_{ij}\}_{(i,j)\in\alpha(r)}$ be a given family of operators, such that $\tilde{A}_{ii} = I$ for all $i \in \mathbb{Z}$. Then

(a)   The band completion problem for $\{\tilde{A}_{ij}\}_{(i,j)\in\alpha(r)}$ is solvable if and only if $[\tilde{A}_{ij}]_{(i,j)\in\gamma\times\gamma} \geq 0$ for all $\gamma \subset \mathbb{Z}$ with $\gamma \times \gamma \subset \beta(r)$.

(b)   There exists a bijective correspondence between the set of the solutions of the band completion problem and the completions of the family $\{\tilde{\Gamma}_{ij}\}_{(i,j)\in\alpha(r)}$ to families $\{\Gamma_{ij} \mid i,j \in \mathbb{Z}, i \leq j\}$ of Schur parameters, such that $\Gamma_{ij} = \tilde{\Gamma}_{ij}$ for $(i,j) \in \alpha(r)$. $\qquad\square$

The same result can be obtained using Theorem 2.2. It is enough to illustrate this with the simplest case of a 3×3 block matrix.

4.10 Example   Let us consider two operators A_{12}, A_{23} and the problem is to determine A_{13} such that the operator

$$A^{(13)} = \begin{bmatrix} I & A_{12} & A_{13} \\ A_{12}^* & I & A_{23} \\ A_{13}^* & A_{23}^* & I \end{bmatrix}$$

is positive in $\mathcal{L}(\oplus_{i=1}^3 \mathcal{H}_i)$. Suppose that A_{12} and A_{23} are contractions. Define $\mathcal{G}_0 = (\mathcal{H}_1 \oplus \mathcal{H}_2)_{A^{(12)}}$ and $\mathcal{G}_1 = (\mathcal{H}_2 \oplus \mathcal{H}_3)_{A^{(23)}}$ to be the Hilbert spaces obtained by renorming $\mathcal{H}_1 \oplus \mathcal{H}_2$ and $\mathcal{H}_2 \oplus \mathcal{H}_3$ with respect to $A^{(12)} = \begin{bmatrix} I & A_{12} \\ A_{12}^* & I \end{bmatrix}$ and, respectively, $A^{(23)} = \begin{bmatrix} I & A_{23} \\ A_{23}^* & I \end{bmatrix}$. Then, denote by $\mathcal{E}_1$ the subspace of $\mathcal{G}_1$ generated by the set $\{(h_2, 0) \mid h_2 \in \mathcal{H}_2\}$ and denote by $\mathcal{F}_0$ the subspace of $\mathcal{G}_0$ generated by the set $\{(0, h_2) \mid h_2 \in \mathcal{H}_2\}$. The map v_0 defined by $v_0(h_2, 0) = (0, h_2)$ for h_2 in $\mathcal{H}_2$ can be extended by continuity to a unitary operator in $\mathcal{L}(\mathcal{E}_1, \mathcal{F}_0)$. Let w_0 in $\mathcal{L}(\mathcal{K}_1, \mathcal{K}_0)$ be a unitary operator extending v_0. The equality

$$T^*T = \begin{bmatrix} I & A_{12} & P_{\mathcal{H}_1}^{\mathcal{K}_0} w_0/\mathcal{H}_3 \\ A_{12}^* & I & A_{23} \\ (P_{\mathcal{H}_1}^{\mathcal{K}_0} w_0/\mathcal{H}_3)^* & A_{23}^* & I \end{bmatrix},$$

where $T = [P_{\mathcal{G}_0}^{\mathcal{K}_0}/\mathcal{H}_1 \quad P_{\mathcal{G}_0}^{\mathcal{K}_0}/\mathcal{H}_2 \quad w_0/\mathcal{H}_3]$, shows that $A_{13} = P_{\mathcal{H}_1}^{\mathcal{K}_0} w_0/\mathcal{H}_3$ is a solution of the considered problem. By Theorem 2.2, there exists a one-to-one correspondence between the solutions of this problem and the set of contractions $\Gamma \in \mathcal{L}(\mathcal{G}_1 \ominus \mathcal{E}_1, \mathcal{G}_0 \ominus \mathcal{F}_0)$. Using the operators defined by (1.2.7) and (1.2.8), it is easy to see that $\mathcal{G}_1 \ominus \mathcal{E}_1$ is identified with $\mathcal{D}_{A_{23}}$ and $\mathcal{G}_0 \ominus \mathcal{F}_0$ is identified with $\mathcal{D}_{A_{12}^*}$. The actual dependence of A_{13} on the parameter Γ is exactly the one indicated in Example 1.5.1. The generalized coresolvent formula can also be used to parametrize the set of solutions of this problem. $\square$

3.5 The Commutant Lifting Method

In this section we show how to use the results concerning the problem of extending families of partial isometries in order to solve norm preserving problems such as the problems of Schur, Nevanlinna-Pick and Nehari, or the four-block problem. For this purpose we prove first a general result known as the *commutant lifting theorem*.

Let $\{T_n\}_{n\in\mathbb{Z}}$ be a family of contractions T_n in $\mathcal{L}(\mathcal{H}_{n+1},\mathcal{H}_n)$ and let $\{W_n\}_{n\in\mathbb{Z}}$ be the Kolmogorov decomposition described by Theorem 1.6.1 of the positive definite kernel A associated to the family $\{T_n\}_{n\in\mathbb{Z}}$ as in Corollary 1.5.9. More precisely, W_n is described by the formula (2.5.5). Moreover, we define the spaces

$$\mathcal{K}_n^+ = \mathcal{H}_n \oplus \oplus_{k\geq n}\mathcal{D}_{T_k} \tag{5.1}$$

and the isometries

$$W_n^+ : \mathcal{K}_{n+1}^+ \longrightarrow \mathcal{K}_n^+ \tag{5.2}$$

$$W_n^+ = W_n/\mathcal{K}_{n+1}^+.$$

We remark that the family $\{W_n^+\}_{n\in\mathbb{Z}}$ has the following dilation properties: for $j > i$,

$$T_i T_{i+1}\ldots T_{j-1} = P_{\mathcal{H}_i} W_i^+ W_{i+1}^+ \ldots W_{j-1}^+/\mathcal{H}_j \tag{5.3}$$

and for $n \in \mathbb{Z}$,

$$\mathcal{K}_n^+ = \mathcal{H}_n \vee W_n^+\mathcal{H}_{n+1} \vee W_n^+ W_{n+1}^+\mathcal{H}_{n+2} \vee \cdots. \tag{5.4}$$

Actually, the properties (5.3) and (5.4) determine the family $\{W_n^+\}_{n\in\mathbb{Z}}$. More exactly, if $\{w_n\}_{n\in\mathbb{Z}}$ is another family of isometries w_n in $\mathcal{L}(\mathcal{G}_{n+1},\mathcal{G}_n)$ such that (5.3) and (5.4) hold, then there exists a family $\{\varphi_n\}_{n\in\mathbb{Z}}$ of unitary operators φ_n in $\mathcal{L}(\mathcal{K}_n^+,\mathcal{G}_n)$ such that $\varphi_n/\mathcal{H}_n = I_{\mathcal{H}_n}$ and $\varphi_n W_n^+ = w_n\varphi_{n+1}$ for all $n \in \mathbb{Z}$. The family $\{W_n^+\}_{n\in\mathbb{Z}}$ will be referred to as the *minimal isometric dilation* of the family $\{T_n\}_{n\in\mathbb{Z}}$.

We consider another family $\{T_n'\}_{n\in\mathbb{Z}}$ of contractions T_n' in $\mathcal{L}(\mathcal{H}_{n+1}',\mathcal{H}_n')$ and define the set:

$$\mathcal{I} = \mathcal{I}(\{T_n\}_{n\in\mathbb{Z}}, \{T_n'\}_{n\in\mathbb{Z}})$$

$$= \{\{X_n\}_{n\in\mathbb{Z}} \mid X_n \in \mathcal{L}(\mathcal{H}_n,\mathcal{H}_n'),\ T_n'X_{n+1} = X_nT_n \quad \text{for all} \quad n \in \mathbb{Z}\}.$$

Let $\{W_n'^+\}_{n\in\mathbb{Z}}$ be the minimal isometric dilation of the family $\{T_n'\}_{n\in\mathbb{Z}}$ and for a family $\{X_n\}_{n\in\mathbb{Z}}$ in $\mathcal{I}$ define the set

$$CID(\{T_n\}_{n\in\mathbb{Z}}, \{T_n'\}_{n\in\mathbb{Z}}; \{X_n\}_{n\in\mathbb{Z}}) = CID(\{X_n\}_{n\in\mathbb{Z}})$$

$$=\{\{Y_n\}_{n\in\mathbb{Z}}\in\mathcal{I}(\{W_n^+\}_{n\in\mathbb{Z}},\{W_n'^+\}_{n\in\mathbb{Z}})|\|Y_n\|\leq 1, P_{\mathcal{H}_n'}Y_n=X_nP_{\mathcal{H}_n} \quad \text{for all} \quad n\in\mathbb{Z}\}.$$

The main result concerning the set $CID(\{X_n\}_{n\in\mathbb{Z}})$ is the following.

5.1 Theorem   *The set $CID(\{X_n\}_{n\in\mathbb{Z}})$ is non-empty if and only if X_n are contractions for all $n \in \mathbb{Z}$.*

Proof The necessity of the condition that all the operators X_n, $n \in \mathbb{Z}$, be contractions is obvious. In order to prove the converse we use unitary couplings as illustrated in Example 2.4.4. Thus, let $\{W_n^+\}_{n\in\mathbb{Z}}$ be the minimal isometric dilation of the family $\{T_n\}_{n\in\mathbb{Z}}$. Define the spaces

$$\mathcal{K}_n^- = \oplus_{k<-n}\mathcal{D}_{T'_{-k}^*} \oplus \mathcal{H}_n' \tag{5.5}$$

and the operators

$$W_n^- : \mathcal{K}_n^- \longrightarrow \mathcal{K}_{n+1}^- \tag{5.6}$$

$$W_n^- = W_n'^*/\mathcal{K}_n^-,$$

where $\{W_n'^+\}_{n\in\mathbb{Z}}$ is the Kolmogorov decomposition described by Theorem 1.6.1 of the positive definite kernel A' associated to the family $\{T_n'\}_{n\in\mathbb{Z}}$ as in Corollary 1.5.9. Finally, define the contractions

$$\widetilde{X}_n : \mathcal{K}_n^+ \longrightarrow \mathcal{K}_n^-$$

$$\widetilde{X}_n = \begin{bmatrix} \vdots & \vdots & \\ 0 & 0 & \cdots \\ X_n & 0 & \cdots \end{bmatrix}$$

and remark that, due to the equality $X_n T_n = T_n' X_{n+1}$, it follows that $\widetilde{X}_n W_n^+ = (W_n^-)^* \widetilde{X}_{n+1}$. Therefore, the pair $(\{W_n^+\}_{n\in\mathbb{Z}}, \{W_n^-\}_{n\in\mathbb{Z}})$ satisfies the coupling property (2.4.7) and the unitary operators

$$v_n : W_n^- \mathcal{K}_n^- \vee \mathcal{K}_{n+1}^+ \longrightarrow \mathcal{K}_n^- \vee W_n^+ \mathcal{K}_{n+1}^+ \tag{5.7}$$

$$v_n(W_n^- k^- + k^+) = k^- + W_n^+ k^+, \quad k^+ \in \mathcal{K}_{n+1}^+, k^- \in \mathcal{K}_n^-,$$

of type (2.4.8) can be taken into account. Define $\mathcal{E}_{n+1} = W_n^- \mathcal{K}_n^- \vee \mathcal{K}_{n+1}^+$, $\mathcal{F}_n = \mathcal{K}_n^- \vee W_n^+ \mathcal{K}_{n+1}^+$ which are subspaces of $\mathcal{G}_n = \mathcal{K}_n^- \vee \mathcal{K}_n^+$, and consider the families $\mathbf{E} = \{\mathcal{E}_n\}_{n\in\mathbb{Z}}$, $\mathbf{F} = \{\mathcal{F}_n\}_{n\in\mathbb{Z}}$ of Hilbert spaces, as well as the family $\mathbf{v} = \{v_n\}_{n\in\mathbb{Z}}$ of the unitary operators v_n defined by (5.7).

We show now that every element of $\mathcal{E}(\mathbf{E}, \mathbf{F}; \mathbf{v})$ produces an element in $\mathrm{CID}(\{X_n\}_{n\in\mathbb{Z}})$. For this purpose, let $\mathbf{w} = \{w_n\}_{n\in\mathbb{Z}}$ be a representative of a class in $\mathcal{E}(\mathbf{E}, \mathbf{F}; \mathbf{v})$, where w_n belongs to $\mathcal{L}(\mathcal{S}_{n+1}, \mathcal{S}_n)$. Define the spaces

$$\widetilde{\mathcal{K}}_n' = \mathcal{H}_n' \vee w_n \mathcal{H}_{n+1}' \vee w_n w_{n+1} \mathcal{H}_{n+2}' \vee \ldots \subset \mathcal{S}_n.$$

If we define the operators

$$\widetilde{w}_n' : \widetilde{\mathcal{K}}_{n+1}' \longrightarrow \widetilde{\mathcal{K}}_n' \tag{5.8}$$

$$\widetilde{w}_n' = w_n/\widetilde{\mathcal{K}}_{n+1}',$$

then it is readily checked that $\{\widetilde{w}'_n\}_{n\in\mathbb{Z}}$ satisfies (5.3) and (5.4) with respect to the family $\{T'_n\}_{n\in\mathbb{Z}}$ of contractions. Consequently, there exists a family $\{\varphi_n\}_{n\in\mathbb{Z}}$ of unitary operators φ_n in $\mathcal{L}(\mathcal{K}'^+_n, \widetilde{\mathcal{K}}'_n)$ such that $\varphi_n/\mathcal{H}'_n = I_{\mathcal{H}'_n}$ and $\varphi_n W'_n = \widetilde{w}'_n \varphi_{n+1}$ for all $n \in \mathbb{Z}$. Finally, define

$$Y_n : \mathcal{K}^+_n \longrightarrow \mathcal{K}'^+_n \tag{5.9}$$

$$Y_n = \varphi_n^* P^{\mathcal{S}_n}_{\widetilde{\mathcal{K}}'_n}/\mathcal{K}^+_n$$

and it is a matter of direct computations to verify that $W'^+_n Y_{n+1} = Y_n W^+_n$ and $P_{\mathcal{H}'_n} Y_n = X_n P_{\mathcal{H}_n}$ for all $n \in \mathbb{Z}$. Therefore, $\{Y_n\}_{n\in\mathbb{Z}}$ belongs to $CID(\{X_n\}_{n\in\mathbb{Z}})$. $\qquad\square$

We note the following reformulation of Theorem 5.1.

5.2 Theorem *Let $\{T_n\}_{n\in\mathbb{Z}}$ and $\{T'_n\}_{n\in\mathbb{Z}}$ be two families of contractions and let $\{W^+_n\}_{n\in\mathbb{Z}}$, respectively, $\{W'^+_n\}_{n\in\mathbb{Z}}$ be their minimal isometric dilations. If the family $\{X_n\}_{n\in\mathbb{Z}}$ belongs to $\mathcal{I}(\{T_n\}_{n\in\mathbb{Z}}, \{T'_n\}_{n\in\mathbb{Z}})$, then there exists a family $\{Y_n\}_{n\in\mathbb{Z}}$ of operators Y_n in $\mathcal{L}(\mathcal{K}^+_n, \mathcal{K}'^+_n)$, such that for all $n \in \mathbb{Z}$,*

$$W'^+_n Y_{n+1} = Y_n W^+_n, \quad P_{\mathcal{H}'_n} Y_n = X_n P_{\mathcal{H}_n},$$

and

$$\sup_{n\in\mathbb{Z}} \|Y_n\| = \sup_{n\in\mathbb{Z}} \|X_n\|. \quad\square$$

5.3 Remark We show here that the distance formula (2.2.7) is a consequence of Theorem 5.2. Let T_{11} in $\mathcal{L}(\mathcal{H}_1, \mathcal{H}'_1)$, T_{12} in $\mathcal{L}(\mathcal{H}_2, \mathcal{H}'_1)$ and T_{21} in $\mathcal{L}(\mathcal{H}_1, \mathcal{H}'_2)$ be given operators. Define $T_0 = [I \ \ 0]^\top$, $T'_0 = [I \ \ 0]$, $X_0 = [T_{11} \ \ T_{12}]$, $X_1 = [T_{11} \ \ T_{21}]^\top$ and notice that $T'_0 X_1 = X_0 T_0$ and $W^+_0 = [I \ \ 0]^\top$, $W'^+_0 = I_{\mathcal{H}'_1 \oplus \mathcal{H}'_2}$. By Theorem 5.2, there exists an operator $Y_1 = \begin{bmatrix} T_{11} & T_{12} \\ T' & T'' \end{bmatrix}$ such that $W'^+_0 [T_{11} \ \ T_{21}]^\top = Y_1 W^+_0$ and $\|Y_1\| = \max\{\|X_0\|, \|X_1\|\}$. Therefore, $T' = T_{21}$ and

$$\inf\{\| \begin{bmatrix} T_{11} & T_{12} \\ T_{21} & X \end{bmatrix} \| \mid X \in \mathcal{L}(\mathcal{H}_2, \mathcal{H}'_2)\} = \max\{\|X_0\|, \|X_1\|\}.$$

In fact, we can round the table and obtain another, more direct proof of Theorem 5.1 in a particular, but generic, case. More precisely, consider contractions T_0 in $\mathcal{L}(\mathcal{H}_1, \mathcal{H}_0)$, T'_0 in $\mathcal{L}(\mathcal{H}'_1, \mathcal{H}'_0)$, and X_0 in $\mathcal{L}(\mathcal{H}_0, \mathcal{H}'_0)$, X_1 in $\mathcal{L}(\mathcal{H}_1, \mathcal{H}'_1)$, such that $T'_0 X_1 = X_0 T_0$. In this case,

$$W^+_0 = \begin{bmatrix} T_0 & 0 \\ D_{T_0} & 0 \end{bmatrix} \quad \text{and} \quad W'^+_0 = \begin{bmatrix} T'_0 & 0 \\ D_{T'_0} & 0 \end{bmatrix}$$

and the solution of the problem of deciding whether there exist two contractions

$$Y_0 = \begin{bmatrix} X_0 & 0 \\ X^0_{21} & X^0_{22} \end{bmatrix} \quad \text{and} \quad Y_1 = \begin{bmatrix} X_1 & 0 \\ X^1_{21} & X^1_{22} \end{bmatrix}$$

such that $W_0'^+ Y_1 = Y_0 W_0^+$, appears as a particular case of Theorem 5.1. A solution of this problem can be immediately obtained as a consequence of (2.2.7). Thus, the equality $W_0'^+ Y_1 = Y_0 W_0^+$ implies the equality $D_{T_0'} X_1 = X_{21}^0 T_0 + X_{22}^0 D_{T_0}$. Therefore, we can take $X_{21}^1 = X_{22}^1 = 0$ and in order to determine whether there exist X_{21}^0 and X_{22}^0 with the desired properties, we multiply Y_0 on the right with $R(T_0)$, the elementary rotation of T_0. We obtain

$$Y_0 R(T_0) = \begin{bmatrix} X_0 T_0 & X_0 D_{T_0^*} \\ D_{T_0'} X_1 & X_{21}^0 D_{T_0^*} - X_{22}^0 T_0^* \end{bmatrix}.$$

But $[\, X_0 T_0 \quad X_0 D_{T_0^*} \,]$ and $[\, X_0 T_0 \quad D_{T_0'} X_1 \,]^\top$ are contractions, hence, by (2.2.7), there exists S such that the operator $Y_0 R(T_0) = \begin{bmatrix} X_0 T_0 & X_0 D_{T_0^*} \\ D_{T_0'} X_1 & S \end{bmatrix}$ is a contraction. The two equalities:

$$\begin{cases} D_{T_0'} X_1 = X_{21}^0 T_0 + X_{22}^0 D_{T_0} \\ S = X_{21}^0 D_{T_0^*} - X_{22}^0 T_0^* \end{cases}$$

determine two operators X_{21}^0 and X_{22}^0 with the required properties. This remark and the structure of the minimal isometric dilations can be used in a stepwise procedure to get another proof of Theorem 5.1. $\square$

The first proof of Theorem 5.1, even less direct than the one sketched in Remark 5.3, has the advantage that if the operators X_n, $n \in \mathbb{Z}$, are contractions, then an explicit connection can be established between the set CID($\{X_n\}_{n \in \mathbb{Z}}$ and the set $\mathcal{E}(\mathbf{E}, \mathbf{F}; \mathbf{v})$ studied in Section 2 and Section 3. Thus, consider two representatives $\mathbf{w} = \{w_n\}_{n \in \mathbb{Z}}$, w_n in $\mathcal{L}(\mathcal{S}_{n+1}, \mathcal{S}_n)$ and $\mathbf{u} = \{u_n\}_{n \in \mathbb{Z}}$, u_n in $\mathcal{L}(\mathcal{S}_{n+1}', \mathcal{S}_n')$ of a class in $\mathcal{E}(\mathbf{E}, \mathbf{F}; \mathbf{v})$. For the elements associated to the family $\mathbf{w}$ we continue to use the notation introduced in the proof of Theorem 5.1. Then, there exists a family $\{\widetilde{\varphi}_n\}_{n \in \mathbb{Z}}$ of unitary operators $\widetilde{\varphi}_n$ in $\mathcal{L}(\mathcal{S}_n, \mathcal{S}_n')$, such that $\widetilde{\varphi}_n / \mathcal{G}_n = I_{\mathcal{G}_n}$ and $u_n \widetilde{\varphi}_{n+1} = \widetilde{\varphi}_n w_n$ for all $n \in \mathbb{Z}$. Moreover, there exists another family $\{\varphi_n'\}_{n \in \mathbb{Z}}$ of unitary operators φ_n' in $\mathcal{L}(\mathcal{K}_n'^+, \widetilde{\mathcal{K}}_n'')$, such that $\varphi_n' / \mathcal{H}_n' = I_{\mathcal{H}_n'}$ and $\varphi_n' W_n'^+ = \widetilde{u}_n' \varphi_{n+1}'$ for all $n \in \mathbb{Z}$, where $\widetilde{\mathcal{K}}_n'' = \mathcal{H}_n' \vee u_n \mathcal{H}_{n+1}' \vee u_n u_{n+1} \mathcal{H}_{n+2}' \vee \ldots$ and $\widetilde{u}_n' = u_n / \mathcal{K}_n''$. But, $\widetilde{\varphi}_n \varphi_n = \varphi_n'$, $\widetilde{\mathcal{K}}_n'' = \widetilde{\varphi}_n \widetilde{\mathcal{K}}_n'$ and $\widetilde{\varphi}_n \mathcal{K}_n^+ = \mathcal{K}_n^+$ for all $n \in \mathbb{Z}$, hence

$$\varphi_n'^* P_{\widetilde{\mathcal{K}}_n''}^{\mathcal{S}_n'} / \mathcal{K}_n^+ = \varphi_n^* P_{\widetilde{\mathcal{K}}_n'}^{\mathcal{S}_n} / \mathcal{K}_n^+$$

for all $n \in \mathbb{Z}$. This means that the operators defined by (5.9) do not depend on the representatives of a class in $\mathcal{E}(\mathbf{E}, \mathbf{F}; \mathbf{v})$, therefore the map

$$\rho : \mathcal{E}(\mathbf{E}, \mathbf{F}; \mathbf{v}) \longrightarrow \mathrm{CID}(\{X_n\}_{n \in \mathbb{Z}}) \tag{5.10}$$

$$\rho([\mathbf{w}]) = \{\varphi_n^* P_{\widetilde{\mathcal{K}}_n'}^{\mathcal{S}_n} / \mathcal{K}_n^+\}_{n \in \mathbb{Z}}$$

is well defined. Actually, we have the following result.

5.4 Theorem   *The map ρ defined by (5.10) is a bijection between the set $\mathcal{E}(\mathbf{E}, \mathbf{F}; \mathbf{v})$ and the set $\mathrm{CID}(\{X_n\}_{n \in \mathbb{Z}})$.*

Proof　First we show that ρ is surjective.　Let $\{Y_n\}_{n\in\mathbb{Z}}$ be an element of $\mathrm{CID}(\{X_n\}_{n\in\mathbb{Z}})$ and remark that for $p < k$, $k > 1$, and $g = W_{n-1}^* \dots W_{n-p}^* k$, $k \in \mathcal{K}_{n-p}^+$, we have

$$W_{n-1}'^* \dots W_{n-k}'^* Y_{n-k} W_{n-k} \dots W_{n-1} P_{W_{n-1}^* \dots W_{n-k}^* \mathcal{K}_{n-k}^+} g = W_{n-1}'^* \dots W_{n-p}'^* Y_{n-p} k.$$

Moreover, $\mathcal{K}_n = \dots \vee W_{n-1}^* W_{n-2}^* \mathcal{K}_{n-2}^+ \vee W_{n-1}^* \mathcal{K}_{n-1}^+ \vee \mathcal{K}_n^+$, hence the operator

$$Z_n = s\text{-}\lim_{k\to\infty} W_{n-1}'^* \dots W_{n-k}'^* Y_{n-k} W_{n-k} \dots W_{n-1} P_{W_{n-1}^* \dots W_{n-k}^* \mathcal{K}_{n-k}^+}$$

exists in $\mathcal{L}(\mathcal{K}_n, \mathcal{K}_n')$, $\|Z_n\| = \|Y_n\| \leq 1$ and $W_n' Z_{n+1} = Z_n W_n$ for all $n \in \mathbb{Z}$. Then, for $n \in \mathbb{Z}$, define $\tilde{\mathcal{G}}_n = (\mathcal{K}_n' \oplus \mathcal{K}_n)_{C_n}$, the Hilbert space obtained by renorming $\mathcal{K}_n' \oplus \mathcal{K}_n$ with respect to $C_n = \begin{bmatrix} I & Z_n \\ Z_n^* & I \end{bmatrix}$. Remark that $\tilde{\mathcal{G}}_n = \mathcal{K}_n' \vee \mathcal{K}_n$ and the map

$$U_n : \tilde{\mathcal{G}}_{n+1} \longrightarrow \tilde{\mathcal{G}}_n \tag{5.11}$$

$$U_n(k' + k) = W_n' k' + W_n k, \quad k' \in \mathcal{K}_{n+1}', k \in \mathcal{K}_{n+1},$$

is well defined and gives rise to a unitary operator in $\mathcal{L}(\tilde{\mathcal{G}}_{n+1}, \tilde{\mathcal{G}}_n)$. By the definition of U_n and of the spaces $\tilde{\mathcal{G}}_n$, it follows that $[\{U_n\}_{n\in\mathbb{Z}}]$ belongs to $\mathcal{E}(\mathbf{E}, \mathbf{F}; \mathbf{v})$ and for the elements $k^\pm$ in $\mathcal{K}_n^\pm$,

$$\langle (P_{\mathcal{K}_n'+}^{\tilde{\mathcal{G}}_n} / \mathcal{K}_n^+) k^+, k^- \rangle_{C_n} = \langle k^+, k^- \rangle_{C_n} = \langle Z_n k^+, k^- \rangle = \langle Y_n k^+, k^- \rangle,$$

hence $P_{\mathcal{K}_n'+}^{\tilde{\mathcal{G}}_n} / \mathcal{K}_n^+ = Y_n$ for all $n \in \mathbb{Z}$. This shows that ρ is surjective.

Let $\mathbf{w} = \{w_n\}_{n\in\mathbb{Z}}$ be a representative of a class in $\mathcal{E}(\mathbf{E}, \mathbf{F}; \mathbf{v})$, where w_n belongs to $\mathcal{L}(\mathcal{S}_{n+1}, \mathcal{S}_n)$. We consider the family $\{\Upsilon_n\}_{n\in\mathbb{Z}}$ of unitary operators Υ_n in $\mathcal{L}(\mathcal{S}_n, \mathcal{S}_n)$ defined by:

$$\Upsilon_n = \begin{cases} w_{-1}^* \dots w_n^* & \text{if } n < 0 \\ I_{\mathcal{S}_0} & \text{if } n = 0 \\ w_0 \dots w_{n-1} & \text{if } n > 0. \end{cases} \tag{5.12}$$

Using the minimality of $\mathbf{w}$, it follows that $\mathcal{S}_0 = \vee_{-\infty}^\infty \Upsilon_n(\mathcal{H}_n \vee \mathcal{H}_n')$. Then, for $n \in \mathbb{Z}$, $m \geq n$, $P_{\mathcal{H}_n} w_n w_{n+1} \dots w_{n+m} / \mathcal{H}_{n+m+1}$ is uniquely determined by the family $\{T_n\}_{n\in\mathbb{Z}}$ and $P_{\mathcal{H}_n'} w_n w_{n+1} \dots w_{n+m} / \mathcal{H}_{n+m+1}'$ is uniquely determined by the family $\{T_n'\}_{n\in\mathbb{Z}}$. Furthermore, for h in $\mathcal{H}_{n+m+1}$ and g' in $\mathcal{H}_n'$, $\langle w_n w_{n+1} \dots w_{n+m} h, g' \rangle = \langle Y_n w_n^+ w_{n+1}^+ \dots w_{n+m}^+ h, g' \rangle$, which shows that the operator $P_{\mathcal{H}_n'} w_n w_{n+1} \dots w_{n+m} / \mathcal{H}_{n+m+1}$ is uniquely determined by the family $\{Y_n\}_{n\in\mathbb{Z}}$, with Y_n defined by (5.9). Similarly, $P_{\mathcal{H}_n} w_n w_{n+1} \dots w_{n+m} / \mathcal{H}_{n+m+1}'$ is uniquely determined by the family $\{Y_n\}_{n\in\mathbb{Z}}$. Consequently, if $\mathbf{w} = \{w_n\}_{n\in\mathbb{Z}}$, with w_n in $\mathcal{L}(\mathcal{S}_{n+1}, \mathcal{S}_n)$ and $\mathbf{w}' = \{w_n'\}_{n\in\mathbb{Z}}$, with w_n' in $\mathcal{L}(\mathcal{S}_{n+1}', \mathcal{S}_n')$ produce the same family $\{Y_n\}_{n\in\mathbb{Z}}$ in $\mathrm{CID}(\{X_n\}_{n\in\mathbb{Z}})$, then we can define the map

$$\psi_0 : \mathcal{S}_0 \longrightarrow \mathcal{S}_0'$$

$$\psi_0\left(\sum_{n\in\mathbb{Z}}\Upsilon_n h_n\right) = \sum_{n\in\mathbb{Z}}\Upsilon'_n h_n,$$

where only a finite number of the vectors $h_n \in \mathcal{H}_n \vee \mathcal{H}'_n$ are different from zero and the family $\{\Upsilon'_n\}_{n\in\mathbb{Z}}$ is associated to $\mathbf{w}'$ by the formula (5.12). This map gives rise to a unitary operator, also denoted by ψ_0, such that $\psi_0/\mathcal{G}_0 = I_{\mathcal{G}_0}$ and we can define the operators $\psi_n = w'^*_{n-1}\cdots w'^*_0\psi_0 w_0 \cdots w_{n-1}$ for $n > 0$ and $\psi_n = w'_n \cdots w'_{-1}\psi_0 w^*_{-1}\cdots w^*_n$ for $n < 0$. It is clear that $\{\psi_n\}_{n\in\mathbb{Z}}$ is a family of unitary operators ψ_n in $\mathcal{L}(\mathcal{S}_n,\mathcal{S}'_n)$ with the property $\psi_n/\mathcal{G}_n = I_{\mathcal{G}_n}$ and $w'_n\psi_{n+1} = \psi_n w_n$ for all $n \in \mathbb{Z}$. This means that $[\mathbf{w}] = [\mathbf{w}']$, hence ρ is injective. $\qquad\square$

5.5 Remark Theorem 5.4 shows that it would be desirable to describe the spaces $\mathcal{Q}_n = \mathcal{G}_n \ominus \mathcal{E}_n$, $\mathcal{R}_n = \mathcal{G}_n \ominus \mathcal{F}_n$, $n \in \mathbb{Z}$ introduced in the proof of Theorem 5.1 in terms of the families $\{T_n\}_{n\in\mathbb{Z}}$, $\{T'_n\}_{n\in\mathbb{Z}}$ and $\{X_n\}_{n\in\mathbb{Z}}$. Thus, we can easily verify that $\mathcal{Q}_{n+1}$ can be identified, up to a certain unitary operator, with the space $(\mathcal{D}_{X_n} \oplus \mathcal{D}_{T_n}) \ominus \mathrm{cl}\mathcal{R}([\,D_{X_n}T_n \quad D_{T_n}\,]^\top)$, while the space $\mathcal{R}_n$ can be identified with the space $(\mathcal{D}_{T'^*_n} \oplus \mathcal{D}_{X^*_{n+1}}) \ominus \mathrm{cl}\mathcal{R}([\,D_{T'^*_n} \quad D_{X^*_{n+1}}T'^*_n\,]^\top)$. $\qquad\square$

We have seen in Remark 5.3 that a special case of Theorem 5.1 is essentially equivalent to the distance formula (2.2.7). Another particular case refers to the situation when $X_n = X$, $T_n = T$ and $T'_n = T'$ for all $n \in \mathbb{Z}$, for certain contractions T, T' in $\mathcal{L}(\mathcal{H})$ and, respectively, $\mathcal{L}(\mathcal{H}')$ and an operator X such that $T'X = XT$. Let W^+ in $\mathcal{L}(\mathcal{K}^+)$ and W'^+ in $\mathcal{L}(\mathcal{K}'^+)$ be the minimal isometric dilations of T and, respectively, T'. In this case we define:

$$\mathrm{CID}(T,T';X) = \mathrm{CID}(X) =$$

$$= \{Y \in \mathcal{L}(\mathcal{K}^+,\mathcal{K}'^+) \mid \|Y\| \le 1,\ W'^+Y = YW^+,\ P_{\mathcal{H}'}Y = XP_{\mathcal{H}}\}.$$

The following result is a direct consequence of Theorem 5.1.

5.6 Theorem *The set $\mathrm{CID}(X)$ is nonempty if and only if the operator X is a contraction.* $\qquad\square$

We conclude this section with the presentation of the solutions of the norm completion problems introduced in Section 1. We begin with Problem 1.10 for $\{\widetilde{X}_{ij}\}_{(i,j)\in\alpha(r)}$. There exists a unique family $\{\beta_n\}_{n\in\mathbb{Z}}$ of intervals $[a_n,b_n] \cap \mathbb{Z}$, $a_n \le b_n$, of maximal length, with the property that $\cup_{n\in\mathbb{Z}}\{(i,j) \mid a_n \le i \le j \le b_n\} = \alpha(r)$. Then, define

$$X_n = \begin{bmatrix} \widetilde{X}_{a_n,a_n} & \widetilde{X}_{a_n,a_{n+1}} & & \widetilde{X}_{a_n,b_n} \\ 0 & \widetilde{X}_{a_{n+1},a_{n+1}} & \ddots & \widetilde{X}_{a_{n+1},b_n} \\ & \ddots & \ddots & \\ 0 & & & \widetilde{X}_{b_n,b_n} \end{bmatrix} \qquad (5.13)$$

and $\delta_n = b_n - a_n + 1$. The operators T_n have the $\delta_n \times \delta_{n+1}$ block matrix representation

$$T_n = [T_n^{(ij)} \mid 1 \le i \le \delta_n, 1 \le j \le \delta_{n+1}], \qquad (5.14)$$

$$T_n^{(ij)} = \begin{cases} I & \text{if } j = i+1 \\ 0 & \text{otherwise.} \end{cases}$$

We obtain the following result.

5.7 Corollary *(a)* *The Problem 1.10 for* $\{\widetilde{X}_{ij}\}_{(i,j)\in\alpha(r)}$ *is solvable if and only if the operators* X_n *defined by (5.13) are contractions for all* $n \in \mathbb{Z}$.

(b) *If the Problem 1.10 for* $\{\widetilde{X}_{ij}\}_{(i,j)\in\alpha(r)}$ *is solvable, then there exists a bijective correspondence between the set of its solutions and the set* $CID(\{T_n\}_{n\in\mathbb{Z}}, \{T_n\}_{n\in\mathbb{Z}};$ $\{X_n\}_{n\in\mathbb{Z}})$, *where* T_n *are given by (5.14).*

Proof We remark that $T_n X_{n+1} = X_n T_n$ for all $n \in \mathbb{Z}$ and the minimal isometric dilation of $\{T_n\}_{n\in\mathbb{Z}}$ consists of a family of marking operators as in (2.4.1). An application of Theorem 5.1 and Theorem 5.4 concludes the proof. □

The Nevanlinna-Pick problem for $\{z_k\}_{k=0}^n$ and $\{w_k\}_{k=0}^n$ can be also solved in the following way. Consider the functions $g_k(z) = (1 - z_k^* z)^{-1}$, $k = 0, 1, \ldots, n$, and denote by $\mathcal{H}$ the subspace of the Hardy space H^2, generated by $\{g_k\}_{k=0}^n$. Let S be the unilateral shift on H^2, defined by the formula $Sf(z) = zf(z)$ for f in H^2, and remark that $S^* g_k = z_k^* g_k$ for $k = 0, 1, \ldots, n$. Hence, define

$$T^* = S^*/\mathcal{H} \tag{5.15}$$

and

$$X : \mathcal{H} \longrightarrow \mathcal{H} \tag{5.16}$$

$$X^* g_k = w_k^* g_k, \quad k = 0, 1, \ldots, n.$$

We obtain the following result.

5.8 Corollary *(a)* *The Nevanlinna-Pick problem for* $\{z_k\}_{k=0}^n$ *and* $\{w_k\}_{k=0}^n$ *is solvable if and only if the operator* X *defined by (5.16) is a contraction.*

(b) *If the Nevanlinna-Pick problem for* $\{z_k\}_{k=0}^n$ *and* $\{w_k\}_{k=0}^n$ *is solvable, then there is a bijective correspondence between the set of its solutions and the set* $CID(T, T; X)$, *where* T *is defined by (5.15).*

Proof Note that $TX = XT$ and S is the minimal isometric dilation of T. An application of Theorem 5.6 and Theorem 5.4 concludes the proof. We also remark that X being a contraction means that for any complex numbers a_k, $k = 0, 1, \ldots, n$,

$$0 \le \langle (I - XX^*) \sum_{k=0}^n a_k g_k, \sum_{j=0}^n a_j g_j \rangle = \sum_{j,k=0}^n \frac{1 - w_j w_k^*}{1 - z_j z_k^*} a_j a_k^*,$$

which is exactly the condition obtained in Theorem 1.12 for the solvability of the Nevanlinna-Pick problem. □

A solution of the Hermite-Fejér problem can be obtained as a combination of the constructions in the proofs of Corollary 5.7 and Corollary 5.8. We omit the details. A solution of the Nehari problem for $\{c_n\}_{n=-\infty}^{-1}$ can be obtained in the framework of the commutant lifting in the following way. Define

$$X = \begin{bmatrix} c_0 & c_{-1} & c_{-2} & c_{-3} & \cdots \\ c_{-1} & c_{-2} & c_{-3} & \cdots & \\ c_{-2} & c_{-3} & \cdots & & \\ c_{-3} & \cdots & & & \\ \vdots & & & & \end{bmatrix} \qquad (5.17)$$

on the Hilbert space $\mathcal{H}$ of the square-summable sequences $\{x_n\}_{n\geq 0}$ of complex numbers, with the inner product $\langle x, y \rangle = \sum_{n\geq 0} x_n y_n^*$. Moreover, consider the contractions

$$T : \mathcal{H} \longrightarrow \mathcal{H} \qquad (5.18)$$

$$T(\oplus_{k\geq 0} x_k) = 0 \oplus \oplus_{k\geq 0} x_k$$

and

$$T' = T^*. \qquad (5.19)$$

We obtain the following result.

5.9 Corollary (a) *The Nehari problem for $\{c_n\}_{n=-\infty}^{-1}$ is solvable if and only if the operator X defined by (5.17) is a contraction.*

(b) *If the Nehari problem for $\{c_n\}_{n=-\infty}^{-1}$ is solvable, then there exists a bijective correspondence between the set of its solutions and the set $CID(T, T'; X)$, where T and T' are given by (5.18) and, respectively, (5.19).*

Proof We remark that $T'X = XT$ and the minimal isometric dilation of T' is the operator W'_+ on l^2 defined by the formula $(W'_+ x)_n = x_{n+1}$ for $x = \{x_n\}_{n\in\mathbb{Z}}$ in l^2. The proof can be concluded by an application of Theorem 5.6 and Theorem 5.4. $\qquad\square$

Finally, we analyze here a version of the four-block problem. Thus, consider families $\mathbf{H}_i = \{\mathcal{H}_{i,n}\}_{n\in\mathbb{Z}}$, $\mathbf{H}'_i = \{\mathcal{H}'_{i,n}\}_{n\in\mathbb{Z}}$, $i = 1, 2$, of Hilbert spaces. Define $\mathcal{H}_i = \oplus_{n\in\mathbb{Z}} \mathcal{H}_{i,n}$, $\mathcal{H}'_i = \oplus_{n\in\mathbb{Z}} \mathcal{H}'_{i,n}$, $i = 1, 2$, and suppose that there is given an operator $L = [L_{ij}]_{i,j=1}^2$ in $\mathcal{L}(\mathcal{H}_1 \oplus \mathcal{H}_2, \mathcal{H}'_1 \oplus \mathcal{H}'_2)$. The problem that we consider requires the computation of the number

$$\delta = \inf\{\| \begin{bmatrix} L_{11} - Q & L_{12} \\ L_{21} & L_{22} \end{bmatrix} \| \mid Q \in \mathcal{S}(\mathbf{H}_1, \mathbf{H}'_1)\}.$$

It is seen that the solution of the Problem 1.15 is a consequence of the computation of δ. To that end, define the projections $P_k = P_{\oplus_{p\leq k} \mathcal{H}_{1,p}}$, $P'_k = P_{\oplus_{p\leq k} \mathcal{H}'_{1,p}}$, and consider the operators $X_{-k} = ((I - P'_k) \oplus I_{\mathcal{H}'_2}) L (P_k \oplus I_{\mathcal{H}_2})$. We obtain the following result.

5.10 Corollary *For an operator $L = [L_{ij}]_{i,j=1}^2$ $\mathcal{L}(\mathcal{H}_1 \oplus \mathcal{H}_2, \mathcal{H}'_1 \oplus \mathcal{H}'_2)$, the following formula holds:*

$$\delta = \sup_{k\in\mathbb{Z}} \|X_k\|.$$

Proof For every element Q in $\mathcal{S}(\mathbf{H}_1, \mathbf{H}_1')$,

$$((I - P_k') \oplus I_{\mathcal{H}_2'}) \begin{bmatrix} L_{11} - Q & L_{12} \\ L_{21} & L_{22} \end{bmatrix} (P_k \oplus I_{\mathcal{H}_2}) = X_{-k},$$

therefore, for every $k \in \mathbb{Z}$,

$$\| \begin{bmatrix} L_{11} - Q & L_{12} \\ L_{21} & L_{22} \end{bmatrix} \| \geq \|X_k\|.$$

This implies that $\delta \geq \sup_{k \in \mathbb{Z}} \|X_k\|$. In order to prove the opposite inequality, we use Theorem 5.2. For $k \in \mathbb{Z}$, we define the contractions: $T_{-k} = N_{\mathbf{H}_1}(k) \oplus I_{\mathcal{H}_2}$ and $T_{-k}' = M_{\mathbf{H}_1'}^*(k) \oplus I_{\mathcal{H}_2'}$, where $N_{\mathbf{H}_1}(k)$ and $M_{\mathbf{H}_1'}(k)$ are the marking operators defined by (2.4.1) and (2.4.2). Remark that $T_{-k}' X_{-k+1} = X_{-k} T_{-k}$ and, by Theorem 5.2, there exists $\widetilde{Q}$ in $\mathcal{L}(\mathcal{H}_1 \oplus \mathcal{H}_2, \mathcal{H}_1' \oplus \mathcal{H}_2')$ such that $\|\widetilde{Q}\| = \sup_{k \in \mathbb{Z}} \|X_k\|$ and $\widetilde{Q} - L = \begin{bmatrix} Q & 0 \\ 0 & 0 \end{bmatrix}$ for some Q in $\mathcal{S}(\mathbf{H}_1, \mathbf{H}_1')$. Consequently,

$$\delta \leq \|L - \begin{bmatrix} Q & 0 \\ 0 & 0 \end{bmatrix} \| = \|\widetilde{Q}\| = \sup_{k \in \mathbb{Z}} \|X_k\|,$$

and the proof is complete. $\qquad\qquad\qquad\qquad\qquad\qquad\qquad\qquad\qquad\qquad\square$

3.6 Notes

For the early work on bounded interpolation and moment theory, we mention here some of the original sources: [Ca1-2], [CF], [Her], [Nevan1-2], [Pi], [Ri], [Sc] (see [Herg] for a recent collection of these papers) and the monographs [Ak], [AK], [KN]. For the theory of extensions of symmetric operators, we mention here [Kr2-3], [Neu], [Na1]. The method of extending isometries to unitary operators was employed by M.A. Naimark and M.G. Krein—see [Na1], [Kr2], [Kr3], and also [AAK1], [KrL1-5]. The proof of Theorem 1.12 appears in [KR].

Theorem 2.2 is proved in [Co3-4]. The first versions of Krein's generalized resolvent formula appeared in [Kr2-3]. Detailed studies of the moment problems discussed in Section 4 can be found in [Kr1], [Ak], [AK], [KN]. Recent papers, which also contain generalizations to indefinite metric spaces, are [KrL1-5], [Lan].

Theorem 5.6 was proved in [Sar1] in a particular case and in [Sz.-NF1] for the general case. The result is also related to the dilation theorem for pairs of contractions [An] (see [FoF1] for a detailed discussion). The version in Theorem 5.1, which technically speaking is equivalent to Theorem 5.6 due to the Toeplitz embedding, was proved in [BG1]. The generalization to nest algebras was made in [PP]—see [Da] for a discussion of some dilation results in nest algebras. The proof of Theorem 5.6 indicated in the text follows an idea in [Ar1] which, in its turn, is a development of [AAK1]— for further developments along this line see [CS]. Theorem 5.4 is a version of the results obtained in [AnCF], [ACF], [CF1-2]. Corollary 5.10 gives the solution of the four-block problem and shows its connection with the commutant lifting theorem. This was done in [FF1-2]. There is a large literature devoted to the applications of the interpolation theory to robust control. We mention here the monographs [DFT], [Fr], [Vi], as well as some of the key papers of the field: [CP], [FHZ], [FT2], [Gl2], [GLDKS], [LA], [Ki1-3], [LG], [Ta1-2], [Za1-2], [ZF]. A collection containing most of these papers is [Dor].

Some other versions of the commutant lifting theorem were developed in recent years. Besides the above mentioned generalization to nest algebras, we mention here a nonlinear generalization in [BFHT], some extensions to Krein spaces [BH1], [CG1-2], [Dr1-2], [DR], multi-dimensional versions in [CS], [Pop] and a so-called spectral version–see [Helt5]. Generalizations of the interpolation problems for meromorphic functions are studied in details in [AAK2], [BH1], [KrL1-5]. These developments have another engineering application in connection with the model reduction problem, see [Gl1] and [DV], [KK], [Ve] for time-varying generalizations.

Many other related interpolation problems are formulated and solved in the literature of the recent years. We mention here the following papers and books devoted to this subject: [AAK3], [ABDS], [BaGK1-5], [BH1-2], [BGK], [Dew1-2], [DeD1-3], [DKV], [DMP], [DFK], [Dy1-2], [DG1-2], [Fe1-2], [Fo], [FoF2], [Go1-4], [GGK], [GKW1], [GrS], [GL], [Helt1-5]. In connection with this chapter see also: [At], [[Ar2], [BAG1-3], [Bu1-2], [JNT], [EGL], [K], [Ka3], [KoP], [La], [Le], [Ne], [Nu], [Po], [Pow], [Ros], [RR1-2], [Sa1-2], [Sar2], [Sz.-NK], [TV], [YS], [Yo].

Chapter 4
Displacement Structures

The displacement structure is introduced in this chapter as a general framework for the presentation of several algorithms related to the description of the positive and contractive block matrices. The Gaussian elimination procedure is exploited in order to deduce a so-called generalized Schur algorithm for displacement structures. An associated transmission-line interpretation of the algorithm is also discussed in details, and several examples involving interpolation and linear filtering illustrate the main results.

4.1 Structured Matrices

Several examples of matrices satisfying a certain matrix equation are discussed in this section in order to motivate the introduction of the notion of displacement structure. A key embedding result is proven and its connection with some dilation results in Chapter 1 is emphasized.

1.1 Example   We first consider an example related to the problem of triangular factorization. Let T be a Toeplitz lower triangular contraction,

$$
T = \begin{bmatrix} c_0 & 0 & \cdots & 0 \\ c_1 & c_0 & & \\ \vdots & & \ddots & \vdots \\ c_n & c_{n-1} & \cdots & c_0 \end{bmatrix},
$$

where $c_0, c_1, \ldots, c_n$ are given complex numbers. It is convenient to introduce a special notation for lower triangular Toeplitz matrices by writing $T = L(c)$, where $c = \begin{bmatrix} c_0 & c_1 & \cdots & c_n \end{bmatrix}^{\top}$. We remark that the upper triangular Cholesky factor of a whole class of matrices can be computed using the upper triangular Cholesky factor F of the matrix $I - TT^*$. Thus, if the positive matrix A has the structure

$$
A = L(x)L^*(x) - L(y)L^*(y), \tag{1.1}
$$

where $L(y) = L(x)T$ and x, y are two vectors in $\mathbb{C}^{n+1}$, then $A = L(x)(I - TT^*)L^*(x) = L(x)F^*FL^*(x)$. It follows that the computation of the upper tri-

angular Cholesky factor of A reduces to the computation of the product of two
upper triangular matrices.

Due to this remark, it appears to be of interest to characterize those positive
matrices of the form (1.1). One possible solution explores the property of T that
it intertwines the truncated shift $(n + 1) \times (n + 1)$ matrix

$$
Z = \begin{bmatrix} 0 & 0 & & 0 \\ 1 & 0 & \ddots & \\ & \ddots & \ddots & \\ 0 & & 1 & 0 \end{bmatrix}. \tag{1.2}
$$

Since $TZ = ZT$, it follows that

$$
I - TT^* - Z(I - TT^*)Z^* = I - ZZ^* - T(I - ZZ^*)T^*
$$

$$
= [\, E_0 \quad c\,] \begin{bmatrix} 1 & 0 \\ 0 & -1 \end{bmatrix} [\, E_0 \quad c\,]^*,
$$

where we also use the fact that $I - ZZ^*$ is the 1-dimensional projection onto the
first component of $\mathbb{C}^{n+1}$. The preceding equality shows that if a matrix A is of
the form (1.1), then

$$
A - ZAZ^* = L(x)[\, E_0 \quad c\,] \begin{bmatrix} 1 & 0 \\ 0 & -1 \end{bmatrix} [\, E_0 \quad c\,]^* L^*(x)
$$

$$
= [\, x \quad y\,] \begin{bmatrix} 1 & 0 \\ 0 & -1 \end{bmatrix} [\, x \quad y\,]^*.
$$

Define $G = [\, x \quad y\,]$ and $J = \begin{bmatrix} 1 & 0 \\ 0 & -1 \end{bmatrix}$. We conclude that if A satisfies (1.1), then
it is a solution of the following equation:

$$
A - ZAZ^* = GJG^*. \tag{1.3}
$$

It turns out that the converse is also true. Since $Z^m = 0$ for $m \geq n+1$, one obtains
that if A is a solution of the equation (1.3), then $A = GJG^* + ZGJG^*Z^* + \ldots +
Z^n GHJG^*(Z^n)^*$, hence $A = L(x)L^*(x) - L(y)L^*(y)$. The fact that A is a positive
matrix and an application of Lemma 1.4.1 show that $L(y) = L(x)T$ for a certain
contraction T. Without loss of generality, we can assume $x_0 \neq 0$. Then, it follows
by direct computation that T must be a lower triangular Toeplitz contraction. In
conclusion, we have proven that a positive matrix A satisfies (1.1) if and only if A
is a solution of the equation (1.3).   □

1.2 Example   We show that the positive Toeplitz matrices and their inverses also
satisfy equations of type (1.3). To that end, suppose $A = [A_{j-i}]_{i,j=0}^n$, $A_0 = 1$,
and define the column vectors $a = [\, 1 \quad A_1^* \quad \ldots A_n^* \,]^T$ and $b = [\, 0 \quad A_1^* \quad \ldots A_n^* \,]^T$.
Then, it is easy to see that A satisfies the equation (1.3) for $G = [\, a \quad b\,]$. Sup-
pose A is invertible. We remark that the two Schur complements of the matrix

$\begin{bmatrix} A & Z^* \\ Z & A^{-1} \end{bmatrix}$ are $A - Z^*AZ$ and, respectively, $A^{-1} - ZA^{-1}Z^*$. Since $A - Z^*AZ$ has exactly two nonzero eigenvalues, one positive and one negative, the same is true for $A^{-1} - ZA^{-1}Z^*$. Consequently, there exist two vectors d and f in $\mathbb{C}^{n+1}$ such that $A^{-1} - ZA^{-1}Z^* = HJH^*$, where $H = [\,d \quad f\,]$. Using the result mentioned in Example 1.1, it follows that

$$A^{-1} = L(d)L^*(d) - L(f)L^*(f).$$

This formula for A^{-1} is referred to as the *Gohberg-Semencul formula* for the inverse of a positive Toeplitz matrix. □

1.3 Example We have seen that the solution of the Nevanlinna-Pick problem was expressed in terms of the Pick matrix

$$P^{(0n)} = \left[\frac{1 - w_i w_j^*}{1 - z_i z_j^*}\right]_{i,j=0}^{n-1}.$$

It turns out that this matrix satisfies an equation of type (1.3). To see this, define

$$F = \begin{bmatrix} z_0 & & & \\ & z_1 & 0 & \\ & 0 & \ddots & \\ & & & z_{n-1} \end{bmatrix}, \qquad G = \begin{bmatrix} 1 & w_0 \\ 1 & w_1 \\ \vdots & \vdots \\ 1 & w_{n-1} \end{bmatrix},$$

and it is easy to verify that $P - FPF^* = GJG^*$. □

1.4 Example The last example involves the field of recursive least-squares estimation. Consider a sequence of scalar data points $\{d(i)\}_{i=0}^t$, and a sequence of row vectors (input signals) $\{\mathbf{u}(i)\}_{i=0}^t$, $\mathbf{u}(i) = [\,u_1(i) \quad u_2(i) \quad \dots \quad u_M(i)\,]$ for all $i = 0, 1, \dots, t$. The entries of $\mathbf{u}(i)$ can be regarded as the values of M input channels at time i. The objective is to determine an $M \times 1$ column vector $\mathbf{w}$ (see Figure 4.1), so as to minimize the weighted error sum:

$$\mathcal{E} = \sum_{i=0}^t \lambda^{t-i} |d(i) - \mathbf{u}(i)\mathbf{w}|^2,$$

where λ is a positive scalar, $0 < \lambda \leq 1$, often called the *forgetting factor*, since past inputs are weighted less than the more recent values. The special case $\lambda = 1$ is known as the *growing memory* case, since, as the length t of data grows, the effect of past data is not attenuated. It is easily seen that the optimal solution $\hat{\mathbf{w}}$ satisfies the so-called *normal equations*,

$$D_t^* \Lambda_t D_t \hat{\mathbf{w}} = D_t^* \Lambda_t d_t,$$

where we have defined

$$d_t = \begin{bmatrix} d(0) \\ d(1) \\ \vdots \\ d(t) \end{bmatrix}, \qquad D_t = \begin{bmatrix} u_1(0) & u_2(0) & \dots & u_M(0) \\ u_1(1) & u_2(1) & & \\ \vdots & & \ddots & \\ u_1(t) & & & u_M(t) \end{bmatrix}.$$

and Λ_t is the diagonal matrix with diagonal entries λ^t, λ^{t-1}, ..., λ and 1.

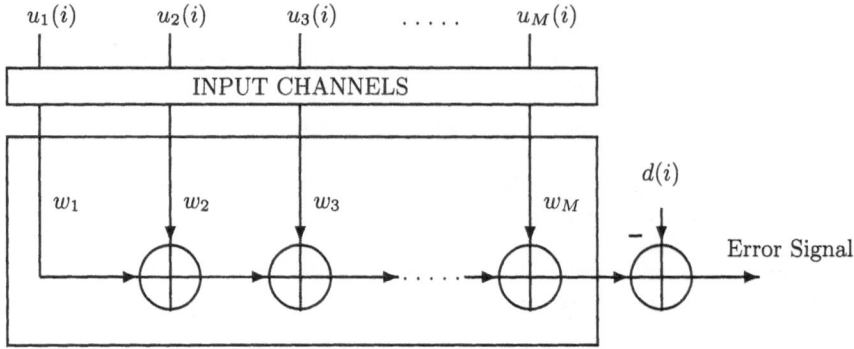

Figure 4.1: Linear least-squares estimation

The normal equations can be compactly rewritten as

$$\Phi(t)\hat{\mathbf{w}} = \vartheta(t),$$

where $\Phi(t) = D_t^* \Lambda_t D_t$ is the *weighted autocorrelation matrix* and $\vartheta(t) = D_t^* \Lambda_t d_t$ is the *weighted cross-correlation vector*. Now, we remark that

$$
\begin{aligned}
\Phi(t) - \lambda\Phi(t-1) &= \sum_{i=0}^{t} \lambda^{t-i}\mathbf{u}^*(i)\mathbf{u}(i) - \lambda\sum_{i=0}^{t-1} \lambda^{t-i-1}\mathbf{u}^*(i)\mathbf{u}(i) \\
&= \sum_{i=0}^{t} \lambda^{t-i}\mathbf{u}^*(i)\mathbf{u}(i) - \sum_{i=0}^{t-1} \lambda^{t-i}\mathbf{u}^*(i)\mathbf{u}(i) \\
&= \mathbf{u}^*(t)\mathbf{u}(t).
\end{aligned}
$$

If we define $F = \sqrt{\lambda}I$, then the preceding equation can be written in the form

$$\Phi(t) - F\Phi(t-1)F^* = \mathbf{u}^*(t)\mathbf{u}(t),$$

which is similar to the equation (1.3).   □

These examples motivate the introduction of the following notion of displacement structure. Consider three families $\{\mathcal{U}(t)\}_{t\in\mathbb{Z}}$, $\{\mathcal{V}(t)\}_{t\in\mathbb{Z}}$ and $\{\mathcal{R}(t)\}_{t\in\mathbb{Z}}$ of Hilbert spaces depending on the parameter $t \in \mathbb{Z}$, two families of bounded linear operators $G(t) = [U(t)\ V(t)]$ in $\mathcal{L}(\mathcal{U}(t) \oplus \mathcal{V}(t), \mathcal{R}(t))$ and $F(t)$ in $\mathcal{L}(\mathcal{R}(t-1), \mathcal{R}(t))$, and define the symmetry $J(t) = I_{\mathcal{U}(t)} \oplus -I_{\mathcal{V}(t)}$.

A family $\{R(t)\}_{t\in\mathbb{Z}}$ of operators in $\mathcal{L}(\mathcal{R}(t))$ is said to have *displacement structure* with respect to the *generators* $\{F(t)\}_{t\in\mathbb{Z}}$, $\{G(t)\}_{t\in\mathbb{Z}}$ if $\{R(t)\}_{t\in\mathbb{Z}}$ is uniformly bounded, viz., there exists $r > 0$ such that $\|R(t)\| \leq r$ for all $t \in \mathbb{Z}$, and $R(t)$ satisfies the *displacement* (or *time-varying Stein-Lyapunov*) equation

$$R(t) - F(t)R(t-1)F^*(t) = G(t)J(t)G^*(t). \tag{1.4}$$

The cardinal number $r(t) = \dim \mathcal{U}(t) + \dim \mathcal{V}(t)$ is called the *displacement rank* of $R(t)$ with respect to the equation (1.4). We say that (1.4) has a *Pick solution* if $R(t)$ is positive for every $t \in \mathbb{Z}$.

We must mention from the very beginning that our main concern in connection with the equation (1.4) will be the study of the effect of the factorization of $R(t)$ at the level of generators. To that end, we will need some convenient assumptions to insure that the equation (1.4) has a unique solution. We introduce the infinite matrices

$$\mathbf{U}(t) = [\ldots \quad F(t)F(t-1)U(t-2) \quad F(t)U(t-1) \quad U(t)],$$

$$\mathbf{V}(t) = [\ldots \quad F(t)F(t-1)V(t-2) \quad F(t)V(t-1) \quad V(t)],$$

and assume that for each $t \in \mathbb{Z}$ and h in $\mathcal{R}(t)$, we have

(H_1) $\quad F^*(t-n)F^*(t-n+1)\ldots F^*(t-1)F^*(t)h \to 0$ as $b \to \infty$,

(H_2) $\quad \mathbf{U}(t)$ and $\mathbf{V}(t)$ are well-defined bounded operators,

(H_3) $\quad \{\mathbf{U}(t)\}_{t \in \mathbb{Z}}$ and $\{\mathbf{V}(t)\}_{t \in \mathbb{Z}}$ are uniformly bounded families.

It is easy to see that the above assumptions imply that the equation (1.4) has a unique uniformly bounded solution given by

$$R(t) = \mathbf{U}(t)\mathbf{U}^*(t) - \mathbf{V}(t)\mathbf{V}^*(t). \tag{1.5}$$

Most of our considerations will be of special interest for the case of finite dimensional Hilbert spaces, so we consider another hypothesis,

(H_4) $\quad \mathcal{U}(t), \mathcal{V}(t)$ and $\mathcal{R}(t)$ are Hilbert spaces of finite dimension for all $t \in \mathbb{Z}$.

The following result plays a key role in our considerations about displacement equations.

1.5 Theorem *(a) Suppose (H_1)-(H_4) hold. If the displacement equation (1.4) has a Pick solution such that all $R(t)$ are invertible, then there exist families $\{H(t)\}_{t \in \mathbb{Z}}$ and $\{K(t)\}_{t \in \mathbb{Z}}$ of operators, $H(t)$ in $\mathcal{L}(\mathcal{R}(t-1), \mathcal{U}(t) \oplus \mathcal{V}(t))$ and $K(t)$ in $\mathcal{L}(\mathcal{U}(t) \oplus \mathcal{V}(t))$, such that the following embedding relation is satisfied*

$$\begin{bmatrix} F(t) & G(t) \\ H(t) & K(t) \end{bmatrix} \begin{bmatrix} R(t-1) & 0 \\ 0 & J(t) \end{bmatrix} \begin{bmatrix} F(t) & G(t) \\ H(t) & K(t) \end{bmatrix}^* = \begin{bmatrix} R(t) & 0 \\ 0 & J(t) \end{bmatrix}. \tag{1.6}$$

(b) If, in addition, the displacement equation (1.4) has a Pick solution $\{R(t)\}_{t \in \mathbb{Z}}$ such that

$$(H) \quad R(t) > \varepsilon I > 0 \text{ for a constant } \varepsilon \text{ and for all } t \in \mathbb{Z},$$

then there exist uniformly bounded families $\{H(t)\}_{t\in\mathbb{Z}}$ and $\{K(t)\}_{t\in\mathbb{Z}}$ satisfying the embedding relations (1.6).

Proof (a) Since we have assumed that all $R(t)$ are strictly invertible, the displacement equation (1.4) can be rewritten as follows (we use the notation $R^{-1/2}(t)$ for $(R(t))^{-1/2}$):

$$I - R^{-1/2}(t)F(t)R^{1/2}(t-1)R^{1/2}(t-1)F^*(t)R^{-1/2}(t)$$
$$= R^{-1/2}(t)G(t)J(t)G^*(t)R^{-1/2}(t).$$

This equality shows that the following claim will lead to a construction of two families satisfying (1.6). Thus, suppose there are given the finite dimensional Hilbert spaces $\mathcal{E}$, $\mathcal{H}$, $\mathcal{H}'$, and the operators F in $\mathcal{L}(\mathcal{H}',\mathcal{H})$ and G in $\mathcal{L}(\mathcal{E},\mathcal{H})$. Suppose that $I - FF^* = GJG^*$, where J is a symmetry (*i.e.* $J^* = J^{-1} = J$) on $\mathcal{E}$. Then, there exist operators H in $\mathcal{L}(\mathcal{H}',\mathcal{H})$ and K in $\mathcal{L}(\mathcal{E})$ such that

$$\begin{bmatrix} F & G \\ H & K \end{bmatrix}\begin{bmatrix} I & 0 \\ 0 & J \end{bmatrix}\begin{bmatrix} F & G \\ H & K \end{bmatrix}^* = \begin{bmatrix} I & 0 \\ 0 & J \end{bmatrix}. \tag{1.7}$$

First we remark that if H and K satisfy (1.7), then $\Theta = K + H(\tau - F)^{-1}G$ is a J-unitary operator (*i.e.* $\Theta J\Theta^* = J = \Theta^*J\Theta$) in $\mathcal{L}(\mathcal{E})$ for every unitary operator τ with the property that $\tau - F$ is invertible. This remark suggests that, if operators H and K are found such that they verify the system of equations

$$\begin{cases} HF^* + KJG^* = 0 \\ K + H(\tau - F)^{-1}G = \Theta, \end{cases} \tag{1.8}$$

where τ is a unitary operator such that $\tau - F$ is invertible and Θ is an arbitrary J-unitary operator, then H and K satisfy (1.7). It is easy to see that if $K + H(\tau - F)^{-1}G$ is J-unitary, then $KJK^* = J - HH^*$. Now, the system (1.8) can be easily solved and one obtains

$$H = \Theta JG^*(I - \tau F^*)^{-1}(\tau - F), \qquad K = \Theta[I - JG^*(I - \tau F^*)^{-1}G],$$

since $I - \tau F^*$ is invertible together with $\tau - F$. In conclusion, the following choices for $H(t)$ and $K(t)$ satisfy the embedding relation (1.6):

$$\begin{aligned} H(t) = \Theta(t)J(t)G^*(t)[R^{1/2}(t) \\ -\tau(t)R^{1/2}(t-1)F^*(t)]^{-1}[\tau(t)R^{-1/2}(t-1) - R^{-1/2}(t)F(t)], \end{aligned} \tag{1.9}$$

$$K(t) = \Theta(t)\{I - J(t)G^*(t)[R^{1/2}(t) - \tau(t)R^{1/2}(t-1)F^*(t)]^{-1}R^{-1/2}(t)G(t)\} \tag{1.10}$$

for an arbitrary $J(t)$-unitary operator $\Theta(t)$ and an arbitrary unitary operator $\tau(t)$, whenever the inverse of $[R^{1/2}(t) - \tau(t)R^{1/2}(t-1)F^*(t)]$ exists.

(b) It is shown that $\Theta(t)$ and $\tau(t)$ can be adequately chosen so as to guarantee the uniform boundedness of the families $\{H(t)\}_{t\in\mathbb{Z}}$ and $\{K(t)\}_{t\in\mathbb{Z}}$ defined by (1.9) and (1.10). This is possible due to the supplementary hypothesis (H). Actually, it is proven that $\tau(t)$ can be found such that

$$[R^{1/2}(t) - \tau(t)R^{1/2}(t-1)F^*(t)][R^{1/2}(t) - \tau(t)R^{1/2}(t-1)F^*(t)]^* \geq \varepsilon'I > 0$$

for some $\varepsilon' > 0$. If $F(t) = 0$, this is obvious. Otherwise, define the operators $A(t) = R^{-1/2}(t-1)F^*(t)R^{1/2}(t)$ and write $A(t) = A_1(t) \oplus 0$ with respect to the decompositions $\mathcal{R}(A^*(t)) \oplus \ker A(t)$ and $\mathcal{R}(A(t)) \oplus \ker A^*(t)$ of $\mathcal{R}(t)$ and, respectively, $\mathcal{R}(t-1)$. Then define $\tau(t) = -A_1^*(t)(A_1(t)A_1^*(t))^{-1/2} \oplus B(t)$, with respect to the above decompositions, and for an arbitrary unitary operator $B(t)$. Since $A_1(t)$ is invertible, it readily follows that $[\tau^*(t) - A(t)][\tau^*(t) - A(t)]^* \geq I$. Therefore, $[R^{1/2}(t) - \tau(t)R^{1/2}(t-1)F^*(t)]$ is invertible and the family

$$\{[R^{1/2}(t) - \tau(t)R^{1/2}(t-1)F^*(t)]^{-1}\}_{t \in \mathbb{Z}}$$

is uniformly bounded. Taking $\Theta(t) = I$ in (1.9) and (1.10) for all $t \in \mathbb{Z}$, leads to uniformly bounded families $\{H(t)\}_{t \in \mathbb{Z}}$ and $\{K(t)\}_{t \in \mathbb{Z}}$. $\square$

1.6 Remark   In order to emphasize the dilation theoretic nature of Theorem 1.5, we show here its explicit connection with the elementary rotation of a contraction. Suppose T is a contraction in $\mathcal{L}(\mathcal{H})$, $\mathcal{H}$ being a finite dimensional Hilbert space. Since $I - TT^* = D_{T^*}D_{T^*}$, it follows from Theorem 1.5 that there exist operators H and K such that $\begin{bmatrix} T & D_{T^*} \\ H & K \end{bmatrix}$ is a unitary operator. Moreover, by formulae (1.9) and (1.10), we get

$$H = \Theta D_{T^*}(I - \tau T^*)^{-1}(\tau - T), \qquad K = \Theta[I - D_{T^*}(I - \tau T^*)^{-1}D_{T^*}],$$

where Θ is a unitary operator and τ is another unitary operator, chosen so that $I - \tau T^*$ is invertible. The previous formulae give rise to the elementary rotation of T if we choose $\Theta = -T^* + D_T(\tau - T)^{-1}D_{T^*}$.

Another case of interest is to consider T a strict contraction and to notice that the following equality holds: $I - D_{T^*}^{-2} = -TD_T^{-2}T^*$. Then, by Theorem 1.5, there exist H and K such that $\begin{bmatrix} D_{T^*}^{-1} & -TD_T^{-1} \\ H & K \end{bmatrix}$ is a $[I \oplus (-I)]$-unitary operator. The formulae (1.9) and (1.10) give

$$H = \Theta D_T^{-1}T^*(I - \tau D_{T^*}^{-1})^{-1}(\tau - D_{T^*}^{-1}), \qquad K = \Theta[I - D_T^{-1}T^*(I - \tau D_{T^*}^{-1})^{-1}TD_T^{-1}], \tag{1.11}$$

where Θ and τ are unitary operators (since $\|D_{T^*}^{-1}\| < 1$, $I - \tau D_{T^*}^{-1}$ is invertible for every unitary operator τ). For instance, for $\tau = I$ and $\Theta = -I$, we obtain the $[I \oplus (-I)]$-unitary operator

$$S(T) = \begin{bmatrix} D_{T^*}^{-1} & -TD_T^{-1} \\ -T^*D_{T^*}^{-1} & D_T^{-1} \end{bmatrix} = \begin{bmatrix} I & -T \\ -T^* & I \end{bmatrix}\begin{bmatrix} D_{T^*}^{-1} & 0 \\ 0 & D_T^{-1} \end{bmatrix}. \tag{1.12}$$

There is a system-theoretic interpretation of the connection between the elementary rotation $R(T)$ and the operator $S(T)$ introduced by (1.12). This can be explained by using so-called *signal flow diagrams*. For instance, the action of a block matrix $T = \begin{bmatrix} T_{11} & T_{12} \\ T_{21} & T_{22} \end{bmatrix}$ can be depicted by the *transfer representation* diagram in Figure 4.2.

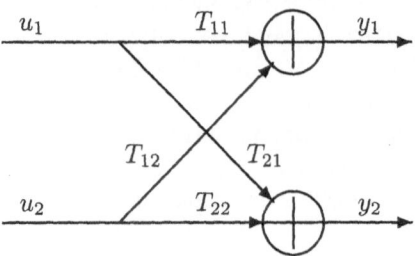

Figure 4.2: Transfer representation

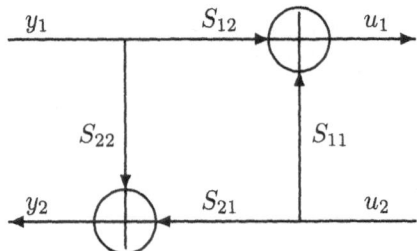

Figure 4.3: Scattering representation

Sometimes it is also useful to describe the dependence $\begin{bmatrix} u_2 \\ y_1 \end{bmatrix} \longrightarrow \begin{bmatrix} u_1 \\ y_2 \end{bmatrix}$, which leads to the *scattering representation* in Figure 4.3.

The connection between the two representations is given by the so-called *Mason rules:*

$$S_{11} = -T_{11}^{-1}T_{12}, \quad S_{21} = T_{22} - T_{21}T_{11}^{-1}T_{12}, \quad S_{12} = T_{11}^{-1}, \quad S_{22} = T_{21}T_{11}^{-1},$$
$$(1.13)$$

which can be easily checked by direct computations, provided the inverse of T_{11} exists. Using the rules (1.13) we see that $R(T)$ and $S(T)$ are the scattering representation, respectively, the transfer representation of the same signal flow. $\square$

4.2   Generalized Schur Algorithm

We use in this section the embedding result of Theorem 1.5 to derive an algorithm related to the lower/upper triangular factorization of the matrices of a Pick solution. Thus, if the classical Gaussian reduction procedure is applied to the elements of a Pick solution $\{R(t)\}_{t \in \mathbb{Z}}$ of a displacement equation, then a convenient algorithm is obtained for the elements of the generator family $\{G(t)\}_{t \in \mathbb{Z}}$. It must be said that this works only if the generators $\{F(t)\}_{t \in \mathbb{Z}}$ have a lower triangular structure, which is the case for many applications.

Let $\{R(t)\}_{t\in\mathbb{Z}}$ be a Pick solution of the displacement equation (1.4) and assume the following conditions:

(H_5) There exist decompositions $\mathcal{R}(t) = \oplus_{i=0}^{n-1} \mathcal{R}_i(t)$ such that $\dim \mathcal{R}_i(t)$ are all equal and finite for $t \in \mathbb{Z}$ and $i = 0, 1, \ldots, n-1$.

(H_6) $\{F(t)\}_{t\in\mathbb{Z}}$ is a uniformly bounded family of lower triangular operators with stable families of diagonal entries $\{f_i(t)\}_{i=0}^{n-1}$.

A family $\{f(t)\}_{t\in\mathbb{Z}}$ is called *stable* if there exists $c_f > 0$ such that $\|f(t)\| \le c_f < 1$ for all $t \in \mathbb{Z}$. It is easily seen that (H_3)-(H_4) together with (H_5)-(H_6) imply (H_1)-(H_2). We now describe the Gaussian elimination.

2.1 Procedure   Due to (H_5), $R(t)$ has a block matrix representation $R(t) = [R_{ij}(t)]_{i,j=0}^{n-1}$. Suppose that $R(t)$ are invertible for all $t \in \mathbb{Z}$. If $l_0(t)$ and $d_0(t)$ stand for the first column and the $(0,0)$ entry of $R(t)$ $(d_0(t) = R_{00}(t))$, then we define for $0 \le i \le n-2$,

$$R_0(t) = R(t),$$

$$\widetilde{R}_{i+1}(t) = \begin{bmatrix} 0 & 0 \\ 0 & R_{i+1}(t) \end{bmatrix} = R_i(t) - l_i(t)d_i^{-1}(t)l_i^*(t),$$

$$l_{i+1}(t) = \text{the first column of } R_{i+1}(t),$$

$$d_{i+1}(t) = \text{the } (0,0) \text{ entry of } R_{i+1}(t). \qquad \qquad \Box$$

This procedure leads to the lower/upper triangular factorization of $R(t)$. Indeed, we obtain

$$R(t) = l_0(t)d_0^{-1}(t)l_0^*(t) + \begin{bmatrix} 0 \\ l_1(t) \end{bmatrix} d_1^{-1}(t) \begin{bmatrix} 0 \\ l_1(t) \end{bmatrix}^* + \ldots = L(t)D^{-1}(t)L^*(t),$$

where $D(t) = \oplus_{i=0}^{n-1} d_i(t)$ and the (nonzero parts of the) columns of the lower triangular matrix $L(t)$ are $l_0(t), l_1(t), \ldots, l_{n-1}(t)$.

The key result about displacement equations is that the triangular factors at time $(t-1)$, viz., $L(t-1)$, can be time-updated to $L(t)$ via a recursive procedure on the generator $G(t)$. Set $F_0(t) = F(t)$ and denote by $F_i(t)$ the submatrix of $F(t)$ obtained after deleting the first (block) row and column of $F_{i-1}(t)$.

2.2 Theorem   *(a) Suppose (H_3)-(H_6) hold and the displacement equation (1.4) has a Pick solution such that all $R(t)$ are invertible. Then the Schur complements $R_i(t)$ satisfy the displacement equation*

$$R_i(t) - F_i(t)R_i(t-1)F_i^*(t) = G_i(t)J(t)G_i^*(t),$$

where $G_i(t)$ satisfies the following generator recursion: $G_0(t) = G(t)$,

$$\begin{bmatrix} 0 \\ G_{i+1}(t) \end{bmatrix} = F_i(t)l_i(t-1)h_i^*(t)J(t) + G_i(t)J(t)k_i^*(t)J(t), \qquad (2.1)$$

where $g_i(t)$ is the top row (block) of $G_i(t)$, and $h_i(t)$, $k_i(t)$ are chosen so as to satisfy the embedding relation:

$$\begin{bmatrix} f_i(t) & g_i(t) \\ h_i(t) & k_i(t) \end{bmatrix} \begin{bmatrix} d_i(t-1) & 0 \\ 0 & J(t) \end{bmatrix} \begin{bmatrix} f_i(t) & g_i(t) \\ h_i(t) & k_i(t) \end{bmatrix}^* = \begin{bmatrix} d_i(t) & 0 \\ 0 & J(t) \end{bmatrix} \qquad (2.2)$$

(b) If, in addition, the displacement equation (1.4) has a Pick solution $\{R(t)\}_{t \in \mathbb{Z}}$ such that (H) holds, then there exist uniformly bounded families $\{h_i(t)\}_{t \in \mathbb{Z}}$, $\{k_i(t)\}_{t \in \mathbb{Z}}$ such that (2.1) and (2.2) hold.

Proof (a) We prove the result for $i = 0$. The same computations are valid for $i \geq 1$. It follows from (1.4) that $l_0(t) - F(t)l_0(t-1)f_0^*(t) = G(t)J(t)g_0^*(t)$, hence

$$\widetilde{R}_1(t) - F(t)\widetilde{R}_1(t-1)F^*(t) = G(t)J(t)[J(t) - g_0^*(t)d_0^{-1}(t)g_0(t)]J(t)G^*(t)$$
$$- G(t)J(t)g_0^*(t)d_0^{-1}(t)f_0(t)l_0^*(t-1)F^*(t)$$
$$- F(t)l_0(t-1)f_0^*(t)d_0^{-1}(t)g_0(t)J(t)G^*(t)$$
$$+ F(t)l_0(t-1)[d_0^{-1}(t-1) - f_0^*(t)d_0^{-1}(t)f_0(t)]l_0^*(t-1)F^*(t).$$

Since $\{d_0(t)\}_{t \in \mathbb{Z}}$ is the Pick solution of the displacement equation

$$d_0(t) - f_0(t)d_0(t-1)f_0^*(t) = g_0(t)J(t)g_0^*(t),$$

it follows from Theorem 1.5(a) that there exist two families $\{h_0(t)\}_{t \in \mathbb{Z}}$, $\{k_0(t)\}_{t \in \mathbb{Z}}$ of operators such that

$$\begin{bmatrix} f_0(t) & g_0(t) \\ h_0(t) & k_0(t) \end{bmatrix} \begin{bmatrix} d_0(t-1) & 0 \\ 0 & J(t) \end{bmatrix} \begin{bmatrix} f_0(t) & g_0(t) \\ h_0(t) & k_0(t) \end{bmatrix}^* = \begin{bmatrix} d_0(t) & 0 \\ 0 & J(t) \end{bmatrix}.$$

Taking into account the assumption that $\dim \mathcal{H}_0(t) = \alpha_0$ for all $t \in \mathbb{Z}$, it follows that

$$\begin{bmatrix} f_0(t) & g_0(t) \\ h_0(t) & k_0(t) \end{bmatrix}^* \begin{bmatrix} d_0^{-1}(t) & 0 \\ 0 & J(t) \end{bmatrix} \begin{bmatrix} f_0(t) & g_0(t) \\ h_0(t) & k_0(t) \end{bmatrix} = \begin{bmatrix} d_0^{-1}(t-1) & 0 \\ 0 & J(t) \end{bmatrix}.$$

Consequently,

$$d_0^{-1}(t-1) - f_0^*(t)d_0^{-1}(t)f_0(t) = h_0^*(t)J(t)h_0(t),$$

$$k_0^*(t)J(t)k_0(t) = J(t) - g_0^*(t)d_0^{-1}(t)g_0(t), \qquad k_0^*(t)J(t)h_0(t) = g_0^*(t)d_0^{-1}(t)f_0(t),$$

and then,

$$\widetilde{R}_1(t) - F(t)\widetilde{R}_1(t-1)F^*(t) = G(t)J(t)k_0^*(t)J(t)k_0(t)J(t)G^*(t)$$
$$+ G(t)J(t)k_0^*(t)J(t)h_0(t)l_0^*(t-1)F^*(t)$$
$$+ F(t)l_0(t-1)h_0^*(t)J(t)k_0(t)J(t)G^*(t)$$
$$+ F(t)l_0(t-1)h_0^*(t)J(t)h_0(t)l_0^*(t-1)F^*(t)$$
$$= \widetilde{G}_1(t)J(t)\widetilde{G}_1^*(t),$$

where $\tilde{G}_1(t) = F(t)l_0(t-1)h_0^*(t)J(t) + G(t)J(t)k_0^*(t)J(t)$. Since $F(t)$ is lower triangular, it follows that

$$\tilde{R}_1(t) - F(t)\tilde{R}_1(t-1)F^*(t) = 0 \oplus [R_1(t) - F_1(t)R_1(t-1)F_1^*(t)],$$

hence $\tilde{G}_1(t) = \begin{bmatrix} 0 \\ G_1(t) \end{bmatrix}$ and (2.1) holds for $i = 0$.

(b) We prove first that there exist real numbers b_d, c_d and c_v (independent of t) such that

$$0 < b_d I < d_i(t) < c_d I, \qquad \|g_i(t)\| < c_v$$

for all $t \in \mathbb{Z}$. It is clear that $\{d_0(t)\}_{t\in\mathbb{Z}}$ is uniformly bounded above since $\{f_0(t)\}_{t\in\mathbb{Z}}$ is stable and $\{g_0(t)J(t)g_0^*(t)\}_{t\in\mathbb{Z}}$ is uniformly bounded. A similar argument shows that $\{l_0(t)\}_{t\in\mathbb{Z}}$ is also uniformly bounded. Moreover, $\{R(t)\}_{t\in\mathbb{Z}}$ is uniformly bounded below, hence the sequence $\{d_0(t)\}_{t\in\mathbb{Z}}$ is obviously uniformly bounded below. By Theorem 1.5(b), uniformly bounded sequences $\{h_0(t)\}_{t\in\mathbb{Z}}$ and $\{k_0(t)\}_{t\in\mathbb{Z}}$ can be chosen so as to satisfy the embedding relation (2.2). It is easy now to conclude that both sequences $\{d_i(t)\}_{t\in\mathbb{Z}}$ and $\{g_i(t)\}_{t\in\mathbb{Z}}$ are uniformly bounded above for $i = 0, 1, \ldots, n-1$. To show that the sequence $\{d_i(t)\}_{t\in\mathbb{Z}}$, $0 < i < n-1$, is bounded below, it is proven by induction that $R_i(t) > \varepsilon_i I$ for some $\varepsilon_i > 0$ independent of t. As a consequence of Procedure 2.1, $R_i(t) = A_i(t)[d_i(t) \oplus R_{i+1}(t)]A_i^*(t)$, where $A_i(t) = \begin{bmatrix} l_i(t)d_i^{-1}(t) & 0 \\ & I_{n-i-1} \end{bmatrix}$. Since $A_i(t)$ is lower triangular and $\{l_i(t)\}_{t\in\mathbb{Z}}$ is uniformly bounded, it follows that $\{A_i^{-1}(t)\}_{t\in\mathbb{Z}}$ is uniformly bounded. This implies that $\begin{bmatrix} d_i(t) & 0 \\ 0 & R_{i+1}(t) \end{bmatrix} \geq \varepsilon_{i+1}I$ for some $\varepsilon_{i+1} > 0$, uniformly with respect to $t \in \mathbb{Z}$. Hence, $R_{i+1}(t) \geq \varepsilon_{i+1}I$ for all $t \in \mathbb{Z}$ and then $\{d_i(t)\}_{t\in\mathbb{Z}}$ is uniformly bounded below. $\square$

2.3 Corollary *Consider the setting of Theorem 2.2. Then*

$$\begin{bmatrix} l_i(t) & 0 \\ & G_{i+1}(t) \end{bmatrix} = [\,F_i(t)l_i(t-1) \quad G_i(t)\,] \begin{bmatrix} f_i^*(t) & h_i^*(t)J(t) \\ J(t)g_i^*(t) & J(t)k_i^*(t)J(t) \end{bmatrix}. \square$$

2.4 Remark Suppose $\mathcal{R}(t) = \oplus_{i=0}^{\infty}\mathcal{R}_i(t)$ and $\dim \mathcal{R}_i(t) = 1$. Suppose $R(t) = [R_{ij}(t)]_{i,j=0}^{\infty}$, $t \in \mathbb{Z}$, is an array of complex numbers such that every finite section, namely every $P_{\oplus_{i=0}^{n-1}\mathcal{R}_i(t)}R(t)/\oplus_{i=0}^{n-1}\mathcal{R}_i(t)$, $n \geq 1$, is a positive matrix. We see that the computation of the entries of $R_i(t)$ in Procedure 2.1 requires a finite number of operations with complex numbers, therefore we can extend those rules for the array $R(t)$. The significance of Procedure 2.1 in this more general situation will be explained later in Chapter 5. Here we remark that since the operators $F(t)$ have an upper triangular matrix representation with respect to the decomposition $\mathcal{R}(t) = \oplus_{i=0}^{\infty}\mathcal{R}_i(t)$, then the displacement equation $R(t) - F(t)R(t-1)F^*(t) = G(t)J(t)G^*(t)$ makes sense if $R(t)$ are arrays as above for $t \in \mathbb{Z}$, and $G(t)$ are also some specified arrays. The generator recursion (2.1) makes sense and a generalized Schur algorithm is described in this way for arrays of complex numbers with positive finite sections and satisfying a displacement equation.   □

We can now explain the connection of Theorem 2.2 with a classical algorithm of I. Schur. Thus, let f in S be a Schur function such that $\|f\|_\infty < 1$, and let $f(z) = \sum_{n=0}^\infty c_n z^n$ be its power series representation about the origin. Define the Toeplitz operator $T_{\widetilde{f}}$, where $\widetilde{f}(z) = (f(z^*))^*$ for $z \in \mathbb{D}$, and remark that the operator $T = \Psi T_{\widetilde{f}} \Psi^*$ (Ψ defined by (1.3.6)), has the upper triangular matrix representation

$$
T = \begin{bmatrix} c_0^* & c_1^* & c_2^* & \\ 0 & c_0^* & c_1^* & \\ 0 & 0 & c_0^* & \ddots \\ & & \ddots & \ddots \end{bmatrix}
\tag{2.3}
$$

with respect to the basis $\{E_n\}_{n\geq 0}$ of the Hilbert space $\mathcal{R}$ of the square-summable sequences $x = \{x_n\}_{n\geq 0}$ of complex numbers with the inner product $\langle x, y \rangle = \sum_{n\geq 0} x_n y_n^*$. Define $R = I - T^*T$ and one sees that R is a strictly positive matrix satisfying the displacement equation $R - FRF^* = GJG^*$, with

$$
F = \begin{bmatrix} 0 & 0 & 0 & \\ 1 & 0 & 0 & \\ 0 & 1 & 0 & \ddots \\ & & \ddots & \ddots \end{bmatrix}, \quad
G = \begin{bmatrix} 1 & c_0 \\ 0 & c_1 \\ 0 & c_2 \\ \vdots & \vdots \end{bmatrix} \quad \text{and} \quad
J = \begin{bmatrix} 1 & 0 \\ 0 & -1 \end{bmatrix}.
\tag{2.4}
$$

2.5 Theorem *The generator recursion for the displacement equation $R - FRF^* = GJG^*$ with generators given by (2.4), has the following simplified, array form:*

$$
\begin{bmatrix} 0 \\ G_{i+1} \end{bmatrix} = FG_i S(\gamma_i) \begin{bmatrix} 1 & 0 \\ 0 & 0 \end{bmatrix} + G_i S(\gamma_i) \begin{bmatrix} 0 & 0 \\ 0 & 1 \end{bmatrix},
$$

for some complex numbers γ_i, $i = 0, 1, \ldots$, with $|\gamma_i| < 1$.

Proof Write $g_i = [\, u_i \;\; v_i \,]$ and due to the strictly lower triangular matrix representation of F, it follows that

$$
d_i = g_i J g_i^* = u_i u_i^* - v_i v_i^* > 0.
$$

Therefore, there exists a uniquely determined contraction γ_i, $|\gamma_i| < 1$, such that $v_i = u_i \gamma_i$. Using the definition (1.12) in order to introduce the J-unitary matrix $S(\gamma_i)$, we remark that $g_i S(\gamma_i) = [\delta_i \;\; 0]$, where $|\delta_i|^2 = d_i$. Using Theorem 1.5, it is possible to choose

$$
h_i = S^{-1}(\gamma_i) J g_i^* d_i^{-1} = \begin{bmatrix} \delta_i^{-1} \\ 0 \end{bmatrix}
$$

and

$$
k_i = S^{-1}(\gamma_i)(I - J g_i^* d_i^{-1} g_i) = \begin{bmatrix} 0 & 0 \\ 0 & 1 \end{bmatrix} S^{-1}(\gamma_i).
$$

Using Corollary 2.3, it follows that

$$
\begin{bmatrix} 0 \\ l_i \\ G_{i+1} \end{bmatrix} = \begin{bmatrix} F_i l_i & G_i \end{bmatrix} \begin{bmatrix} 0 & [(\delta_i^*)^{-1} & 0] \\ S(\gamma_i)\begin{bmatrix} \delta_i^* \\ 0 \end{bmatrix} & S(\gamma_i)\begin{bmatrix} 0 & 0 \\ 0 & 1 \end{bmatrix} \end{bmatrix},
$$

hence $l_i = G_i S(\gamma_i)\begin{bmatrix} \delta_i^* & 0 \end{bmatrix}^T$ and then

$$
\begin{bmatrix} 0 \\ G_{i+1} \end{bmatrix} = F G_i S(\gamma_i) \begin{bmatrix} 1 & 0 \\ 0 & 0 \end{bmatrix} + G_i S(\gamma_i) \begin{bmatrix} 0 & 0 \\ 0 & 1 \end{bmatrix}.
$$

Schematically, the mapping from G_i to G_{i+1} has the following array interpretation

$$
G_i = \begin{bmatrix} * & * \\ * & * \\ * & * \\ \vdots & \vdots \end{bmatrix} \longrightarrow \begin{bmatrix} * & 0 \\ * & * \\ * & * \\ \vdots & \vdots \end{bmatrix} \longrightarrow \begin{bmatrix} 0 & 0 \\ * & * \\ * & * \\ \vdots & \vdots \end{bmatrix} = \begin{bmatrix} 0 \\ G_{i+1} \end{bmatrix},
$$

where the first arrow represents the multiplication with $S(\gamma_i)$ and the second one is to shift down the first column. $\quad\square$

Write $G_i = \begin{bmatrix} u_{i0} & v_{i0} \\ u_{i1} & v_{i1} \\ u_{i2} & v_{i2} \\ \vdots & \vdots \end{bmatrix}$, where $u_{i0} = u_i$ and $v_{i0} = v_i$, and define the formal

power series $u_i(z) = \sum_{k \geq 0} u_{ik} z^k$, $v_i(z) = \sum_{k \geq 0} v_{ik} z^k$. The generator recursion of Theorem 2.5 can be rewritten in the form

$$
\begin{bmatrix} u_{i+1}(z) & v_{i+1}(z) \end{bmatrix} = \frac{1}{(1 - |\gamma_i|^2)^{1/2}} \begin{bmatrix} u_i(z) & v_i(z) \end{bmatrix} \begin{bmatrix} 1 & -\gamma_i z^{-1} \\ -\gamma_i^* & z^{-1} \end{bmatrix}. \tag{2.5}
$$

Since $u_0(z) = 1$ and $v_0(z) = f(z)$ for all $z \in \mathbb{D}$, it follows that u_i and v_i are analytic functions in $\mathbb{D}$, for all $i \geq 1$. Define $v_i'(z) = z v_i(z)$, then

$$
|u_{i+1}(z)|^2 - |v_{i+1}'(z)|^2 = |u_i(z)|^2 - |v_i(z)|^2, \quad z \in \mathbb{D}.
$$

It follows by induction on i that $u_i(z) \neq 0$ for $z \in \mathbb{D}$ and that $f_i = v_i / u_i$ are all Schur functions such that $f_i(0) = \gamma_i$. Equation (2.5) can be rewritten in the form of the following *classical Schur algorithm*: $f_0 = f$ and for $i \geq 1$,

$$
\gamma_i = f_i(0), \quad \text{and} \quad f_{i+1}(z) = \frac{f_i(z) - \gamma_i}{z(1 - \gamma_i^* f_i(z))}. \tag{2.6}
$$

We can round the table and view (2.6) as a continued fraction algorithm describing the structure of the Schur functions. Thus, let f be a Schur function. As a consequence of Schwartz's lemma, all the functions f_i, $i \geq 1$, defined by (2.6) belong to $\mathcal{S}$ and $|\gamma_i| \leq 1$ for all $i \geq 0$. In addition, it is a consequence of the maximum modulus principle the fact that if $|\gamma_{i_0}| = 1$ for some $i_0 \geq 1$, then

$f_{i_0}(z) = \gamma_{i_0}$ for $|z| < 1$, hence $f_i \equiv 0$ and $\gamma_i = 0$ for $i > i_0$. The elements of the set $\{\gamma_i\}_{i \geq 0}$ are referred to as the *Schur parameters of* f. We notice the following result which shows that the Schur algorithm applied to the function f is consistent with the structure of f.

2.6 Theorem   *A Schur function f is uniquely determined by its Schur parameters. Furthermore, there exists $i_0 \geq 1$ such that $|\gamma_{i_0}| = 1$ if and only if f is a finite Blaschke product.*

Proof It follows from (2.6) that, for $i \geq 0$,

$$f_i(z) = \frac{\gamma_i + z f_{i+1}(z)}{1 + z \gamma_i^* f_{i+1}(z)}, \tag{2.7}$$

therefore

$$f(z) = \frac{\mathcal{A}_i(z) + z \mathcal{B}_i^{\#}(z) f_{i+1}(z)}{\mathcal{B}_i(z) + z \mathcal{A}_i^{\#}(z) f_{i+1}(z)} \tag{2.8}$$

for some polynomials $\mathcal{A}_i$ and $\mathcal{B}_i$ of degree n, referred to as the *Schur polynomials* associated to f. Note that $\mathcal{A}_0(z) = \gamma_0$, $\mathcal{B}_0(z) = 1$ and

$$\mathcal{A}_{i+1}(z) = \mathcal{A}_i(z) + z \gamma_{i+1} \mathcal{B}_i^{\#}(z), \tag{2.9}$$

$$\mathcal{B}_{i+1}(z) = \mathcal{B}_i(z) + z \gamma_{i+1} \mathcal{A}_i^{\#}(z) \tag{2.10}$$

for $i \geq 0$. From these relations, it is deduced that the polynomials $\mathcal{A}_i$ and $\mathcal{B}_i$ are uniquely determined by the Schur parameters $\gamma_0, \gamma_1, \ldots, \gamma_i$. Moreover, (2.8) shows that f is a finite Blaschke product if and only if $|\gamma_{i_0}| = 1$ for a certain $i_0 \geq 1$, and in this case

$$f(z) = \frac{\mathcal{A}_{i_0-1}(z) + z \mathcal{B}_{i_0-1}^{\#}(z) \gamma_{i_0}}{\mathcal{B}_{i_0-1}(z) + z \mathcal{A}_{i_0-1}^{\#}(z) \gamma_{i_0}}. \tag{2.11}$$

We show now that if $|\gamma_i| < 1$ for all $i \geq 0$, then the polynomials $\mathcal{B}_i$, $i \geq 0$, have no zero in the closed unit disc and the rational functions $\mathcal{A}_i \mathcal{B}_i^{-1}$ approximate the function f uniformly on the compact subsets of the unit disc. In order to prove these assertions, it is useful to rewrite the recurrence formulae (2.9) and (2.10) in the following matrix form:

$$\begin{bmatrix} \mathcal{B}_{i+1}(z) & \mathcal{A}_{i+1}(z) \\ \mathcal{A}_{i+1}^{\#}(z) & \mathcal{B}_{i+1}^{\#}(z) \end{bmatrix} = \begin{bmatrix} 1 & z\gamma_{i+1} \\ \gamma_{i+1}^* & z \end{bmatrix} \begin{bmatrix} \mathcal{B}_i(z) & \mathcal{A}_i(z) \\ \mathcal{A}_i^{\#}(z) & \mathcal{B}_i^{\#}(z) \end{bmatrix}. \tag{2.12}$$

It is seen that

$$\begin{bmatrix} 1 & z\gamma_{i+1} \\ \gamma_{i+1}^* & z \end{bmatrix} J \begin{bmatrix} 1 & z\gamma_{i+1} \\ \gamma_{i+1}^* & z \end{bmatrix}^* \geq (1 - |\gamma_i|^2) J$$

for $|z| \leq 1$. This inequality and (2.12), show that for $i \geq 0$ and $|z| \leq 1$,

$$\begin{bmatrix} \mathcal{B}_i(z) & \mathcal{A}_i(z) \\ \mathcal{A}_i^{\#}(z) & \mathcal{B}_i^{\#}(z) \end{bmatrix} J \begin{bmatrix} \mathcal{B}_i(z) & \mathcal{A}_i(z) \\ \mathcal{A}_i^{\#}(z) & \mathcal{B}_i^{\#}(z) \end{bmatrix}^* \geq \prod_{k=0}^{i} (1 - |\gamma_k|^2) J.$$

Consequently, $|\mathcal{B}_i(z)|^2 - |\mathcal{A}_i(z)|^2 \geq \prod_{k=0}^{i}(1 - |\gamma_k|^2)$ for $|z| \leq 1$, which shows that $\mathcal{B}_i$ has no zero in the closed unit disc. Taking the determinants of the both sides in (2.12), it follows that

$$\mathcal{B}_i(z)\mathcal{B}_i^{\#}(z) - \mathcal{A}_i(z)\mathcal{A}_i^{\#}(z) = z^i \prod_{k=0}^{i}(1 - |\gamma_k|^2)$$

and then, for $|z| < 1$,

$$f(z) - \frac{\mathcal{A}_i(z)}{\mathcal{B}_i(z)} = z^{i+1} \prod_{k=0}^{i}(1 - |\gamma_k|^2) \frac{f_{i+1}(z)}{\mathcal{B}_i(z)(\mathcal{B}_i(z) + z\mathcal{A}_i^{\#}(z)f_{i+1}(z))}. \qquad (2.13)$$

This equality shows that f is approximated by $\mathcal{A}_i\mathcal{B}_i^{-1}$ uniformly on the compact subsets of the unit disc. Since $\mathcal{A}_i$ and $\mathcal{B}_i$ are uniquely determined by $\{\gamma_k\}_{k=0}^{i}$ for every $i \geq 0$, it follows that f is uniquely determined by its Schur parameters. $\qquad\square$

In the remainder of this section, we render explicit the connection between the Schur algorithm, the structure of the triangular contractions as described in Theorem 2.2.1 and the orthogonal polynomials on the unit circle.

2.7 Proposition *If f is a Schur function, if f_i, $i > 0$, are the Schur functions associated to f by the formula (2.6) and if $\{\gamma_i\}_{i\geq 0}$ is the set of the Schur parameters of f, then*

(a) *The Schur parameters of f_i are exactly γ_i, γ_{i+1}, γ_{i+2}, ...*

(b) *The Schur parameters of $\mathcal{A}_i\mathcal{B}_i^{-1}$ are exactly $\gamma_0, \gamma_1, \ldots, \gamma_i, 0, 0, \ldots$.*

(c) *The Schur parameters of f coincide with complex conjugates of the Schur parameters associated by the procedure described in Theorem 2.2.1 to the Toeplitz upper triangular contraction T defined by (2.3).*

Proof Remark that the Schur algorithm applied to the function f_i will produce the sequence of Schur functions f_{i+1}, f_{n+2}, ... Hence, the set of the Schur parameters of f_i is $\{\gamma_m\}_{m\geq i}$. In order to prove (b), let us denote by g_i the Schur function associated to the Schur parameters $\{\gamma_0, \gamma_1, \ldots, \gamma_i, 0, 0, \ldots\}$. Moreover, denote by $\mathcal{A}_k(g_i)$, $\mathcal{B}_k(g_i)$, $k \geq 0$, the Schur polynomials of g_i. Then, by (2.10) and (2.11), $\mathcal{A}_k(g_i) = \mathcal{A}_k$ and $\mathcal{B}_k(g_i) = \mathcal{B}_k$ for $k = 0, 1, \ldots, i$. When we apply the Schur algorithm to the function g_i, the $(i+1)$-th Schur function produced by the algorithm is identically zero, hence $g_i = \mathcal{A}_i\mathcal{B}_i^{-1}$ and the function $\mathcal{A}_i\mathcal{B}_i^{-1}$ has the Schur parameters $\gamma_0, \gamma_1, \ldots, \gamma_i, 0, 0, \ldots$.

Then, let $\{c_i\}_{i\geq 0}$ be the set of the Taylor coefficients of f about the origin. By the formula (2.2.3) and Proposition 2.2.3, we know that

$$c_i^* = a_i + \prod_{k=0}^{i-1}(1 - |\Gamma_k|^2)\Gamma_i, \qquad (2.14)$$

where a_i is uniquely determined by the Schur parameters Γ_k, $k = 0, \ldots i - 1$ associated by the procedure in Theorem 2.2.1 to the Toeplitz operator T. On the other hand, the formula (2.8) shows that

$$c_i = a_i' + \prod_{k=0}^{i-1}(1 - |\gamma_k|^2)\gamma_i, \qquad (2.15)$$

where a_i' is uniquely determined by the Schur parameters γ_k, $k = 0, \ldots, n-1$, of f. Since in both formulae (2.17) and (2.18), the correspondences $\{T_i\}_{i \geq 0} \leftrightarrow \{\Gamma_i\}_{i \geq 0}$ and, respectively, $\{T_i\}_{i \geq 0} \leftrightarrow \{\gamma_i\}_{i \geq 0}$, are bijective, it follows that $\gamma_i^* = \Gamma_i$ for all $i \geq 0$. $\qquad\square$

We show now the connection between the Schur polynomials and the Szegö polynomials. Let μ be a probability measure on $[0, 2\pi)$ and let $\{\Gamma_i'\}_{i \geq 0}$ be the set of the Schur parameters of the positive definite kernel A associated to the Fourier coefficients of μ as in Example 1.3.3. Let $\{\psi_i\}_{i \geq 0}$ and $\{\hat{\psi}_i\}_{i \geq 0}$ be the sets of the monic orthogonal polynomials of first and second kind of μ. Let f be the Schur function corresponding to the set $\{\Gamma_i'\}_{i \geq 0}$ by the Schur algorithm. Let $\{\mathcal{A}_i\}_{i \geq 0}$, $\{\mathcal{B}_i\}_{i \geq 0}$ be the sets of the Schur polynomials associated to f.

2.8 Proposition   For $i \geq 1$, $\psi_i = \mathcal{B}_i^\# - \mathcal{A}_i^\#$ and $\hat{\psi}_i = \mathcal{B}_i^\# + \mathcal{A}_i^\#$

Proof   Since $\Gamma_0' = 0$, then $\psi_1(z) = z - \Gamma_1'^*$, $\mathcal{A}_1^\#(z) = \Gamma_1'^*$ and $\mathcal{B}_1^\#(z) = z$. Consequently, $\psi_1 = \mathcal{B}_1^\# - \mathcal{A}_1^\#$ and $\hat{\psi}_1 = \mathcal{B}_1^\# + \mathcal{A}_1^\#$. The general statement follows by induction, using the relations (2.9), (2.10) and (1.6.19). $\qquad\square$

2.9 Remark   Using this relation between the monic orthogonal polynomials and the Schur polynomials, (2.8) shows that we can choose $a = \hat{\psi}_N^\#$, $b = \psi_N^\#$, $c = \hat{\psi}_N$ and $d = \psi_N$ in formula (4.4.1). $\qquad\square$

4.3   Discrete Transmission-Line Models

In this section we show that the Schur algorithm is related to a discrete transmission-line model which appears in connection with the discretization of the differential equations describing the propagation of signals in lossless nonuniform media.

Let f be a Schur function and remark that the linearized form (2.5) of the Schur algorithm can be rewritten as

$$\begin{bmatrix} u_{i+1}(z) \\ v_{i+1}(z) \end{bmatrix} = \frac{1}{(1 - |\gamma_i|^2)^{1/2}} \begin{bmatrix} 1 & -\gamma_i^* \\ -\gamma_i z^{-1} & z^{-1} \end{bmatrix} \begin{bmatrix} u_i(z) \\ v_i(z) \end{bmatrix} =$$

$$= \frac{1}{(1 - |\gamma_i|^2)^{1/2}} \begin{bmatrix} 1 & -\gamma_i^* \\ -\gamma_i z^{-1} & z^{-1} \end{bmatrix} \cdots \frac{1}{(1 - |\gamma_0|^2)^{1/2}} \begin{bmatrix} 1 & -\gamma_0^* \\ -\gamma_0 z^{-1} & z^{-1} \end{bmatrix} \begin{bmatrix} u_0(z) \\ v_0(z) \end{bmatrix},$$

which describes the cascade composition of the first $i+1$ layers of the *transmission-line model* in transfer representation as shown in Figure 4.4. We set $e_k = (1 - |\gamma_k|^2)^{1/2}$.

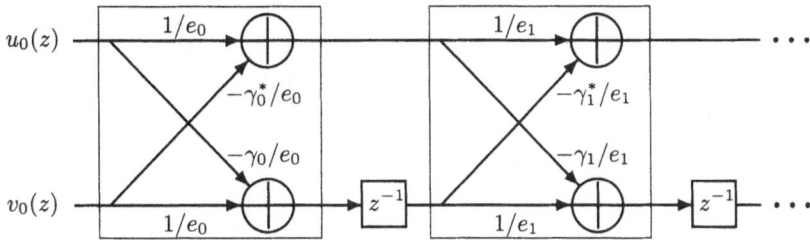

Figure 4.4: Transmission-line model in transfer representation

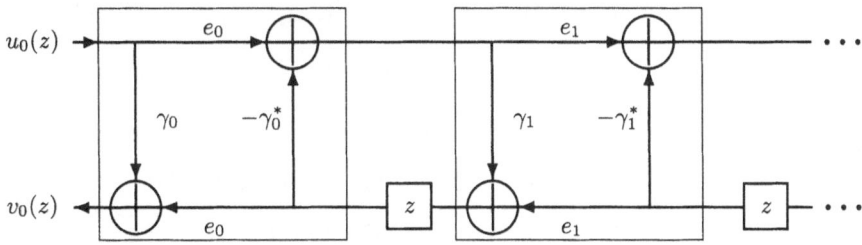

Figure 4.5: Transmission-line model in scattering representation

The scattering representation of the same transmission-line is shown in Figure 4.5.

It is interesting to point out that the formula (2.6) incorporates the evolution of two distinct processes. First, it is the evolution of the "residuals" f_i defined by (2.6). We write this as the following procedure.

3.1 Procedure *Let f be a Schur function. Then its Schur parameters are computed by the recursions: $v_0 = f$, $u_0 = 1$, $\gamma_0 = v_0(0)$ and for $i \geq 1$,*

$$\begin{bmatrix} u_{i+1}(z) \\ v_{i+1}(z) \end{bmatrix} = \frac{1}{(1 - |\gamma_i|^2)^{1/2}} \begin{bmatrix} 1 & -\gamma_i^* \\ -\gamma_i z^{-1} & z^{-1} \end{bmatrix} \begin{bmatrix} u_i(z) \\ v_i(z) \end{bmatrix}$$

$$\gamma_{i+1} = v_{i+1}(0)u_{i+1}^{-1}(0). \ \square$$

We could say that Procedure 3.1 consists in a *layer identification and peeling* process applied to the data described by the Schur function f. But, (2.6) also gives rise to the rational functions $g_i = \mathcal{A}_i \mathcal{B}_i^{-1}$. This time, the formulae (2.9) and (2.10) have the nature of an *adjoining identified layers* process, since at the step $i + 1$ we need γ_{i+1}. A convenient way to get γ_{i+1} from the data, is to use the equality (2.13). Thus, it follows that

$$f(z)\mathcal{B}_i(z) = \mathcal{A}_i(z) + z^{i+1} \prod_{k=0}^{i}(1 - |\gamma_k|^2)\frac{f_{i+1}(z)}{\mathcal{B}_i(z) + z\mathcal{A}_i^{\#}(z)f_{i+1}(z)}$$

and identifying the coefficients of the power z^{i+1} in the both sides, one obtains

$$c_1b_{ii} + c_2b_{i,i-1} + \ldots + c_ib_{i1} + c_{i+1} = \gamma_{i+1}\prod_{k=0}^{i}(1 - |\gamma_k|^2),$$

where $\mathcal{B}_i(z) = 1 + b_{i1}z + \ldots + b_{ii}z^i$. We can summarize this analysis by writing the following procedure, known as the *Levinson algorithm*.

3.2 Procedure   *Let f be a Schur function. Then its Schur parameters are computed by the recursion: $\gamma_0 = f(0)$, $\mathcal{A}_0 = \gamma_0$, $\mathcal{B}_0 = 1$, $\Delta_1 = c_1$, $\gamma_1 = c_1/(1 - |\gamma_0|^2)$ and for $i \geq 1$,*

$$\mathcal{A}_i(z) = \mathcal{A}_{i-1}(z) + z\gamma_i\mathcal{B}_{i-1}^{\#}(z)$$

$$\mathcal{B}_i(z) = 1 + b_{i1}z + \ldots + b_bz^i = \mathcal{B}_{i-1}(z) + z\gamma_i\mathcal{A}_{i-1}^{\#}(z)$$

$$\Delta_{i+1} = c_1b_b + c_2b_{i,i-1} + \ldots + c_ib_{i1} + c_{i+1}$$

$$\gamma_{i+1} = \Delta_{i+1}\prod_{k=0}^{i}(1 - |\gamma_k|^2)^{-1}. \quad \Box$$

We conclude this section by showing that the transmission-line models introduced in Figure 4.4 and Figure 4.5 also appear in connection with the discretization of some differential equations describing the propagation of signals in lossless nonuniform media. The differential equations describing this propagation are so-called *telegrapher's equations*,

$$\begin{cases} \dfrac{\partial}{\partial x}v(x,t) = -Z(x)\dfrac{\partial}{\partial t}i(x,t) \\[2mm] \dfrac{\partial}{\partial x}i(x,t) = -Z^{-1}(x)\dfrac{\partial}{\partial t}v(x,t), \end{cases} \tag{3.1}$$

where $v(x,t)$ is the voltage at point $x \geq 0$ on the line, at time t, $i(x,t)$ is the current at point $x \geq 0$ at time t and $Z(x)$ is the local impedance at the point $x \geq 0$. By assumption, Z is a strictly positive function and set $Z(0) = 1$. Moreover, all the functions are supposed sufficiently smooth. We first normalize the voltage and current variables by defining the new variables $V(x,t) = v(x,t)Z^{-1/2}(x)$ and $I(x,t) = i(x,t)Z^{1/2}(x)$, and the equations (3.1) become:

$$\begin{cases} \dfrac{\partial}{\partial x}V(x,t) + \dfrac{\partial}{\partial t}I(x,t) = -k(x)V(x,t) \\[2mm] \dfrac{\partial}{\partial x}I(x,t) + \dfrac{\partial}{\partial t}V(x,t) = k(x)I(x,t), \end{cases} \tag{3.2}$$

where $k(x) = \frac{1}{2}\frac{d}{dx}lnZ(x)$ is the so-called *reflectivity parameter*. It is also convenient to introduce the *wave variables*, $W_R(x,t) = \frac{1}{2}(V(x,t) + I(x,t))$, $W_L(x,t) = \frac{1}{2}(V(x,t) - I(x,t))$ and the evolution equations (3.2) become:

$$\begin{cases} \dfrac{\partial}{\partial x}W_R(x,t) + \dfrac{\partial}{\partial t}W_R(x,t) = -k(x)W_L(x,t) \\[2mm] \dfrac{\partial}{\partial x}W_L(x,t) - \dfrac{\partial}{\partial t}W_L(x,t) = k(x)W_R(x,t). \end{cases} \tag{3.3}$$

We now suppose that the impedance is constant on the intervals of length $\frac{1}{2}$, *i.e.* $Z(x) = Z_{n-1}$ for $x \in [\frac{n-1}{2}, \frac{n}{2})$. It follows that $k(x) = 0$ on this interval, hence the left wave and the right wave satisfy the equations:

$$\begin{cases} \dfrac{\partial}{\partial x}W_R(x,t) + \dfrac{\partial}{\partial t}W_R(x,t) = 0 \\[2mm] \dfrac{\partial}{\partial x}W_L(x,t) - \dfrac{\partial}{\partial t}W_L(x,t) = 0 \end{cases} \tag{3.4}$$

for $x \in [\frac{n-1}{2}, \frac{n}{2})$. Consequently, $W_R(x,t) = W_R(x - t)$ and $W_L(x,t) = W_L(x + t)$ for $x \in [\frac{n-1}{2}, \frac{n}{2})$ and the whole process is described by the evolution of the vectors:

$$\begin{bmatrix} W_R(n,t) \\ W_L(n,t) \end{bmatrix} = \begin{bmatrix} W_R(x,t) \\ W_L(x,t) \end{bmatrix}_{x=\frac{n}{2}+} = \lim_{x\downarrow\frac{n}{2}} \begin{bmatrix} W_R(x,t) \\ W_L(x,t) \end{bmatrix}.$$

Note that by the continuity of the functions v and i,

$$\begin{bmatrix} v(x,t) \\ i(x,t) \end{bmatrix}_{x=\frac{n}{2}+} = \begin{bmatrix} v(x,t) \\ i(x,t) \end{bmatrix}_{x=\frac{n}{2}-} = \lim_{x\uparrow\frac{n}{2}} \begin{bmatrix} v(x,t) \\ i(x,t) \end{bmatrix}. \tag{3.5}$$

On the other hand, note that

$$\begin{bmatrix} W_R(x,t) \\ W_L(x,t) \end{bmatrix} = \frac{1}{2}\begin{bmatrix} Z^{-1/2}(x) & Z^{1/2}(x) \\ Z^{-1/2}(x) & -Z^{1/2}(x) \end{bmatrix}\begin{bmatrix} v(x,t) \\ i(x,t) \end{bmatrix}$$

and it follows from (3.5) that

$$\begin{bmatrix} W_R(n,t) \\ W_L(n,t) \end{bmatrix} = \frac{1}{2}\begin{bmatrix} Z_n^{-1/2} & Z_n^{1/2} \\ Z_n^{-1/2} & -Z_n^{1/2} \end{bmatrix} (\frac{1}{2}\begin{bmatrix} Z_{n-1}^{-1/2} & Z_{n-1}^{1/2} \\ Z_{n-1}^{-1/2} & -Z_{n-1}^{1/2} \end{bmatrix})^{-1}\begin{bmatrix} W_R(x,t) \\ W_L(x,t) \end{bmatrix}_{x=\frac{n}{2}-}$$

$$= \frac{Z_n + Z_{n-1}}{2(Z_n Z_{n-1})^{1/2}}\begin{bmatrix} 1 & -\dfrac{Z_n - Z_{n-1}}{Z_n + Z_{n-1}} \\[3mm] -\dfrac{Z_n - Z_{n-1}}{Z_n + Z_{n-1}} & 1 \end{bmatrix}\begin{bmatrix} W_R(n-1, t - \frac{1}{2}) \\[2mm] W_L(n-1, t + \frac{1}{2}) \end{bmatrix}.$$

Define for $n \geq 0$ the *reflection coefficients* $k_n = (Z_n - Z_{n-1})(Z_n + Z_{n-1})^{-1}$. Since $Z_n > 0$ for all $n \geq 0$, it follows that $|k_n| < 1$ for all $n \geq 0$. Define the *time delay operator* by

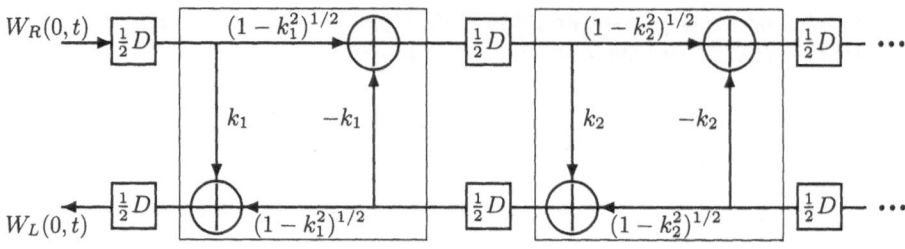

Figure 4.6: The current-voltage propagation in scattering representation

the formula $(\frac{1}{2}D)f(t) = f(t - \frac{1}{2})$ and the evolution of the vectors $\begin{bmatrix} W_R(n,t) \\ W_L(n,t) \end{bmatrix}$, $n \geq 0$, is described by the recursion:

$$\begin{bmatrix} W_R(n,t) \\ W_L(n,t) \end{bmatrix} = S(k_i) \begin{bmatrix} \frac{1}{2}D & 0 \\ 0 & (\frac{1}{2}D)^{-1} \end{bmatrix} \begin{bmatrix} W_R(n-1,t) \\ W_L(n-1,t) \end{bmatrix}, \tag{3.6}$$

hence

$$\begin{bmatrix} W_R(n,t) \\ W_L(n,t) \end{bmatrix} = S(k_i) \begin{bmatrix} \frac{1}{2}D & 0 \\ 0 & (\frac{1}{2}D)^{-1} \end{bmatrix} \cdots S(k_1) \begin{bmatrix} \frac{1}{2}D & 0 \\ 0 & (\frac{1}{2}D)^{-1} \end{bmatrix} \begin{bmatrix} W_R(0,t) \\ W_L(0,t) \end{bmatrix},$$

which is a cascade composition of the first n layers of the transmission-line model in scattering representation showed in Figure 4.6.

By the cumulate effect of the two delays on each section, the transmission-line in Figure 4.6 is essentially the same as the (time domain representation) of the transmission-line in Figure 4.5.

4.4   Displacement Structure and Completion Problems

In this section we show how to use displacement structures in order to solve completion problems as those introduced in Chapter 3. The main idea is to use a system theoretic interpretation of Theorem 1.5 and Theorem 2.2. The main advantage of this method is that it produces a Schur type algorithm for the construction of a particular solution. We also discuss the connection with some other methods used to deal with completion problems.

Let $\{R(t)\}_{t \in \mathbb{Z}}$ be a Pick solution of the displacement equation (1.4) and suppose (H_2)-(H_4) hold. In addition, suppose that dim $\mathcal{R}(t)$ are all equal for $t \in \mathbb{Z}$ and $\{F(t)\}_{t \in \mathbb{Z}}$ is a stable family. Let $\{H(t)\}_{t \in \mathbb{Z}}$ and $\{K(t)\}_{t \in \mathbb{Z}}$ be two families of operators given by Theorem 1.5(b) and consider the system

$$\Omega \begin{cases} x(t) = F^*(t)x(t+1) + H^*(t)J(t)u(t) \\ y(t) = J(t)G^*(t)x(t+1) + J(t)K^*(t)J(t)u(t), \quad t \in \mathbb{Z}. \end{cases}$$

Denote the transfer map of Ω by $\mathbf{T}$ and define the symmetry $\mathbf{J} = \oplus_{t \in \mathbb{Z}} J(t)$.

4.1 Lemma *The transfer map $\mathbf{T}$ of the system Ω is a J-unitary bounded operator.*

Proof By (H_3) and the assumption that $\{F(t)\}_{t \in \mathbb{Z}}$ is a stable family, it follows that $\widetilde{G} = \oplus_{t \in \mathbb{Z}} G(t)$ and $\widetilde{F} = \oplus_{t \in \mathbb{Z}} F(t)$ are bounded operators. Moreover, by Theorem 1.5(b), one obtains that $\widetilde{H} = \oplus_{t \in \mathbb{Z}} H(t)$ and $\widetilde{K} = \oplus_{t \in \mathbb{Z}} K(t)$ are also bounded operators. Taking into account once again the fact that $\{F(t)\}_{t \in \mathbb{Z}}$ is stable, one deduces that the matrix $[L_{ij}]_{i,j \in \mathbb{Z}}$, where

$$L_{ij} = \begin{cases} F^*(i) & \text{if } j = i+1 \\ 0 & \text{otherwise,} \end{cases}$$

represents a bounded operator L, $\|L\| < 1$. Consequently, $\mathbf{J}\widetilde{K}^*\mathbf{J} + \mathbf{J}\widetilde{G}^*(I - L)^{-1}\widetilde{H}^*\mathbf{J}$ is a bounded operator and it is readily checked that this operator coincides with $\mathbf{T}$. Since $\dim \mathcal{H}(t) = \alpha < \infty$ for all $t \in \mathbb{Z}$, it follows from (1.6) that

$$\begin{bmatrix} F(t) & G(t) \\ H(t) & K(t) \end{bmatrix}^* \begin{bmatrix} R^{-1}(t) & 0 \\ 0 & J(t) \end{bmatrix} \begin{bmatrix} F(t) & G(t) \\ H(t) & K(t) \end{bmatrix} = \begin{bmatrix} R^{-1}(t-1) & 0 \\ 0 & J(t) \end{bmatrix} \quad (4.1)$$

and this equality helps proving that $\mathbf{T}$ is J-unitary. Indeed, the entries of $\mathbf{TJT}^*$ and $\mathbf{T}^*\mathbf{JT}$ can be easily computed. For instance, the t-th element (denoted by λ_{tt}) of the main diagonal of $\mathbf{TJT}^*$ is given by

$$\lambda_{tt} = J(t)(K^*(t)J(t)K(t) + G^*(t)H^*(t+1)J(t+1)H(t+1)G(t) + \ldots)J(t),$$

and by (4.1) it follows that

$$R^{-1}(t) = H^*(t+1)J(t+1)H(t+1) + F^*(t+1)H^*(t+2)J(t+2)H(t+2)F(t+1) + \ldots,$$

hence $\lambda_{tt} = J(t)$. Similar computations can be carried out in order to determine all the entries of $\mathbf{TJT}^*$ and $\mathbf{T}^*\mathbf{JT}$, and to see that $\mathbf{TJT}^* = \mathbf{T}^*\mathbf{JT} = \mathbf{J}$. $\square$

We further partition the entries T_{ij} of $\mathbf{T}$ accordingly with $J(j)$ and $J(i)$,

$$T_{ij} = \begin{bmatrix} T_{11}^{ij} & T_{12}^{ij} \\ T_{21}^{ij} & T_{22}^{ij} \end{bmatrix}$$

and consider the upper triangular operators $\mathbf{T}_{kl} = [T_{kl}^{ij}]_{i,j \in \mathbb{Z}}$, $k, l = 1, 2$.

4.2 Lemma *$\mathbf{T}_{22}$ is an invertible operator such that $\mathbf{T}_{22}^{-1}$ is upper triangular and $\mathbf{T}_{12}\mathbf{T}_{22}^{-1}$ is a strict contraction which belongs to the Schur class $\mathcal{S}(\{\mathcal{V}(t)\}_{t \in \mathbb{Z}}, \{\mathcal{U}(t)\}_{t \in \mathbb{Z}})$.*

Proof Since $\mathbf{T}$ is J-unitary, it is easily seen that

$$\mathbf{T}_{22}\mathbf{T}_{22}^* > I \quad \text{and} \quad \mathbf{T}_{22}^*\mathbf{T}_{22} > I. \quad (4.2)$$

This implies that $\mathbf{T}_{22}$ is invertible and that $\|\mathbf{T}_{22}^{-1}\| < 1$. Now one defines the operator $X(t) = P_{\oplus_{k \leq t} \mathcal{V}(k)} \mathbf{T}_{22} / \oplus_{k \leq t} \mathcal{V}(k)$ and it follows from (4.2) that $X^*(t)X(t) > I$. Moreover, if $\mathbf{T}(t) = P_{\oplus_{k \leq t} (\mathcal{U}(t) \oplus \mathcal{V}(t))} \mathbf{T} / \oplus_{k \leq t} (\mathcal{U}(k) \oplus \mathcal{V}(k))$ and $\mathbf{J}_t = \oplus_{k \leq t} J(k)$, then the embedding relation (1.6) implies that

$$\mathbf{J}_t - \mathbf{T}(t) \mathbf{J}_t \mathbf{T}^*(t) = [\ldots \quad F(t)G(t-1) \quad G(t)]^* R^{-1}(t) [\ldots \quad F(t)G(t-1) \quad G(t)] \geq 0.$$

Hence, $X(t)X^*(t) \geq I$ and it may be concluded that $X(t)$ is invertible for every $t \in \mathbb{Z}$ and the family $\{X^{-1}(t)\}_{t \in \mathbb{Z}}$ is uniformly bounded by one. Define the operators $\widetilde{X}(t) = X(t) \oplus 0$ on the same space as $\mathbf{T}_{22}$, and remark that

$$\widetilde{X}(t+1) = \begin{bmatrix} \widetilde{X}(t) & * \\ 0 & * \end{bmatrix},$$

where $*$ denotes irrelevant entries. This shows that the sequence $\{\widetilde{X}(t)\}_{t \geq 0}$ strongly converges to a bounded operator $\widetilde{X}$ as $t \to \infty$. It is easily checked that $\widetilde{X}$ is upper triangular and that it actually coincides with $\mathbf{T}_{22}^{-1}$. The fact that $\mathbf{T}_{12} \mathbf{T}_{22}^{-1}$ is an upper triangular strict contraction in $\mathcal{S}(\{\mathcal{V}(t)\}_{t \in \mathbb{Z}}, \{\mathcal{U}(t)\}_{t \in \mathbb{Z}})$ is a consequence of Lemma 4.1.   □

The $\mathbf{J}$-unitary operator $\mathbf{T}$ with the property that $\mathbf{T}_{22}^{-1}$ is an upper triangular operator is called $\mathbf{J}$-*inner*. The following result emphasizes the so-called *blocking property* of $\mathbf{T}$.

4.3 Lemma   *The operator* $\mathbf{T}$ *satisfies the identity:*

$$[\ldots \quad F(t)G(t-1) \quad G(t) \quad 0] \mathbf{T} = [0 \quad *],$$

where $*$ *denotes the irrelevant k-th entries, $k > t$.*

Proof For $s \leq t$, the s-th entry $y(s)$ of the row matrix

$$[\ldots \quad F(t)G(t-1) \quad G(t) \quad 0] \mathbf{T}$$

can be easily computed as follows:

$$\begin{aligned}
y(s) &= G(s)T_{ss} + F(s)G(s-1)T_{s-1,s} + F(s)F(s-1)G(s-2)T_{s-2,s} + \ldots \\
&= G(s)J(s)K^*(s)J(s) + F(s)G(s-1)J(s-1)G^*(s-1)H^*(s)J(s) + \ldots \\
&= F(s)(-R(s-1) + G(s-1)J(s-1)G^*(s-1) + \ldots)H^*(s)J(s) = 0,
\end{aligned}$$

and the proof is concluded.   □

We are now in a position to prove the following result.

4.4 Theorem   *Suppose (H_2)-(H_4) hold. In addition, suppose* $\dim \mathcal{R}(t) = \alpha < \infty$ *for* $t \in \mathbb{Z}$, *$\{F(t)\}_{t \in \mathbb{Z}}$ is a stable family and the nondegeneracy condition* $\mathbf{U}(t)\mathbf{U}^*(t) > \mu I > 0$ *holds for all $t \in \mathbb{Z}$. Then the displacement equation (1.4) has a*

Pick solution satisfying (H) if and only if there exists S in $\mathcal{S}(\{\mathcal{V}(t)\}_{t\in\mathbb{Z}}, \{\mathcal{U}(t)\}_{t\in\mathbb{Z}})$ such that $\|S\| < 1$ and

$$\mathbf{V}(t) = \mathbf{U}(t) P_{\oplus_{k\le t}\mathcal{U}(k)} S / \oplus_{k\le t} \mathcal{V}(k), \quad t \in \mathbb{Z}. \tag{4.3}$$

Proof One implication is immediate. If there exists a strict contraction S in the Schur class $\mathcal{S}(\{\mathcal{V}(t)\}_{t\in\mathbb{Z}}, \{\mathcal{U}(t)\}_{t\in\mathbb{Z}})$ such that (4.3) holds, then one defines $S(t) = P_{\oplus_{k\le t}\mathcal{U}(k)} S / \oplus_{k\le t} \mathcal{V}(k)$ and it is remarked that

$$R(t) = \mathbf{U}(t)(I - S(t)S^*(t))\mathbf{U}^*(t).$$

Since S is a strict contraction, there exists a number $\delta > 0$ such that $I - SS^* \ge \delta I$, hence $I - S(t)S^*(t) \ge \delta I$ for all $t \in \mathbb{Z}$. Consequently, $R(t) \ge \delta \mathbf{U}(t)\mathbf{U}^*(t)$ and by the nondegeneracy condition, $R(t) \ge \delta \mu I > 0$ for all $t \in \mathbb{Z}$.

Conversely, assume that $\{R(t)\}_{t\in\mathbb{Z}}$ is a Pick solution of the equation (1.4) that satisfies (H). By Theorem 1.5(b), there exist uniformly bounded families $\{H(t)\}_{t\in\mathbb{Z}}$ and $\{K(t)\}_{t\in\mathbb{Z}}$ of operators such that (1.6) holds. We are led to consider the linear system

$$\Omega \begin{cases} x(t) = F^*(t)x(t+1) + H^*(t)J(t)u(t) \\ \dot{y}(t) = J(t)G^*(t)x(t+1) + J(t)K^*(t)J(t)u(t), \quad t \in \mathbb{Z}. \end{cases} \tag{4.4}$$

By Lemma 4.1, the transfer map $\mathbf{T}$ of this system **J**-inner. By Lemma 4.2, $S = -\mathbf{T}_{12}\mathbf{T}_{22}^{-1}$ belongs to $\mathcal{S}(\{\mathcal{V}(t)\}_{t\in\mathbb{Z}}, \{\mathcal{U}(t)\}_{t\in\mathbb{Z}})$ and $\|S\| < 1$. By Lemma 4.3, it follows that S satisfies (4.3) and the proof is complete. $\square$

The connection with the completion problems studied in Chapter 3 is illustrated by the following example. More examples will be seen in the next section.

4.5 Example Consider the displacement equation of the form

$$R - FRF^* = GJG^*, \tag{4.5}$$

and assume that $\mathcal{R}(t) = \mathbb{C}^n, \mathcal{U}(t) = \mathbb{C}^p$ and $\mathcal{V}(t) = \mathbb{C}^q$ for all $t \in \mathbb{Z}$. By the Jordan structure theorem, there exists an invertible matrix W such that

$$WFW^{-1} = F_0 = \begin{bmatrix} F_0^{(0)} & 0 & & \cdots & 0 \\ 0 & F_0^{(1)} & 0 & \cdots & 0 \\ \vdots & & & \ddots & \\ 0 & \cdots & & & F_0^{(m-1)} \end{bmatrix} \tag{4.6}$$

and $F_0^{(j)}, j = 0, 1, \ldots, m-1$, are $r_j \times r_j$ Jordan blocks of the form

$$F_0^{(j)} = \begin{bmatrix} z_j & 0 & \cdots & 0 \\ 1 & z_j & \cdots & \\ \vdots & & \ddots & \\ 0 & \cdots & 1 & z_j \end{bmatrix} \tag{4.7}$$

with $|z_j| < 1$ for $j = 0, 1, \ldots, m - 1$. Note that if R is an invertible Pick solution of the displacement equation (4.5), then $R_0 = W R W^{-1}$ is an invertible Pick solution of the displacement equation

$$R_0 - F_0 R_0 F_0^* = G_0 J G_0^*, \tag{4.8}$$

where $G_0 = WG = [\,U_0 \quad V_0\,]$. We write U_0 and V_0 with respect to the decomposition of F_0 in Jordan blocks, as follows:

$$U_0 = \begin{bmatrix} U_0^{(0)} \\ U_0^{(1)} \\ \vdots \\ U_0^{(m-1)} \end{bmatrix} \quad \text{and} \quad V_0 = \begin{bmatrix} V_0^{(0)} \\ V_0^{(1)} \\ \vdots \\ V_0^{(m-1)} \end{bmatrix}. \tag{4.9}$$

Supposing that the equation (4.5) has an invertible Pick solution, it follows from Theorem 4.4 that there exists an upper triangular strict contraction S such that (4.3) holds. Even more, S is actually a Toeplitz operator, as it easily follows by an inspection of the proof of Theorem 4.4 and let f in $\mathcal{S}(\mathbb{C}^p, \mathbb{C}^q)$ be its symbol. Writing $U_0^{(i)}$ and $V_0^{(i)}$ as columns of row vectors,

$$U_0^{(i)} = \begin{bmatrix} u_1^{(i)} \\ u_2^{(i)} \\ \vdots \\ u_{r_i}^{(i)} \end{bmatrix} \quad \text{and} \quad V_0^{(i)} = \begin{bmatrix} v_1^{(i)} \\ v_2^{(i)} \\ \vdots \\ v_{r_i}^{(i)} \end{bmatrix}, \tag{4.10}$$

where $u_j^{(i)}$ and $v_j^{(i)}$ are $1 \times p$ and $1 \times q$ row vectors respectively, it follows that

$$\begin{bmatrix} v_1^{(i)} & \cdots & v_{r_i}^{(i)} \end{bmatrix} = \begin{bmatrix} u_1^{(i)} & \cdots & u_{r_i}^{(i)} \end{bmatrix} \mathcal{H}_f^{r_i}(z_i), \qquad 0 \leq i \leq m - 1, \tag{4.11}$$

where $\mathcal{H}_f^{r_i}(z_i)$ was defined by (3.1.7). In conclusion, it follows from Theorem 4.4 that the displacement equation (4.8) has an invertible Pick solution if and only if there exists a function f in $\mathcal{S}(\mathbb{C}^p, \mathbb{C}^q)$, $\|F\| < 1$, which is the solution of the Hermite-Fejér problem (4.11). $\square$

We now explain the connection between the generalized Schur algorithm and Theorem 4.4. Let $\{R(t)\}_{t \in \mathbb{Z}}$ be the Pick solution of the displacement equation

$$R(t) - F(t) R(t - 1) F^*(t) = G(t) J(t) G^*(t).$$

Suppose (H_3)-(H_6) and (H) hold. Then, it follows from Theorem 2.2 that the families $\{R_i(t)\}_{t \in \mathbb{Z}}$, $i = 0, 1, \ldots, n - 1$, of Schur complements satisfy the displacement equations

$$R_i(t) - F_i(t) R_i(t - 1) F_i^*(t) = G_i(t) J(t) G_i^*(t), \tag{4.12}$$

with generators given by the generalized Schur algorithm (2.1). At each step, the elements $f_i(t)$, $g_i(t)$, $h_i(t)$ $k_i(t)$ satisfy the embedding relation (2.2). Due to the lower triangular structure of $F(t)$, the relations in Corollary 2.3 can be rewritten in the following form:

$$[F_i(t) \quad G_i(t)] = A_i(t) \begin{bmatrix} 1 & 0 & 0 \\ 0 & F_{i+1}(t) & G_{i+1}(t) \end{bmatrix}$$
$$\times \begin{bmatrix} f_i(t) & 0 & g_i(t) \\ 0 & I_{n-i-1} & 0 \\ h_i(t) & 0 & k_i(t) \end{bmatrix} \begin{bmatrix} A_i(t-1) & 0 \\ 0 & I_{p+q} \end{bmatrix}^{-1}, \quad (4.13)$$

where $A_i(t) = \begin{bmatrix} l_i(t)d_i^{-1}(t) & 0 \\ & I_{n-i-1} \end{bmatrix}$. This formula suggests to define

$$[H_{n-1} \quad K_{n-1}] = [h_{n-1} \quad k_{n-1}]$$

and then, for $0 \le i < n-1$, the matrices $[H_i \quad K_i]$ are introduced such that

$$\begin{bmatrix} F_i(t) & G_i(t) \\ H_i(t) & K_i(t) \end{bmatrix} = \begin{bmatrix} A_i(t) & 0 \\ 0 & I_{p+q} \end{bmatrix} \begin{bmatrix} 1 & 0 & 0 \\ 0 & F_{i+1}(t) & G_{i+1}(t) \\ 0 & H_{i+1}(t) & K_{i+1}(t) \end{bmatrix}$$
$$\times \begin{bmatrix} f_i(t) & 0 & g_i(t) \\ 0 & I_{n-i-1} & 0 \\ h_i(t) & 0 & k_i(t) \end{bmatrix} \begin{bmatrix} A_i(t-1) & 0 \\ 0 & I_{p+q} \end{bmatrix}^{-1}. \quad (4.14)$$

Since $R_i(t) = A_i(t)[d_i(t) \oplus R_{i+1}(t)]A_i^*(t)$, it follows from (4.14) that the families $\{H_i(t)\}_{t \in \mathbb{Z}}$, $\{K_i(t)\}_{t \in \mathbb{Z}}$, $i = 0, 1, \ldots, n-1$, satisfy the embedding relation (1.6) for the displacement equations (4.12). Finally, consider for $i = 0, 1, \ldots, n-1$, the systems

$$\Omega_i \begin{cases} x_i(t) = f_i^*(t)x_i(t+1) + h_i^*(t)J(t)u_i(t) \\ y_i(t) = J(t)g_i^*(t)x_i(t+1) + J(t)k_i^*(t)J(t)u_i(t), \quad t \in \mathbb{Z}, \end{cases} \quad (4.15)$$

and denote by $\mathbf{T}_i$ the transfer map of Ω_i.

4.6 Lemma *The operator* $\mathbf{T} = \mathbf{T}_0\mathbf{T}_1 \ldots \mathbf{T}_{n-1}$ *is the transfer map of the system*

$$\Omega_i \begin{cases} x(t) = F^*(t)x(t+1) + H_0^*(t)J(t)u(t) \\ y(t) = J(t)G^*(t)x(t+1) + J(t)K_0^*(t)J(t)u(t), \quad t \in \mathbb{Z}. \end{cases}$$

Proof It follows immediately from 4.14. □

We are now in a position to prove the main result of this section.

4.7 Theorem *Suppose (H_3)-(H_6) hold, as well as the additional nondegeneracy condition $\mathbf{U}(t)\mathbf{U}^*(t) > \mu I > 0$ for all $t \in \mathbb{Z}$. Then the solution of the displacement equation (1.4) is a Pick solution satisfying (H) if and only if there exists S in $\mathcal{S}(\{\mathcal{V}(t)\}_{t \in \mathbb{Z}}, \{\mathcal{U}(t)\}_{t \in \mathbb{Z}})$ such that $\|S\| < 1$ and (4.3) holds. Moreover, a particular* *solution can be given by the formula* $S = -\mathbf{T}_{12}\mathbf{T}_{22}^{-1}$, *where* $\mathbf{T} = \mathbf{T}_0\mathbf{T}_1\ldots\mathbf{T}_{n-1}$ *and* $\mathbf{T}_i$ *are the transfer maps of the systems (4.15).*

Proof This is a consequence of Theorem 4.4 and Lemma 4.6. $\square$

One more proof of Theorem 4.4 can be obtained by exploiting the connection between displacement equations and the lifting of commutants method described in Section 3.5. In fact, we notice the following generalization of Theorem 4.4.

4.8 Theorem *Suppose* (H_1)-(H_3) *hold. Then the displacement equation (1.4) has a Pick solution if and only if there exists* S *in* $\mathcal{S}(\{\mathcal{V}(t)\}_{t\in\mathbb{Z}}, \{\mathcal{U}(t)\}_{t\in\mathbb{Z}})$ *such that (4.3) holds.*

Proof One implication is immediate. If an upper triangular contraction S exists such that (4.3) holds, then the solution of (1.4) given by (1.5) is a Pick solution. Conversely, assume that $\{R(t)\}_{t\in\mathbb{Z}}$ is a Pick family, then $R(t) = U(t)\mathbf{U}^*(t) - \mathbf{V}(t)\mathbf{V}^*(t)$ are positive operators for all $t \in \mathbb{Z}$. By Lemma 1.4.1, there exist contractions $\bar{S}(t)$ in $\mathcal{L}(\oplus_{j\le t}\mathcal{V}(j), \mathrm{cl}\mathcal{R}(\mathbf{U}^*(t)))$ such that $\mathbf{V}(t) = \mathbf{U}(t)\bar{S}(t)$. Moreover, remark that $\mathbf{U}(t)N_{\{\mathcal{U}(s)\}_{s\in\mathbb{Z}}}(t) = F(t)\mathbf{U}(t-1)$ and $\mathbf{V}(t)N_{\{\mathcal{V}(s)\}_{s\in\mathbb{Z}}}(t) = F(t)\mathbf{V}(t-1)$, where $N_{\{\mathcal{U}(s)\}_{s\in\mathbb{Z}}}(t)$ and $N_{\{\mathcal{V}(s)\}_{s\in\mathbb{Z}}}(t)$ are marking operators defined by (2.4.2). Hence,

$$
\begin{aligned}
N^*_{\{\mathcal{V}(s)\}_{s\in\mathbb{Z}}}(t)\bar{S}^*(t)\mathbf{U}^*(t) &= N^*_{\{\mathcal{V}(s)\}_{s\in\mathbb{Z}}}(t)\mathbf{V}^*(t) = \mathbf{V}^*(t-1)F^*(t) \\
&= \bar{S}^*(t-1)\mathbf{U}^*(t-1)F^*(t) \\
&= \bar{S}^*(t-1)N^*_{\{\mathcal{U}(s)\}_{s\in\mathbb{Z}}}(t)\mathbf{U}^*(t).
\end{aligned}
$$

By Theorem 3.5.1, there exist contractions $\hat{S}(t)$ in $\mathcal{L}(\oplus_{j\le t}\mathcal{V}(k), \oplus_{j\le t}\mathcal{U}(j))$ such that $\bar{S}^*(t) = \hat{S}^*(t)/\mathrm{cl}\mathcal{R}(\mathbf{U}^*(t))$ and $\hat{S}(t)N_{\{\mathcal{V}(s)\}_{s\in\mathbb{Z}}}(t) = N_{\{\mathcal{U}(s)\}_{s\in\mathbb{Z}}}(t)\hat{S}(t-1)$. It follows that the family $\{\hat{S}(t)\}_{t\in\mathbb{Z}}$ induces an upper triangular contraction S such that (4.3) holds. $\square$

4.9 Remark Theorem 4.8 shows the connection between Theorem 4.4 and the commutant lifting theorem, and a parametrization of all the contractions S satisfying (4.3) can be obtained using the methods introduced in Chapter 3. However, it is worth mentioning a more direct approach to the parametrization that works in the conditions of Theorem 4.4. Thus, it is obtained that all the strict contractions S satisfying (4.3), are given by the formula:

$$S = \mathbf{T}[K] = -(\mathbf{T}_{11}K + \mathbf{T}_{12})(\mathbf{T}_{21}K + \mathbf{T}_{22})^{-1}, \tag{4.16}$$

for arbitrary K in $\mathcal{S}(\{\mathcal{V}(t)\}_{t\in\mathbb{Z}}, \{\mathcal{U}(t)\}_{t\in\mathbb{Z}})$ with $\|K\| < 1$, and $\mathbf{T} = [\mathbf{T}_{ij}]_{i,j=1,2}$ the transfer map of the system (4.4). One implication is immediate. Consider a K as above. By Lemma 4.1 and Lemma 4.2, it follows that $S = -(\mathbf{T}_{11}K + \mathbf{T}_{12})(\mathbf{T}_{21}K + \mathbf{T}_{22})^{-1}$ is an upper triangular strict contraction. Set $S_1 = \mathbf{T}_{11}K + \mathbf{T}_{12}$ and $S_2 = \mathbf{T}_{21}K + \mathbf{T}_{22}$, then $\begin{bmatrix} S_1 \\ S_2 \end{bmatrix} = T\begin{bmatrix} K \\ I \end{bmatrix}$ and Lemma 4.3 implies that S is a solution of (4.3). The converse implication is based on the following remark. If Up denotes the set of upper triangular operators in $\mathcal{L}(\oplus_{t\in\mathbb{Z}}\mathcal{V}(t), \oplus_{t\in\mathbb{Z}}(\mathcal{U}(t) \oplus \mathcal{V}(t)))$, then

$$\mathbf{T}Up = \{X \in Up \mid \mathbf{G}(t)X(t) = 0, \quad t \in \mathbb{Z}\},$$

where $\mathbf{G}(t) = [\ldots \quad F(t)F(t-1)G(t-2) \quad F(t)G(t-1) \quad G(t)]$ and for X in Up, $X(t) = P_{\oplus_{j\leq t}\mathcal{U}(j)\oplus\mathcal{V}(j)}X/\oplus_{j\leq t}\mathcal{V}(j)$. Indeed, take Y in Up, then $\mathbf{G}(t)(\mathbf{T}Y)(t) = \mathbf{G}(t)\mathbf{T}(t)Y(t) = 0$ by Lemma 4.3. Conversely, take X in Up such that $\mathbf{G}(t)X(t) = 0$ for all $t \in \mathbb{Z}$. Define $Y = \mathbf{T}^{-1}X = \mathbf{J}\mathbf{T}^*\mathbf{J}X$, and due to the structure of the entries of $\mathbf{T}$, it is readily checked that all the entries of Y under the main diagonal are zero. That is, Y belongs to Up and the claim is proved.

If S is a strict contraction such that (4.3) holds, then it follows by the previous claim that there exists $\begin{bmatrix} K_1 \\ K_2 \end{bmatrix}$ in Up such that $\begin{bmatrix} S \\ I \end{bmatrix} = T\begin{bmatrix} K_1 \\ K_2 \end{bmatrix}$. We deduce that K_2 is invertible and K_2^{-1} is upper triangular. Set $K = K_1K_2^{-1}$, then K belongs to $\mathcal{S}(\{\mathcal{V}(t)\}_{t\in\mathbb{Z}}, \{\mathcal{U}(t)\}_{t\in\mathbb{Z}})$ and $\|K\| < 1$. Since $S = (\mathbf{T}_{11}K + \mathbf{T}_{12})K_2$ and $I = (\mathbf{T}_{21}K + \mathbf{T}_{22})K_2$, it follows that $S = (\mathbf{T}_{11}K + \mathbf{T}_{12})(\mathbf{T}_{21}K + \mathbf{T}_{22})^{-1}$. $\square$

4.5 Other Applications

In this section we present some other applications of the theory of displacement equations. We begin with a so-called *time-variant Hermite-Fejér problem* generalizing Problem 3.1.13. For a certain simplicity we discuss here only the matrix case.

We introduce a time-variant *tangential evaluation* along the direction defined by a uniformly bounded family $\{u(t)\}_{t\in\mathbb{Z}}$ of $1 \times r(t)$ vectors. Thus, introduce the symmetric functions $s_k^{(n)}$ of n variables, $s_0^{(n)} = 1$ and

$$s_k^{(n)}(x_1, x_2, \ldots, x_n) = \sum_{1\leq i_1 < \ldots < i_k \leq n} x_{i_1}x_{i_2}\ldots x_{i_k}.$$

Let $T = [T_{ij}]_{i,j\in\mathbb{Z}}$ be an upper triangular bounded operator with T_{ij} being $r(i) \times r(j)$ matrices and let $\{f(t)\}_{t\in\mathbb{Z}}$ be a stable family of complex numbers. Then, for $p \geq 0$, define

$$u(t) \bullet \frac{1}{p!}T^{(p)}(f(t)) = \sum_{m=0}^{\infty} s_m^{(m+p)}(f(t), f(t-1), \ldots, f(t-m-p-1))u(t-m-p)T_{t-m-p,t}.$$

More generally, we write

$$[u_1(t) \quad u_2(t) \quad \ldots \quad u_r(t)] \bullet \mathcal{H}_T^r(f(t)) = [u_1(t) \quad u_2(t) \quad \ldots \quad u_r(t)]$$

$$\bullet \begin{bmatrix} T(f(t)) & \frac{1}{1!}T^{(1)}(f(t)) & \frac{1}{2!}T^{(2)}(f(t)) & & \frac{1}{(r-1)!}T^{(r-1)}(f(t)) \\ & T(f(t)) & \frac{1}{1!}f^{(1)}(z) & \ddots & \frac{1}{(r-2)!}T^{(r-2)}(f(t)) \\ & & \ddots & \ddots & \\ 0 & & & T(f(t)) & \frac{1}{1!}f^{(1)}(z) \\ & & & & T(f(t)) \end{bmatrix}$$

$$= \begin{bmatrix} u_1(t) \bullet T(f(t)) & u_1(t) \bullet \frac{1}{1!}T^{(1)}(f(t)) + u_2(t) \bullet T(f(t)) & \cdots \end{bmatrix}.$$

We now state the time-variant Hermite-Fejér problem treated in this section. Consider m stable families $\{\alpha_i(t)\}_{t\in\mathbb{Z}}$, $i = 0,1,\ldots,m-1$, of complex numbers and associated to each family a positive integer $r_i \geq 1$ and uniformly bounded row vectors $\mathbf{a}_i(t)$ and $\mathbf{b}_i(t)$ partitioned as follows

$$\mathbf{a}_i(t)=\begin{bmatrix} u_1^{(i)}(t) & u_2^{(i)}(t) & \cdots & u_{r_i}^{(i)}(t) \end{bmatrix} \quad \text{and} \quad \mathbf{b}_i(t)=\begin{bmatrix} v_1^{(i)}(t) & v_2^{(i)}(t) & \cdots & v_{r_i}^{(i)}(t) \end{bmatrix},$$

where $u_j^{(i)}(t)$ and $v_j^{(i)}(t)(j = 1,\ldots r_i)$ are $1 \times p(t)$ and $1 \times q(t)$ row vectors, respectively.

5.1 Problem   *Given m stable families $\{\alpha_i(t)\}_{t\in\mathbb{Z}}$ and associated uniformly bounded data $\mathbf{a}_i(t)$ and $\mathbf{b}_i(t)$, it is required to find conditions for the existence of an upper triangular (strict) contraction S such that*

$$\mathbf{b}_i(t) = \mathbf{a}_i(t)\mathcal{H}_S^{r_i}(\alpha_i(t)), \quad 0 \leq i \leq m-1, t \in \mathbb{Z}.$$

A solution of this problem can be obtained in the following way. Define $J(t) = I_{p(t)} \oplus (-I_{q(t)})$ and to each $\alpha_i(t)$ we associate a Jordan block $\bar{F}_i(t)$ of size $r_i \times r_i$,

$$\bar{F}_i(t) = \begin{bmatrix} \alpha_i(t) & & & \\ 1 & \alpha_i(t) & & \mathbf{0} \\ & \ddots & \ddots & \\ & & 1 & \alpha_i(t)cr \end{bmatrix},$$

and two $r_i \times p(t)$, $r_i \times q(t)$ matrices $U_i(t)$ and $V_i(t)$, respectively,

$$U_i(t) = \begin{bmatrix} u_1^{(i)}(t) \\ u_2^{(i)}(t) \\ \vdots \\ u_{r_i}^{(i)}(t) \end{bmatrix} \quad \text{and} \quad V_i(t) = \begin{bmatrix} v_1^{(i)}(t) \\ v_2^{(i)}(t) \\ \vdots \\ v_{r_i}^{(i)}(t) \end{bmatrix}.$$

Then $F(t) = \oplus_{i=0}^{m-1}\bar{F}_i(t)$ and

$$G(t) = \begin{bmatrix} U_0(t) & V_0(t) \\ U_1(t) & V_1(t) \\ \vdots & \vdots \\ U_{m-1}(t) & V_{m-1}(t) \end{bmatrix} = \begin{bmatrix} U(t) & V(t) \end{bmatrix}.$$

5.2 Corollary   (*a*) *The time-variant Hermite-Fejér problem is solvable if and only if the solution of the displacement equation with* $\{F(t)\}_{t\in\mathbb{Z}}$ *and* $\{G(t)\}_{t\in\mathbb{Z}}$ *defined as above is a Pick solution.*
(*b*) *Under the nondegeneracy condition* $\mathbf{U}(t)\mathbf{U}^*(t) > \mu I > 0$ *for all* $t \in \mathbb{Z}$, *the Hermite-Fejér problem is solvable with an* S *such that* $\|S\| < 1$ *if and only if the solution of the displacement equation with* $\{F(t)\}_{t\in\mathbb{Z}}$ *and* $\{G(t)\}_{t\in\mathbb{Z}}$ *defined as above is a Pick solution satisfying* (H).

Proof We easily check that the interpolation condition in the Hermite-Fejér problem and the relation (4.3) are equivalent.   □

The formula (1.5) renders the above result explicit. We mention here several particular cases. For instance, if $m = 1$, $\alpha_0(t) = 0$, $r_0 = n$, $p(t) = 1$, $q(t) = 1$, $\mathbf{a}_0(t) = \begin{bmatrix} 1 & 0 & \dots & 0 \end{bmatrix}$, $\mathbf{b}_0(t) = \begin{bmatrix} \beta_0(t) & \beta_1(t) & \dots & \beta_{n-1}(t) \end{bmatrix}$, then we obtain a *time-variant Schur problem*. We remark that this is exactly the scalar case of Problem 3.1.10. Using Corollary 5.2, the same solution as that described by Corollary 3.5.7 is obtained. In addition, a time-variant analogue of the array form of the Schur algorithm can be described as follows:

$$\begin{bmatrix} 0 & 0 \\ G_{i+1}(t) \end{bmatrix} = ZG_i(t-1)S(\gamma_i(t-1))\begin{bmatrix} 1 & 0 \\ 0 & 0 \end{bmatrix} + G_i(t)S(\gamma_i(t))\begin{bmatrix} 0 & 0 \\ 0 & 1 \end{bmatrix}, \quad (5.1)$$

where $\gamma_i(t) = v_i(t)/u_i(t)$, $g_i(t) = \begin{bmatrix} u_i(t) & v_i(t) \end{bmatrix}$ and Z is given by (1.2). In this case, a time-variant transmission-line model is represented in Figure 4.7.

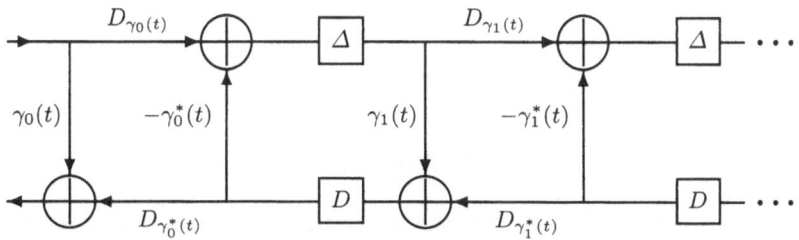

Figure 4.7: Time-variant transmission-line model

Moreover, the connection with the structure of triangular contractions as described in Section 2.1 can be rendered explicitly. For instance, in Figure 4.8 it is represented the transmission-line corresponding to a 3×4 contraction $T = [T_{ij}]$, which follows from both formula (5.1) and from Remark 2.2.5. (for clarity, it is also represented the elementary rotation of T).

Another particular case of Problem 5.1 is the following *time-variant Nevanlinna-Pick problem*: $m = n$, $\{\alpha_i(t)\}_{t\in\mathbb{Z}}$ stable, $r_i = 1$, $p(t) = q(t) = 1$, $\mathbf{a}_i(t) = 1$, $\mathbf{b}_i(t) = \beta_i(t)$. In this case, we are interested to finding upper triangular (strict) contractions S such that $S(\alpha_i(t)) = \beta_i(t)$ for $i = 0, 1, \dots, n-1$. Following the

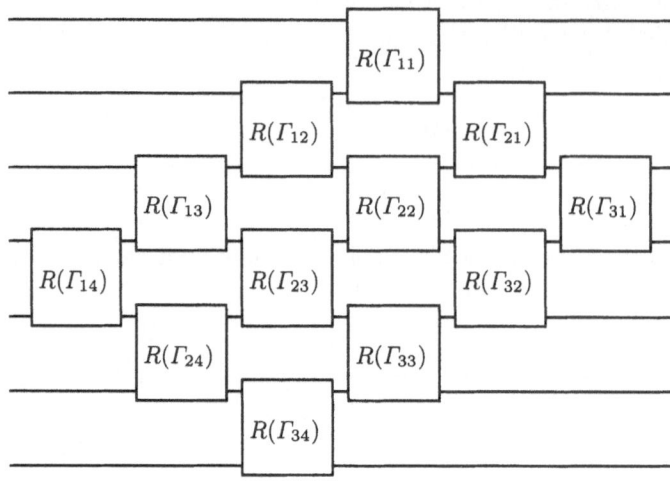

Figure 4.8: The structure of a 3×4 contraction and of its elementary rotation

construction given above, we get

$$F(t) = \begin{bmatrix} \alpha_0(t) & & & \\ & \alpha_1(t) & & 0 \\ & & \ddots & \\ 0 & & & \alpha_{n-1}(t) \end{bmatrix}, \quad G(t) = \begin{bmatrix} 1 & \beta_0(t) \\ 1 & \beta_1(t) \\ \vdots & \vdots \\ 1 & \beta_{n-1}(t) \end{bmatrix}.$$

5.3 Corollary *The time-variant Nevanlinna-Pick problem has solutions if and only if*

$$[\{1 - \beta_i(t)\beta_j^*(t)\} \bullet N_{\alpha_j^*}(\alpha_i(t))]_{i,j=0}^{n-1} \geq 0, \quad t \in \mathbb{Z},$$

where, for a stable family $\{\alpha(t)\}$, the upper triangular operator N_α is defined by $(N_\alpha)_{tt} = 1$, and $(N_\alpha)_{t-j,t} = \alpha(t-j+1)\alpha(t-j+2)\ldots\alpha(t)$ for $j \geq 1$ (the stability of $\{\alpha(y)\}$ assures that N_α is a well-defined bounded operator). □

We now consider another application that arises in the problem of *model (in)validation*. Roughly speaking, the problem is to decide whether a postulated nominal model is consistent with measured input-output data. A simple example is that of a postulated model of the form $M_0 + WS$, where S is constrained to be an upper triangular (strict) contraction and M_0, W are upper triangular operators. The associated model validation problem can be stated as follows: assume an input sequence $\{u_0, u_1, \ldots, u_{n-1}\}$ is applied to the physical system, where u_i is a $1 \times p$ row vector, and measure the associated output sequence $\{v_0, v_1, \ldots, v_{n-1}\}$, where v_i is a $1 \times q$ row vector. The question is to verify whether there exists a model for the plant of the form $M_0 + WS$ that maps the given input sequence to the measured output sequence $\begin{bmatrix} v_0 & v_1 & \cdots & v_{n-1} \end{bmatrix} = \begin{bmatrix} u_0 & u_1 & \cdots & u_{n-1} \end{bmatrix}(M_0 + WS)$. This reduces to checking whether there exists an upper triangular (strict) contraction S matching the sequence $\begin{bmatrix} v_0 & v_1 & \cdots & v_{n-1} \end{bmatrix} - \begin{bmatrix} u_0 & u_1 & \cdots & u_{n-1} \end{bmatrix} M_0$ with the sequence $\begin{bmatrix} u_0 & u_1 & \cdots & u_{n-1} \end{bmatrix} W$. This is seen to be a *tangential time-variant Schur problem* of the form: given data points $\{u_i(t), v_i(t)\}_{t \in \mathbb{Z}}$, $i = 0, 1, \ldots, n-1$, it is required to find conditions for the existence of an upper triangular (strict) contraction $S = [S_{ij}]$ such that

$$\begin{bmatrix} u_0(t-n+1) & \cdots & u_{n-1}(t) \end{bmatrix} \begin{bmatrix} S_{t-n+1,t-n+1} & \cdots & & S_{t-n+1,t} \\ & \ddots & & \\ 0 & & S_{t-1,t-1} & S_{t-1,t} \\ & & & S_{tt} \end{bmatrix}$$
$$= \begin{bmatrix} v_0(t-n+1) & \cdots & v_{n-1}(t) \end{bmatrix}.$$

The solution of this problem is obtained as a particular case of Theorem 5.2. Thus, set

$$U(t) = \begin{bmatrix} u_0(t) \\ \vdots \\ u_{n-1}(t) \end{bmatrix}, \quad V(t) = \begin{bmatrix} v_0(t) \\ \vdots \\ v_{n-1}(t) \end{bmatrix},$$

$F(t) = Z$, given by (1.2), and $G(t) = \begin{bmatrix} U(t) & V(t) \end{bmatrix}$.

5.4 Corollary   *The tangential time-variant Schur problem is solvable if and only if the solution of the displacement equation associated to the above $\{F(t)\}_{t \in \mathbb{Z}}$, $\{G(t)\}_{t \in \mathbb{Z}}$ is a Pick solution.* $\square$

Also in the realm of the time-variant Hermite-Fejér problem, there is the following so-called *strong Parrott problem*: Given matrices B_{ij}, $1 \le j \le i \le n$, $S = \begin{bmatrix} S_1 & S_2 & \cdots & S_n \end{bmatrix}$ and $T = \begin{bmatrix} T_1 & T_2 & \cdots & T_n \end{bmatrix}$, it is required to find conditions for the existence of contractions B of the form

$$B = \begin{bmatrix} B_{11} & & & \\ B_{21} & B_{22} & & ? \\ \vdots & & \ddots & \\ B_{n1} & B_{n2} & \cdots & B_{nn} \end{bmatrix},$$

such that $SB = T$, where ? denotes unspecified entries. To put this problem into our framework, define

$$U(t) = \begin{bmatrix} 0 \\ I \\ 0 \\ \vdots \\ 0 \end{bmatrix}, \quad 1 \le t \le n-1, \quad U(t) = \begin{bmatrix} S_{n+t} \\ I \\ 0 \\ \vdots \\ 0 \end{bmatrix}, \quad -n+1 \le t \le 0,$$

$$
V(0) = \begin{bmatrix} T_1 \\ B_{n1} \\ B_{n-1,1} \\ \vdots \\ B_{11} \end{bmatrix}, \; V(1) = \begin{bmatrix} T_2 \\ 0 \\ B_{n2} \\ \vdots \\ B_{22} \end{bmatrix}, \; V(2) = \begin{bmatrix} T_3 \\ 0 \\ 0 \\ B_{n3} \\ \vdots \\ B_{33} \end{bmatrix}, \; \dots \; V(n-1) = \begin{bmatrix} T_n \\ 0 \\ 0 \\ \vdots \\ 0 \\ B_{nn} \end{bmatrix},
$$

$$
F(t) = \begin{bmatrix} I \\ 0 & 0 \\ & I & 0 \\ & & I & 0 \\ & & & \ddots & \ddots \\ & & & & I & 0 \end{bmatrix}, \; G(t) = [\, U(t) \quad V(t) \,], \text{ for } -n+1 \le t \le n-1,
$$

and all the elements equal to zero for the other time indices. We then have the following result.

5.5 Corollary   *The strong Parrott problem problem has solutions if and only if the solution of the displacement equation associated with the above data has a Pick solution.* □

We conclude this section with a discussion of Example 1.4. We were faced with the recursive solution of the normal equation

$$
\Phi(t)\mathbf{w}(t) = \vartheta(t) \tag{5.2}
$$

and we remarked that $\{\Phi(t)\}_{t\in\mathbb{Z}}$ satisfies the displacement equation $\Phi(t) - F\Phi(t-1)F^* = \mathbf{u}^*(t)\mathbf{u}(t)$ with $F = \sqrt{\lambda}I$, and it is easy to see that $\vartheta(t) - F\vartheta(t-1)F^* = \mathbf{u}^*(t)d(t)$. In order to use the generalized Schur algorithm in this context, we define the $2n \times 2n$ matrix

$$
R(t) = \begin{bmatrix} \Phi(t) & \vartheta(t) & 0 \\ I_n & & 0 \end{bmatrix}.
$$

It follows from the displacement structures of $\{\Phi(t)\}_{t\in\mathbb{Z}}$ and $\{\vartheta(t)\}_{t\in\mathbb{Z}}$ that $\{R(t)\}_{t\in\mathbb{Z}}$ is a solution of the equation

$$
R(t) - \widetilde{F}(t)R(t-1)\widetilde{F}^*(t) = \begin{bmatrix} \mathbf{u}^*(t) \\ 0 \end{bmatrix} [\, \mathbf{u}(t) \quad d(t) \quad 0 \,],
$$

which is a non-Hermitian time-variant displacement equation with $\widetilde{F}(t) = F(t) \oplus I_n$. Although $R(t)$ is non-Hermitian, its leading $n \times n$ is Hermitian (equal to $\Phi(t)$) and hence, we obtain the following result as a consequence of Theorem 2.2.

5.6 Corollary   *The solution of the normal equations (5.2) can be recursively updated as follows: form the prearray of numbers on the left hand-side of the equation*

$$
\begin{bmatrix} \sqrt{\lambda}\Phi^{1/2}(t-1) & \mathbf{u}^*(t) \\ \sqrt{\lambda}\vartheta^*(t-1)\Phi^{-1/2}(t-1) & d(t) \\ \frac{1}{\sqrt{\lambda}}\Phi^{-1/2}(t-1) & 0 \end{bmatrix} \Theta(t) = \begin{bmatrix} \Phi^{1/2}(t) & 0 \\ \vartheta^*(t)\Phi^{-1/2}(t) & \bar{e}(t) \\ \Phi^{-1/2}(t) & \Delta\mathbf{w}(t) \end{bmatrix};
$$

choose any unitary matrix $\Theta(t)$ that produces the zero block in the postarray; the other entries in the postarray can then be shown to be the quantities needed for the next time-instant plus the entries denoted by $\bar{e}(t)$ and $\Delta\mathbf{w}(t)$ which are used in updating the solution, $\mathbf{w}(t) = \mathbf{w}(t-1) - \Delta\mathbf{w}(t)\bar{e}^*(t)$. □

4.6 Notes

The concept of displacement structure was formalized in [KKM] and since then an important number of applications appeared. For the history and evolution of this subject see [Ka3-5]. Among the applications, we mention here inverse scattering [BK], root distribution of polynomials [LBK], triangular factorization [LAK3-4], modeling of nonstationary processes [Lev], [LAK1-2], adaptive filtering [SLK], [SK1-2]. The recent thesis [Say] contains an overview of old and new in displacement structure theory. The results mentioned in Example 1.1 appear in [LAK2]. The Gohberg-Semencul formula appears in [GS]. Theorem 1.5 is noted in [GDKDM]. This is a dilation theoretic result, as explained in Remark 1.6. Even more, it is not only that the elementary rotation of a contraction is a particular case of that construction, but, in fact, the equality (1.7) is a consequence of the existence of an elementary rotation for any bounded operator on an indefinite metric space (according to [ACG]). In the proof of Theorem 4.4 we have seen that Theorem 1.5 plays a key role in exploring the interpolation properties encoded in the state-space representation. This idea is used systematically in [BGR], [Gl1], [GLDKS], and the existence result about J-conjugation in [Ki3] amounts to the same thing. Applications are also indicated in [BaGK1], [LAK4], [Say], [SCK], [SLK], [Ve]. Theorem 2.2 is the key result in the study of matrices with displacement structure—see [Ka3-5], [KKM], [Lev], [Say], [SCK], [SLK]. For transmission-line models, see [BK], [FoF], [HM], [IS], [KB], [MG], [Re], [Va], [YJN], [Wo]. The role of the generalized Schur algorithm in emphasizing the connection between displacement structure equations and interpolation was noticed in [CSK1], [Say], [SCK], [SKLC]. The first paper dealing with time-varying interpolation is [Arv]. Theorem 4.7 was noticed in several places, [CSK1], [DeD3], [Ve] (in the last two references the language of time-varying Z transform introduced in [ADD2] is systematically used). In [CSK1], it is pointed out that the result itself is a simple consequence of the commutant lifting theorem (for time-invariant case, this is shown in [Ros]), therefore the connection between the two methods of dealing with completion problems is made explicit. For the parametrization in Remark 4.9, we followed [BGR] and [BaGK1]. In fact, we can note a rather general picture of the connections between most of the methods used in interpolation. We have seen that the commutant lifting can be dealt with in the framework of extending families of isometries analyzed in Chapter 3. Theorem 4.8 shows that Theorem 4.4 is a consequence of the commutant lifting theorem. The usefulness of the band method developed in [DG1] comes from the remark that most of the interpolation problems described in Chapter 3 (including the Nehari problem) can be reduced to the band completion problem. In its turn, this can be also solved using Theorem 4.4 and a slight extension of the construction in Example 1.2 (see [SCK] for details). We already noted the connection between displacement structures, the J-conjugation approach in [Ki1-3] and the state-space method employed in [Gl1] and [BaGK]. Formula (4.14) is very close to explain the connection between displacement structure and Potapov's factorization theory (see [GDKDM] and [LAK4] for more details), and some continuous time displacement equations were also used in interpolation in [Sa1-2]. Finally, the connection between displacement equations and the reproducing kernel space approach to interpolation is explained in [Dy2]. Problem 5.1 is addressed in [DeD3] and [SCK] - see also [Ko]. The time-varying Nevanlinna-Pick is addressed in [BG1], [BaGK1-2], [CSK1], [Dew2], [DeD3], [SCK], [Ve]. The application of Theorem 4.7 to the model validation problem was indicated in [SCK]. The problem itself is discussed in [PKTKN], [SD]. The strong Parrott problem was addressed in [FT1] (some other solutions were indicated later, for instance in [BW1]). The connection with Theorem 4.7 was shown in [CSK1] and [SCK]. For a detailed discussion about Corollary 5.6, see [SLK]. In connection with this chapter see also: [Ak], [AK], [AD], [ACF], [BC2], [Bo], [Br1], [Bu1-2], [ByL], [BLM], [Cl], [Co2], [Cy], [DM], [DI], [DGK2], [DD], [Dy1], [DM], [Fo], [FoF1], [GeK], [GKKL], [GKO], [Hay], [HM], [KO], [Levi], [LY], [Nel], [Po], [Sa1-2], [Sc].

Chapter 5

Factorization of Positive Definite Kernels

In this chapter we extend the Cholesky factorization for positive block matrices to the case of positive definite kernels. The basic construction exploits the Kolmogorov decomposition and the role of the Schur parameters is also emphasized. Several particular cases are discussed in some details. The connections between Szegö polynomials and factorization are presented in Section 3, together with the asymptotic properties of the orthogonal polynomials and properties of the associated Schur parameters. In the last section we explore the connection between the spectral factorization and the maximum entropy principle.

5.1 Spectral Factors

In this section we present a general result concerning the factorization of positive definite kernels. We already discussed in Section 1.5 the Cholesky factorization for positive block matrices. Apparently unrelated, there is the following classical result of F. Riesz and L. Fejér.

1.1 Theorem *Every positive trigonometric polynomial $P = \sum_{k=0}^{n}(a_k\nu_{-k} + a_k^*\nu_k)$, $a_n \neq 0$, admits the factorization $P = |Q|^2$, where Q is a polynomial of degree n which has no zero in the unit disc.*

Proof Remark that $\nu_n P$ extends to a polynomial of degree $2n$ and, for $|\zeta| = 1$,

$$P(\zeta) = a_n\zeta^{-n}\prod_{k=1}^{p}(\zeta - z_k)(\zeta - \frac{1}{z_k^*})\prod_{j=1}^{m}(\zeta - \xi_j),$$

where $|z_k| > 1$ for $k = 1, 2, \ldots p$, $|\xi_j| = 1$ for $j = 1, 2, \ldots m$ and $2p + m = 2n$. Set $2s = 2n - 2p = m$, then

$$P(\zeta) = A|\prod_{k=1}^{p}(\zeta - z_k)|^2|\prod_{j=1}^{s}(\zeta - \xi_j)|^2$$

for a certain constant $A \geq 0$ and the polynomial

$$Q(z) = A^{1/2} \prod_{k=1}^{p} (\zeta - z_k) \prod_{j=1}^{s} (\zeta - \xi_j)$$

has the required properties.   □

It turns out that a factorization result for positive definite kernels can be formulated in such a way to contain both the Cholesky factorization and Theorem 1.1 as particular cases. The proof of this result exploits the connection between the Cholesky factorization and the Kolmogorov decomposition already emphasized in Example 1.6.5.

In fact, we derive first a generalization of the upper-lower triangular factorization (1.5.15). Consider two families $\mathbf{H} = \{\mathcal{H}_n\}_{n \in \mathbb{Z}}$ and $\mathbf{L} = \{\mathcal{L}_n\}_{n \in \mathbb{Z}}$ of Hilbert spaces. By a *lower triangular array* we mean here an array G with elements $G_{ij} = G(i,j)$ in $\mathcal{L}(\mathcal{H}_j, \mathcal{L}_i)$ such that $G_{ij} = 0$ for $i < j$ and for all $i \in \mathbb{Z}$, $col_i G$ belongs to $\mathcal{L}(\mathcal{H}_i, \oplus_{k \in \mathbb{Z}} \mathcal{L}_k)$ ($col_i G$ denotes the i-th column of the array G). We denote by $\bar{\mathcal{H}}^2(\mathbf{H}, \mathbf{L})$ the set of the lower triangular arrays. Moreover, a lower triangular array G is called *outer* if

$$\bigvee_{i \geq k} (col_i G) \mathcal{H}_i = \oplus_{i \geq k} \mathcal{L}_i \tag{1.1}$$

for all $k \in \mathbb{Z}$. If G is an element in $\bar{\mathcal{H}}^2(\mathbf{H}, \mathbf{L})$, then we define a kernel A_G by the formula:

$$A_G(i,j) = (col_i G)^* (col_j G) \tag{1.2}$$

and it is easy to see that

$$\sum_{i,j \in \mathbb{Z}} \langle A_G(i,j) h_j, h_i \rangle = \sum_{i,j \in \mathbb{Z}} \langle (col_j G) h_j, (col_i G) h_i \rangle$$

$$= \sum_{i \in \mathbb{Z}} \| (col_i G) h_i \|^2 \geq 0$$

for all sequences $\{h_n\}_{n \in \mathbb{Z}}$ in $\oplus_{n \in \mathbb{Z}} \mathcal{H}_n$ with finite support, therefore A_G is a positive definite kernel. An order is introduced on the set of positive definite kernels by the rule: $A_1 \leq A_2$ if $A_2 - A_1$ is a positive definite kernel.

1.2 Theorem   *Let A be a positive definite kernel. Then there exist a family $\mathbf{L}$ of Hilbert spaces and an outer array G_A in $\bar{\mathcal{H}}^2(\mathbf{H}, \mathbf{L})$, such that*

(a) $A_{G_A} \leq A$.

(b)   *If there is another family $\mathbf{L}'$ of Hilbert spaces and another lower triangular array G' in $\bar{\mathcal{H}}^2(\mathbf{H}, \mathbf{L}')$ satisfying $A_{G'} \leq A$, then $A_{G'} \leq A_{G_A}$.*

Proof   Let A be a positive definite kernel and suppose, without loss of generality, that $A_{nn} = I_{\mathcal{H}_n}$ for all $n \in \mathbb{Z}$. Consider $\{\Gamma_{ij} \mid i,j \in \mathbb{Z}, \ i \leq j\}$ the set of the Schur parameters of A and let V be the Kolmogorov decomposition associated to A by Theorem 1.6.1. For $n \in \mathbb{Z}$, consider the Hilbert spaces

$$\mathcal{K}_n^+ = \mathcal{H}_n \oplus \mathcal{D}_n, \qquad (1.3)$$

and the isometries

$$W_n^+ : \mathcal{K}_{n+1}^+ \longrightarrow \mathcal{K}_n^+ \qquad (1.4)$$
$$W_n^+ = W_n / \mathcal{K}_{n+1}^+,$$

where the operators W_n are defined by the formula (1.6.3). Taking into account the Wold-von Neumann decomposition of the family $\{W_n^+\}_{n \in \mathbb{Z}}$ of isometries, we introduce the spaces $\mathcal{L}_n = \mathcal{K}_n^+ \ominus W_n^+ \mathcal{K}_{n+1}^+$, $n \in \mathbb{Z}$, and due to the structure of the operators W_n, it follows that $\mathcal{L}_n = W_n(\ldots \oplus 0 \oplus \mathcal{D}_{n,*} \oplus 0_{\mathcal{H}_{n+1}} \oplus \ldots)$. Define $\mathbf{L} = \{\mathcal{L}_n\}_{n \in \mathbb{Z}}$ and the array $G_A = [G_{ij}]_{i,j \in \mathbb{Z}}$ is introduced by the following formula:

$$(G_A)_{ij} = \begin{cases} P_{\mathcal{L}_i}^{\mathcal{K}_i} W_{i-1}^* \ldots W_j^* / \mathcal{H}_j & \text{if } j < i \\ P_{\mathcal{L}_i}^{\mathcal{K}_i} / \mathcal{H}_i & \text{if } j = i \\ 0 & \text{if } j > i. \end{cases} \qquad (1.5)$$

Denote by P_n the orthogonal projection of $\mathcal{K}_n$ onto the space

$$\ldots \oplus W_{n-1}^* \mathcal{L}_{n-1} \oplus \mathcal{L}_n \oplus W_n \mathcal{L}_{n+1} \oplus W_n W_{n+1} \mathcal{L}_{n+2} \oplus \ldots$$

and, since

$$\mathcal{K}_n = (\ldots \oplus W_{n-1}^* \mathcal{L}_{n-1} \oplus \mathcal{L}_n \oplus W_n \mathcal{L}_{n+1} \oplus W_n W_{n+1} \mathcal{L}_{n+2} \oplus \ldots) \oplus \mathcal{R}_n, \quad (1.6)$$

it follows that $W_{n-1} P_n = P_{n-1} W_{n-1}$. Taking into account the structure of W_n, one infers that $P_n / \mathcal{H}_n = (P_{\mathcal{L}_n} + W_n P_{\mathcal{L}_{n+1}} W_n^* + \ldots) / \mathcal{H}_n$ for all $n \in \mathbb{Z}$. Consequently,

$$(col_i G_A)^* (col_j G_A) = P_{\mathcal{H}_i}^{\mathcal{K}_i} W_i \ldots W_{j-1} P_j / \mathcal{H}_j = P_{\mathcal{H}_i}^{\mathcal{K}_i} P_i W_i \ldots W_{j-1} / \mathcal{H}_j$$

for $i \leq j$. It follows that $col_i G_A$ are contractions for $i \in \mathbb{Z}$ and it is easy to check that G_A is outer, hence G_A is an outer lower triangular array in $\bar{\mathcal{H}}^2(\mathbf{H}, \mathbf{L})$.

Then, for a sequence $\{h_n\}_{n \in \mathbb{Z}}$ in $\oplus_{n \in \mathbb{Z}} \mathcal{H}_n$ with finite support,

$$\sum_{i,j \in \mathbb{Z}} \langle A_{G_A}(i,j) h_j, h_i \rangle = \sum_{i \leq j} \langle P_{\mathcal{H}_i}^{\mathcal{K}_i} P_i W_i \ldots W_{j-1} h_j, h_i \rangle$$

$$+ \sum_{j < i} \langle h_j, P_{\mathcal{H}_j}^{\mathcal{K}_j} P_j W_j \ldots W_{i-1} h_i \rangle$$

$$\leq \sum_{i \leq j} \langle P_{\mathcal{H}_i}^{\mathcal{K}_i} W_i \ldots W_{j-1} h_j, h_i \rangle$$

$$+ \sum_{j < i} \langle h_j, P_{\mathcal{H}_j}^{\mathcal{K}_j} W_j \ldots W_{i-1} h_i \rangle$$

$$= \sum_{i,j \in \mathbb{Z}} \langle A(i,j) h_j, h_i \rangle,$$

which means that $A_{G_A} \leq A$. The property (a) is proven for G_A.

If G' in $\bar{\mathcal{H}}^2(\mathbf{H}, \mathbf{L}')$ is another array such that $A_{G'} \leq A$, then it follows by the minimality property of the Kolmogorov decomposition that $\mathcal{K}_n^+ = \mathcal{H}_n \vee \vee_{k=1}^{\infty} W_n W_{n+1} \ldots W_{n+k} \mathcal{H}_{n+k+1}$ for all $n \in \mathbb{Z}$. This suggests to define the map

$$X_n : \mathcal{K}_n^+ \longrightarrow \oplus_{k \geq n} \mathcal{L}'_k,$$

$$X_n(\sum_{k=n}^{m} W_n \ldots W_{k-1} h_k) = \sum_{k=n}^{m} (col_k G') h_k,$$

where $m \geq n$ is an arbitrary integer. Since

$$\| X_n(\sum_{k=n}^{m} W_n \ldots W_{k-1} h_k) \|^2 = \sum_{j,k=n}^{m} \langle A_{G'}(j,k) h_k, h_j \rangle$$

$$\leq \sum_{j,k=n}^{m} \langle A(j,k) h_k, h_j \rangle$$

$$= \| \sum_{k=n}^{m} W_n \ldots W_{k-1} h_k \|^2,$$

it follows that the map X_n extends to a contraction, also denoted by X_n, from $\mathcal{K}_n^+$ into the space $\oplus_{k \geq n} \mathcal{L}'_k$. Moreover, it is remarked that

$$X_n \mathcal{R}_n = X_n(\cap_{p \geq n} W_n \ldots W_{p-1} \mathcal{K}_p^+)$$

$$\subseteq \cap_{p \geq n} X_n W_n \ldots W_{p-1} \mathcal{K}_p^+$$

$$\subseteq \cap_{p \geq n} (col_p G')(\oplus_{q \geq n} \mathcal{L}'_q) = 0,$$

which implies that $X_n P_n = X_n$. Using this relation, it follows that for $m \leq n$ and h_k in $\mathcal{H}_k$, $m \leq k \leq n$,

$$\sum_{j,k=n}^{m} \langle A_{G'}(j,k) h_k, h_j \rangle = \| \sum_{k=m}^{n} (col_k G') h_k \|^2$$

$$= \| X_m(\sum_{k=m}^{n} W_m \ldots W_{k-1} h_k) \|^2$$

$$= \| X_m P_m(\sum_{k=m}^{n} W_m \ldots W_{k-1} h_k) \|^2$$

$$\leq \| P_m(\sum_{k=m}^{n} W_m \ldots W_{k-1} h_k) \|^2$$

$$= \sum_{j,k=m}^{n} \langle A_{G_A}(j,k) h_k, h_j \rangle$$

i.e. $A_{G'} \leq A_{G_A}$. The property (*b*) of G_A is proven. □

1.3 Remark It is easy now to derive the lower/upper triangular factorization of the positive definite kernel A. For a family $\mathbf{H} = \{\mathcal{H}_n\}_{n \in \mathbb{Z}}$ of Hilbert spaces, define the unitary operator

$$\widetilde{\mathcal{I}} : \oplus_{n \in \mathbb{Z}} \mathcal{H}_n \longrightarrow \oplus_{n \in \mathbb{Z}} \mathcal{H}_n$$

$$\widetilde{\mathcal{I}}(\oplus_{n \in \mathbb{Z}} h_n) = \oplus_{n \in \mathbb{Z}} h_{-n}, \quad \oplus_{n \in \mathbb{Z}} h_n \in \oplus_{n \in \mathbb{Z}} \mathcal{H}_n,$$

which plays the same role as $\mathcal{I}$ defined by (1.5.14). Suppose $A = [A_{ij}]_{n \in \mathbb{Z}}$ is a positive definite kernel such that A_{ij} belongs to $\mathcal{L}(\mathcal{H}_j, \mathcal{H}_i)$. Then, for $m \leq n$, define $B^{(mn)} = \mathcal{I} A^{(mn)} \mathcal{I} = [B_{ij}]_{m \leq i,j \leq n}$, where $A^{(mn)} = [A_{ij}]_{m \leq i,j \leq n}$. It follows that $B = [B_{ij}]_{i,j \in \mathbb{Z}}$ is a positive definite kernel and let G_B in $\bar{\mathcal{H}}^2(\mathbf{H}, \widetilde{\mathbf{L}})$ be the outer lower triangular array associated to B according to the construction in Theorem 1.2. Then $F_A = \widetilde{\mathcal{I}} G_B \widetilde{\mathcal{I}}$ is an upper triangular array such that $col_i F_A$ belongs to $\mathcal{L}(\mathcal{H}_i, \oplus_{k \in \mathbb{Z}} \widetilde{\mathcal{L}}_k)$ for all $i \in \mathbb{Z}$ and $\vee_{i \leq k} (col_i F_A) \mathcal{H}_i = \oplus_{i \leq k} \widetilde{\mathcal{L}}_i$ for all $k \in \mathbb{Z}$ (we say that F_A is an *upper triangular outer array* and denote by $\mathcal{H}^2(\mathbf{H}, \widetilde{\mathbf{L}})$ the set of all upper triangular arrays). Besides, for arbitrary integers m, n such that $m \leq n$ and vectors h_i in $\mathcal{H}_i$, $m \leq i \leq n$, one obtains that

$$\sum_{i,j=m}^{n} \langle A_{F_A}(i,j) h_j, h_i \rangle = \sum_{i,j=m}^{n} \langle (col_i F_A)^* (col_j F_A) h_j, h_i \rangle$$

$$= \sum_{i,j=m}^{n} \langle (col_i G_B)^* (col_j G_B) \widetilde{\mathcal{I}} h_j, \widetilde{\mathcal{I}} h_i \rangle$$

$$\leq \sum_{i,j=m}^{n} \langle B(i,j) \widetilde{\mathcal{I}} h_j, \widetilde{\mathcal{I}} h_i \rangle = \langle B^{(mn)} \widetilde{\mathcal{I}} h, \widetilde{\mathcal{I}} h \rangle$$

$$= \langle A^{(mn)} h, h \rangle = \sum_{i,j=m}^{n} \langle A(i,j) h_j, h_i \rangle,$$

where $h = [h_m \quad h_{m+1} \quad \ldots \quad h_n]^\top$. Consequently, $A_{F_A} \leq A$ and one can see that for any other array F' in $\mathcal{H}^2(\mathbf{H}, \mathbf{L}')$ satisfying $A_{F'} \leq A$, it follows that $A_{F'} \leq A_{F_A}$. □

In the remainder of this section, we discuss two properties of the array G_A. We begin by showing that G_A is uniquely determined up to a left unitary diagonal factor. Let G' and G'' be two lower triangular arrays, G' in $\bar{\mathcal{H}}^2(\mathbf{H}, \mathbf{L}')$ and G'' in $\bar{\mathcal{H}}^2(\mathbf{L}', \mathbf{L})$, where $\mathbf{H}$, $\mathbf{L}$ and $\mathbf{L}'$ are families of Hilbert spaces. In addition, suppose that G'' is a contraction from $\oplus_{n \in \mathbb{Z}} \mathcal{L}'_n$ to $\oplus_{n \in \mathbb{Z}} \mathcal{L}_n$. In this case, we remark that the product $G'' G'$ makes sense as a lower triangular array in $\bar{\mathcal{H}}^2(\mathbf{H}, \mathbf{L})$. Note now the following result, which is an extension of Lemma 1.4.1.

1.4 Proposition *Let G and G' be two lower triangular arrays, G arbitrary in $\bar{\mathcal{H}}^2(\mathbf{H}, \mathbf{L})$ and G' outer in $\bar{\mathcal{H}}^2(\mathbf{H}, \mathbf{L}')$. If $A_G \leq A_{G'}$, then there exists a lower* triangular contraction G'' in $\bar{\mathcal{H}}^2(\mathbf{L}',\mathbf{L})$ such that $G = G''G'$. Furthermore, if $A_G = A_{G'}$, then G'' is an isometry. If, in addition, G is also outer, then G'' is a unitary diagonal operator.

Proof For $i \in \mathbb{Z}$ and arbitrary $n \geq i$, define $Y_i(\sum_{k=i}^n (col_k G')h_k) = \sum_{k=i}^n (col_k G)h_k$. Then,

$$\|Y_i(\sum_{k=i}^n (col_k G')h_k)\|^2 = \sum_{j,k=i}^n \langle A_G(j,k)h_k, h_j \rangle \leq$$

$$\leq \sum_{j,k=i}^n \langle A'_G(j,k)h_k, h_j \rangle = \|\sum_{k=i}^n (col_k G')h_k\|^2.$$

Since G' is outer, it follows that Y_i extends to a contraction, also denoted by Y_i, from $\oplus_{k\geq i}\mathcal{L}'_k$ into $\oplus_{k\geq i}\mathcal{L}_k$. Moreover, we remark that for all $i \in \mathbb{Z}$, $M_{\mathbf{L}}(i)Y_{i+1} = Y_i M_{\mathbf{L}'}(i)$, where $M_{\mathbf{L}}$ and $M_{\mathbf{L}'}$ are the marking operators associated to the families $\mathbf{L}$ and $\mathbf{L}'$ by the formula (2.4.1). It follows that the family $\{Y_n\}_{n\in\mathbb{Z}}$ induces a lower triangular contraction G'' from $\oplus_{n\in\mathbb{Z}}\mathcal{L}'_n$ into $\oplus_{n\in\mathbb{Z}}\mathcal{L}_n$ such that $G = G''G'$. If $A_G = A_{G'}$, then the operators Y_n are isometries for all $n \in \mathbb{Z}$, hence G'' is an isometry. Finally, if it supposed in addition that G is outer, then Y_n are unitary operators and G'' must be unitary diagonal. □

1.5 Theorem *The array G_A associated to the positive definite kernel A in Theorem 1.2 is uniquely determined up to a left unitary diagonal factor.*

Proof Let G be another outer array satisfying (a) and (b) in Theorem 1.2. It follows that $A_G = A_{G'}$ and then the uniqueness property of G_A is a consequence of Proposition 1.4. □

Hereby, the outer factor G_A associated to the positive definite kernel A by Theorem 1.2 will be referred to as the *lower triangular spectral factor* of A. The dual object F_A introduced in Remark 1.3, will be referred to as the *upper triangular spectral factor* of A. According to Theorem 1.5, G_A is uniquely determined up to a left unitary diagonal factor. As a first consequence of the existence of the spectral factor, let us note here the following *inner-outer factorization* for arbitrary arrays in $\bar{\mathcal{H}}^2(\mathbf{H},\mathbf{H}')$.

1.6 Proposition *Every array G in $\bar{\mathcal{H}}^2(\mathbf{H},\mathbf{H}')$ admits a uniquely determined (up to a unitary diagonal factor) factorization $G = G_iG_o$, where G_o in $\bar{\mathcal{H}}^2(\mathbf{H},\mathbf{H}'')$ is an outer array and G_i in $\bar{\mathcal{H}}^2(\mathbf{H}'',\mathbf{H}')$ is isometric.*

Proof Let A_G be the positive definite kernel associated to G and let G_o in $\bar{\mathcal{H}}^2(\mathbf{H},\mathbf{H}'')$ be the spectral factor of A_G. By the property (a) of G_o, $A_{G_o} \leq A_G$ and by the property (b) of G_o, it follows that $A_G \leq A_{G_o}$. Hence, $A_{G_o} = A_G$ and by Theorem 1.5, there exists an isometric array G_i in $\bar{\mathcal{H}}^2(\mathbf{H}'',\mathbf{H}')$, such that $G = G_iG_o$. Consider $G = G'_iG'_o$ another factorization of G with G'_o an outer array and G'_i an isometric one. Due to the properties of G_o, we get $A_{G'_o} = A_{G_o} = A_G$. By Theorem 1.5, $G'_o = UG_o$, where U is a unitary diagonal array and then $G'_i = G_iU^*$. □

Finally, we mention that there is of interest to know when the equality occurs in Theorem 1.2(a), in which case we say that the positive definite kernel A is *factorable*. A first answer can be easily subtracted from the proof of Theorem 1.2.

1.7 Proposition   *The positive definite kernel A is factorable if and only if all the residual spaces $\mathcal{R}_n$ in the Wold-von Neumann decomposition of the family $\{W_n^+\}_{n\in\mathbb{Z}}$ of isometries defined by (1.4) are trivial.*   □

This criterion is quite intricate. In the next section we discuss a few examples when the factorability can be easier described.

5.2 Examples

The purpose of this section is to describe in some details several particular cases of Theorem 1.2. We begin with the factorization of the semispectral measures on the unit circle.

Let $\mathcal{H}$ be a separable Hilbert space. A *semispectral measure* μ on the unit circle is a linear positive map from $C(\mathbb{T})$ to $\mathcal{L}(\mathcal{H})$. When $\mathcal{H} = \mathbb{C}$, a semispectral measure is simply a positive measure on the unit circle. The *Fourier coefficients* of μ are the operators

$$A_n = \mu(\nu_{-n}), \quad n \in \mathbb{Z}, \tag{2.1}$$

where $\{\nu_n\}_{n\in\mathbb{Z}}$ is the standard orthonormal basis in L^2 introduced in Example 1.1.4. A computation similar to the one in Example 1.3.3 shows that the map A from $\mathbb{Z} \times \mathbb{Z}$ to $\mathcal{L}(\mathcal{H})$ defined by $A(i,j) = A_{j-i}$ for $i,j \in \mathbb{Z}$, is a positive definite Toeplitz kernel. Conversely, given a positive definite Toeplitz kernel, we can define a semispectral measure μ by the formula $\mu(f) = Q^*E(S)Q$ for f in $C(\mathbb{T})$, where the unitary operator S on $\mathcal{H}_A$ and the operator Q in $\mathcal{L}(\mathcal{H}, \mathcal{H}_A)$ are those introduced in Theorem 1.3.2, while E is the spectral measure of the unitary operator S. We also remark that $\mu_1 \leq \mu_2$ if and only if $\mu_1(f) \leq \mu_2(f)$ for every f in $C(\mathbb{T})$. Then, consider the function F in $H^2(\mathcal{L}(\mathcal{H}, \mathcal{H}'))$, where $\mathcal{H}$, $\mathcal{H}'$ are separable Hilbert spaces and the set of the Taylor coefficients of F is $\{F_n\}_{n\geq 0}$. We define the following map:

$$V_F : \mathcal{H} \longrightarrow L^2(\mathcal{H}') \tag{2.2}$$

$$(V_F h)(e^{it}) = \sum_{n=0}^{\infty} e^{int} F_n h, \quad h \in \mathcal{H},$$

and remark that since F belongs to $H^2(\mathcal{L}(\mathcal{H}, \mathcal{H}'))$, then $\|V_F h\|^2_{L^2(\mathcal{H}')} \leq M\|h\|^2$. Consequently, V_F is a bounded operator. Obviously, $\mathcal{R}(V_F) \subset L^2_+(\mathcal{H}')$. The function F is called *outer* if

$$\bigvee_{n=0}^{\infty} T_{\nu_n} V_F \mathcal{H} = L^2_+(\mathcal{H}'). \tag{2.3}$$

Finally, we associate to the function F the semispectral measure μ_F defined by $\mu_F(f) = V_F^* E_{\nu_1}(f) V_F$ for f in $C(\mathbb{T})$, where E_{ν_1} denotes the spectral measure of the unitary operator M_{ν_1}. We can notice now the following result concerning the factorization of semispectral measures.

2.1 Theorem   Let μ be a semispectral measure. Then there exist a separable Hilbert space $\widetilde{\mathcal{L}}$ and an outer function F_μ in $\mathcal{H}^2(\mathcal{L}(\mathcal{H}, \widetilde{\mathcal{L}}))$, such that

(a) $\mu_{F_\mu} \leq \mu$.

(b) If there is another Hilbert space $\mathcal{L}'$ and another function F' in $\mathcal{H}^2(\mathcal{H}, \mathcal{L}')$ satisfying $\mu_{F'} \leq \mu$, then $\mu_{F'} \leq \mu_{F_\mu}$.

Proof Let μ be a semispectral measure and suppose, without loss of generality, that $\mu(\nu_0) = I$. Let A be the positive definite Toeplitz kernel associated to the Fourier coefficients of μ, which means that $A(i, j) = A_{j-i}$ for $i, j \in \mathbb{Z}$, where A_n, $n \in \mathbb{Z}$, are given by (2.1). Consider the family $\mathbf{H} = \{\mathcal{H}_n\}_{n \in \mathbb{Z}}$ with $\mathcal{H}_n = \mathcal{H}$ for all $n \in \mathbb{Z}$ and let F_A in $\mathcal{H}^2(\mathbf{H}, \widetilde{\mathbf{L}})$ be the upper triangular spectral factor of A. By the constructions in Theorem 1.2 and Remark 1.3, we can easily see that $\widetilde{\mathcal{L}}_n = \widetilde{\mathcal{L}}$ for all $n \in \mathbb{Z}$ and that F_A is a Toeplitz upper triangular array. The fact that F_A belongs to $\mathcal{H}^2(\mathbf{H}, \widetilde{\mathbf{L}})$ shows that $F_\mu = \sum_{n=0}^\infty \nu_n F_n$ belongs to $\mathcal{H}^2(\mathcal{L}(\mathcal{H}, \widetilde{\mathcal{L}}))$. It is then readily seen that F_μ is an outer element in $\mathcal{H}^2(\mathcal{L}(\mathcal{H}, \widetilde{\mathcal{L}}))$ (in the sense of (2.3)) and that μ_{F_μ} has the required properties.

Finally, let us mention that the construction in Theorem 1.2 provides an explicit form of F_μ. If B denotes the positive definite Toeplitz kernel associated to A by the construction in Remark 1.3, then it is obvious that B is the matrix transpose of A, i.e. $B(i, j) = A^*(i, j)$. Let $\mu^\top$ be the semispectral measure associated to B. If $\{\Gamma_n\}_{n \geq 0}$ are the Schur parameters of μ then the Schur parameters of $\mu^\top$ are exactly $\{\Gamma_n^*\}_{n \geq 0}$. By the construction in the proof of Theorem 1.2,

$$F_{\mu^\top}(z) = P_{\mathcal{D}_*}(I - zW^*)^{-1}W^*/\mathcal{H}, \qquad |z| < 1, \tag{2.4}$$

where W is the Naimark dilation of μ, as described in Theorem 1.6.1, and

$$\mathcal{D}_* = \mathrm{cl}\mathcal{R}(s\text{-}\lim_{n \to \infty}(D_{\Gamma_1^*} \ldots D_{\Gamma_{n-1}^*} D_{\Gamma_n^*}^2 D_{\Gamma_{n-1}^*} \ldots D_{\Gamma_1^*})). \ \square \tag{2.5}$$

The above result can be further specialized. Let N be a function defined on $[0, 2\pi)$ with values in $\mathcal{L}(\mathcal{H})$, where $\mathcal{H}$ is a separable Hilbert space. Due to the separability of $\mathcal{H}$, it is equivalent that N is strongly or weakly measurable.

2.2 Theorem   Let N be a measurable function on $[0, 2\pi)$ with values positive contractions. Then there exist a separable Hilbert space $\mathcal{L}$ and an outer Schur function F_N in $\mathcal{S}(\mathcal{L}(\mathcal{H}, \mathcal{L}))$, such that

(a) $F_N^*(e^{it})F_N(e^{it}) \leq N^2(t)$ a.e. on $[0, 2\pi)$;

(b) If there exist another Hilbert space $\mathcal{L}'$ and another function F' in $\mathcal{S}(\mathcal{L}(\mathcal{H}, \mathcal{L}'))$ satisfying $F'^*(e^{it})F'(e^{it}) \leq N^2(t)$ a.e. on $[0, 2\pi)$, then $F'^*(e^{it})F'(e^{it}) \leq F_N^*(e^{it})F_N(e^{it})$ a.e. on $[0, 2\pi)$.

Proof Define the semispectral measure

$$\mu_N(f) = \frac{1}{2\pi} \int_0^{2\pi} f(e^{it})N^2(t)dt, \tag{2.6}$$

for f in $C(\mathbb{T})$, and the proof can be concluded by an application of Theorem 2.1.

□

2.3 Remark Let L^1 be the Lebesgue space of the Lebesgue integrable functions on $[0, 2\pi)$ and let f be a positive function in L^1. Suppose that $\log f$ belongs to L^1 and define the *Szegö function*

$$F(z) = \exp(\frac{1}{4\pi} \int_0^{2\pi} \frac{e^{it} + z}{e^{it} - z} \log f(t) dt). \qquad (2.7)$$

The integral is well defined by Jensen's inequality and F is an analytic outer function in the Hardy space H^2. By Fatou's theorem, $f(t) = |F(e^{it})|^2$ a.e. on $[0, 2\pi)$, hence F is the spectral factor of $N = f^{1/2}$. The same construction can be used to produce the spectral factor of a positive measure on the unit circle with the property that $\log \mu'$ belongs to L^1, where μ' denotes the Radon-Nicodym derivative of the measure μ. In this case, we define the *Szegö function*

$$F_\mu(z) = \exp(\frac{1}{4\pi} \int_0^{2\pi} \frac{e^{it} + z}{e^{it} - z} \log \mu'(t) dt), \qquad (2.8)$$

which is an outer function with the property that $|F_\mu(e^{it})|^2 = \mu'(t)$ a.e., hence $|F_\mu|^2 \times$ (Lebesgue measure) $\leq \mu$. Let F be another outer function with the property that $|F|^2 \times$ (Lebesgue measure) $\leq \mu$, then

$$\frac{1}{4\pi} \int_0^{2\pi} \operatorname{Re} \frac{e^{it} + z}{e^{it} - z} |F(e^{it})|^2 dt \leq \frac{1}{4\pi} \int_0^{2\pi} \operatorname{Re} \frac{e^{it} + z}{e^{it} - z} d\mu(t)$$

and by Fatou's theorem, $|F(e^{it})|^2 \leq \mu'(t) = |F_\mu(e^{it})|^2$ a.e. Hence F_μ can be chosen the spectral factor of μ.

□

2.4 Remark The factorability criterion obtained in Proposition 1.7 can be written in a more tractable form for the functions considered in Theorem 2.2. Thus, let N be a measurable function on $[0, 2\pi)$ with values in $\mathcal{L}(\mathcal{H})$ such that $0 \leq N(t) \leq I$ for $t \in [0, 2\pi)$. Consider the semispectral measure introduced by the formula (2.6). Moreover, define the bounded operator

$$Q' : \mathcal{H} \longrightarrow L^2(\mathcal{H})$$

$$(Q'h)(t) = N(t)h, \quad h \in \mathcal{H}.$$

It is readily checked that the equality $\mu_N(f) = Q'^* E_{\nu_1}(f) Q'$ holds for every f in $C(\mathbb{T})$, where E_{ν_1} is the spectral measure of the unitary operator M_{ν_1} on $L^2(\mathcal{H})$. By Theorem 1.3.2, there exists an isometry Φ from $\mathcal{H}_A$ to $L^2(\mathcal{H})$ such that $\Phi Qh = Q'h$ for h in $\mathcal{H}$. Moreover, $M_{\nu_1} \Phi = \Phi S$, where S is the Naimark dilation of μ_N. We remark that

$$\cap_{n \geq 0} S^n \mathcal{K}_+ = \cap_{n \geq 0} \Phi^* M_{\nu_n} \Phi \mathcal{K}_+$$

and

$$\Phi \mathcal{K}_+ = \Phi(\vee_{n\geq 0} S^n Q\mathcal{H}) = \vee_{n\geq 0}\Phi S^n Q\mathcal{H}$$

$$= \vee_{n\geq 0} M_{\nu_n}\Phi Q\mathcal{H} = \vee_{n\geq 0} M_{\nu_n} Q'\mathcal{H} = \mathrm{cl}(NH^2(\mathcal{H})).$$

Since the space $\vee_{n\geq 0} M_{\nu_n}\Phi\mathcal{K}_+$ is contained in the range of Φ, we obtain that N^2 is factorable if and only if $\cap_{n\geq 0} M_{\nu_n}\mathrm{cl}(NH^2(\mathcal{H})) = 0.$ □

With the criterion established in Remark 2.4 we can handle an important class of factorable functions. Thus, for a separable Hilbert space $\mathcal{H}$ define

$$\mathcal{P}L^\infty(\mathcal{L}(\mathcal{H}))=\{N\in L^\infty(\mathcal{L}(\mathcal{H}))|\chi N^*\in H^\infty(\mathcal{L}(\mathcal{H}))\text{for some scalar inner function}\chi\}.$$

2.5 Theorem   If N is a positive function in $\mathcal{P}L^\infty(\mathcal{L}(\mathcal{H}))$, then N is factorable.

Proof   Let N be a positive function in $\mathcal{P}L^\infty(\mathcal{L}(\mathcal{H}))$ and suppose, without loss of generality, that $0 \leq N \leq I$. Let χ be an inner scalar function with the property that χN belongs to $H^\infty(\mathcal{L}(\mathcal{H}))$. Then, one obtains that

$$\cap_{n\geq 0} M_{\nu_n}\mathrm{cl}(NH^2(\mathcal{H})) = M_\chi^*(\cap_{n\geq 0} M_{\nu_n}\mathrm{cl}(M_\chi NH^2(\mathcal{H})))$$

$$\subset M_\chi^*(\cap_{n\geq 0} M_{\nu_n} H^2(\mathcal{H})) = 0.$$

Therefore N^2 is factorable. But, due to the assumption that $0 \leq N \leq I$, it folllows $N^2 \geq N$. Therefore, there exists a contraction

$$X : \mathrm{cl}(NH^2(\mathcal{H})) \longrightarrow \mathrm{cl}(NH^2(\mathcal{H}))$$

$$XN^{\frac{1}{2}}F = NF, \qquad F \in H^2(\mathcal{H}).$$

Clearly, X is one-to-one and

$$X(\cap_{n\geq 0} M_{\nu_n}\mathrm{cl}(N^{\frac{1}{2}}H^2(\mathcal{H}))) \subset \cap_{n\geq 0} M_{\nu_n}\mathrm{cl}(XN^{\frac{1}{2}}H^2(\mathcal{H}))$$

$$= \cap_{n\geq 0} M_{\nu_n} NH^2(\mathcal{H}) = 0,$$

hence N itself is factorable. □

Let us notice as a consequence of this result the following operatorial version of Theorem 1.1.

2.6 Corollary   Let $P = \sum_{k=0}^{n}(A_k\nu_{-k} + A_k^*\nu_k)$ be a positive trigonometric polynomial with A_k in $\mathcal{L}(\mathcal{H})$, $k = 0, 1, \ldots, n$ and $A_n \neq 0$. Then P admits the factorization $P = Q^*Q$, where $Q(z) = \sum_{k=0}^{n} B_k z^k$ is an analytic trigonometric polynomial and B_k belongs to $\mathcal{L}(\mathcal{H})$ for $k = 0, 1, \ldots, n$.

Proof   Since $\nu_n P$ belongs to $H^\infty(\mathcal{L}(\mathcal{H}))$, it follows that P belongs to $\mathcal{P}L^\infty(\mathcal{L}(\mathcal{H}))$. By Theorem 2.5, there exists an outer function Q in $H^\infty(\mathcal{H}, \mathcal{L})$ such that $P = Q^*Q$. Since $\nu_n P$ belongs to $H^\infty(\mathcal{L}(\mathcal{H}))$, it follows that $\nu_n PH^2(\mathcal{H}) \subset H^2(\mathcal{H})$, or $\nu_n Q^*QH^2(\mathcal{H}) \subset H^2(\mathcal{H})$. But Q is outer, therefore $\mathrm{cl}(QH^2(\mathcal{H})) = H^2(\mathcal{L})$. It follows that $\nu_n Q^*H^2(\mathcal{L}) \subset H^2(\mathcal{H})$, i.e. $\nu_n Q^*$ belongs to $H^\infty(\mathcal{L}, \mathcal{H})$. This means that $Q(z) = \sum_{k=0}^{n} B_k z^k$ and since $\dim H \geq \dim \mathcal{L}$, we can view the operators B_k as lying in $\mathcal{L}(\mathcal{H})$. □

5.3   Schur's Algorithm, Szegö's Theory and Spectral Factors

In this section we discuss in more details the connections between Szegö polynomials and spectral factors.

Let μ be a probability measure on $[0, 2\pi)$ and let $\{\Gamma_n\}_{n \geq 0}$ be its Schur parameters. Let $\{\psi_n\}_{n \geq 0}$ be the monic orthogonal polynomials of μ and $\psi_n(0) = -\Gamma_n^*$ are the Szegö coefficients of μ. We say that μ belongs to the *Szegö class* if $\log \mu'$ belongs to L^1. We obtain the following characterization of the Szegö class in terms of the Schur parameters.

3.1 Proposition   *The positive measure μ belongs to the Szegö class if and only if its Schur parameters satisfy the condition:*

$$\sum_{n=1}^{\infty} |\Gamma_n|^2 < \infty. \tag{3.1}$$

Proof   By Remark 2.3, the spectral factor of μ coincides with the Szegö function of μ defined by

$$F_\mu(z) = \exp\left(\frac{1}{4\pi} \int_0^{2\pi} \frac{e^{it} + z}{e^{it} - z} \log \mu'(t) dt\right).$$

On the other hand, by Theorem 2.1, $F_{\mu^\top}(z) = P_{\mathcal{D}_*} W^*(I - zW^*)^{-1}/\mathbb{C}$. Due to the structure of the Naimark dilation, it follows that $F_{\mu^\top}(0) = \prod_{n=1}^{\infty}(1 - |\Gamma_n|^2)^{\frac{1}{2}}$. Since the Szegö coefficients of $\mu^\top$ are $\{-\Gamma_n\}_{n \geq 0}$, it follows that $F_{\mu^\top}(0) = F_\mu(0)$, or

$$\prod_{n=1}^{\infty}(1 - |\Gamma_n|^2) = \exp\left(\frac{1}{2\pi} \int_0^{2\pi} \log \mu'(t) dt\right). \tag{3.2}$$

Therefore, $\log \mu'$ belongs to L^1 if and only if $\sum_{n=1}^{\infty} |\Gamma_n|^2 < \infty$. $\square$

As a consequence of this result, we can clarify a remark preceding Proposition 1.6.8. We noted that the space $\mathcal{D}_*$ is either $\mathbb{C}$ or 0. Using Proposition 3.1, we can see that $\mathcal{D}_* = \mathbb{C}$ if and only if μ belongs to the Szegö class.

We now show a connection between the probability measures on $[0, 2\pi)$ and the Schur functions. Thus, define the function

$$g(z) = \frac{1}{4\pi} \int_0^{2\pi} \frac{e^{it} + z}{e^{it} - z} d\mu(t) \tag{3.3}$$

which is analytic on the unit disc, with positive real part and then,

$$f(z) = \frac{g(z) - g(0)}{g(z) + g(0)} \tag{3.4}$$

is a Schur function. If $\{\gamma_n\}_{n \geq 0}$ are the Schur parameters of f and $\{\Gamma_n\}_{n \geq 0}$ are the Schur parameters of μ, then for all $n \geq 0$ $\Gamma_n = \gamma_n$ (with the mention that $\Gamma_0 = \gamma_0 = 0$). By Fatou's theorem, we have that

$$\mu'(t) = \operatorname{Re} \frac{1 + f(e^{it})}{1 - f(e^{it})} = \frac{1 - |f(e^{it})|^2}{|1 - f(e^{it})|^2} \quad a.e. \tag{3.5}$$

Since $1 - f$ is outer, it follows from (3.5) that μ belongs to the Szegö class if and only if $\log(1 - |f(e^{it})|^2) \in L^1$. If this is the case, then $1 - |f|^2$ belongs to L^1 as well, hence there exists an outer function F_f in H^1 such that $1 - |g|^2 = |F_f|^2$ a.e. on the unit circle. F_f is the spectral factor of $1 - |f|^2$ and by (3.5), it follows that the spectral factor of μ can be expressed as

$$F_\mu = F_f(1 - f)^{-1}. \tag{3.6}$$

We made this connection with the Schur functions because we remarked in Section 4.2 the role played by the rational functions $\mathcal{A}_n\mathcal{B}_n^{-1}$ in the uniform approximation of the function f ($\mathcal{A}_n$ and $\mathcal{B}_n$ are the Schur polynomials associated to f). Due to the simple interpretation (given by Proposition 4.2.7(b)) of this approximation procedure, we expect that it can be used further. For instance, we have seen in Proposition 4.2.7(b) that the Schur polynomials of $g_n = \mathcal{A}_n\mathcal{B}_n^{-1}$ are

$$\mathcal{A}_k(g_n) = \begin{cases} \mathcal{A}_k & \text{if } k \leq n \\ \mathcal{A}_n & \text{if } k > n \end{cases} \quad \text{and} \quad \mathcal{B}_k(g_n) = \begin{cases} \mathcal{B}_k & \text{if } k \leq n \\ \mathcal{B}_n & \text{if } k > n. \end{cases}$$

Consequently, for $k > n$, $\mathcal{A}_k^\#(g_n) = \nu^{k-n}\mathcal{A}_n^\#$, $\mathcal{B}_k^\#(g_n) = \nu^{k-n}\mathcal{B}_n^\#$, hence

$$\lim_{k \to \infty} \mathcal{A}_k^\#(g_n) = 0 \quad \text{and} \quad \lim_{k \to \infty} \mathcal{B}_k^\#(g_n) = 0$$

uniformly on the compact subsets of the unit disc. We use the symbol "$\Longrightarrow$" to denote this kind of convergence. Let us compute now the spectral factor of $1 - |g_n|^2$. Using formulae established in the proof of Theorem 4.2.6, we obtain for $|z| < 1$, that

$$\exp(\frac{1}{4\pi} \int_0^{2\pi} \frac{e^{it} + z}{e^{it} - z} \log(1 - |g_n(e^{it})|^2)dt) =$$

$$= \exp(\frac{1}{4\pi} \int_0^{2\pi} \frac{e^{it} + z}{e^{it} - z} \log \frac{|\mathcal{B}_n(e^{it})|^2 - |\mathcal{A}_n(e^{it})|^2}{|\mathcal{B}_n(e^{it})|^2} dt) = \prod_{k=1}^n (1 - |\gamma_k|^2)^{1/2} \frac{1}{\mathcal{B}_n(z)}.$$

Since $g_n \Longrightarrow f$, all these remarks suggest that we have the following result.

3.2 Theorem Let f be a Schur function such that $\log(1 - |f|^2)$ belongs to L^1. Let $\{\mathcal{A}_n\}_{n \geq 0}$ and $\{\mathcal{B}_n\}_{n \geq 0}$ be the Schur polynomials of f and let $\{\gamma_n\}_{n \geq 0}$ be the Schur parameters of f. Then

$$\prod_{k=1}^n (1 - |\gamma_k|^2)^{1/2} \frac{1}{\mathcal{B}_n} \Longrightarrow F_f \tag{3.7}$$

and

$$\mathcal{A}_n^\# \Longrightarrow 0, \qquad \mathcal{B}_n^\# \Longrightarrow 0. \tag{3.8}$$

Proof Using the formula (4.2.8) we obtain that a.e. on $|\zeta| = 1$,

$$1 - |f(\zeta)|^2 = \frac{\prod_{k=1}^n (1 - |\gamma_k|^2)}{|\mathcal{B}_n(\zeta) + \zeta\mathcal{A}_n^\#(\zeta)f_{n+1}(\zeta)|^2}(1 - |f_{n+1}(\zeta)|^2) \quad . \tag{3.9}$$

Using the relations established in the proof of Theorem 4.2.6, it follows that $\mathcal{A}_n^{\#}\mathcal{B}_n^{-1}$ and $\prod_{k=1}^{n}(1 - |\gamma_k|^2)^{1/2}\mathcal{B}_n^{-1}$ are Schur functions. Since $\mathcal{B}_n$ has no zero inside the closed unit disc, $\mathcal{B}_n + \nu_1\mathcal{A}_n^{\#}f_{n+1} = \mathcal{B}_n(1 + \nu_1(\mathcal{A}_n^{\#}\mathcal{B}_n^{-1}))f_{n+1}$ is an outer function. It follows from (3.9) that

$$F_f(\mathcal{B}_n + \nu_1\mathcal{A}_n^{\#}f_{n+1}) = \prod_{k=1}^{n}(1 - |\gamma_k|^2)^{1/2}F_{f_{n+1}}. \qquad (3.10)$$

Evaluating (3.10) at $z = 0$, it follows that $F_{f_n}(0) \to 1$ for $n \to \infty$, and since F_{f_n} belongs to $\mathcal{S}$, we get $F_{f_n} \Longrightarrow 1$. We deduce that

$$\mathcal{B}_n + \nu_1\mathcal{A}_n^{\#}f_{n+1} \Longrightarrow \prod_{n=1}^{\infty}(1 - |\gamma_n|^2)^{1/2}\frac{1}{F_f}, \qquad (3.11)$$

and by (3.11) and (4.2.8), it follows that

$$\mathcal{A}_n + \nu_1\mathcal{B}_n^{\#}f_{n+1} \Longrightarrow \prod_{n=1}^{\infty}(1 - |\gamma_n|^2)^{1/2}\frac{f}{F_f}. \qquad (3.12)$$

By Jensen inequality,

$$F_{f_{n+1}}^2(0) \le 1 - \frac{1}{2\pi}\int_0^{2\pi}|f_{n+1}(e^{it})|^2dt \le 1,$$

hence $f_n \to 0$ in H^2, therefore $f_n \Longrightarrow 0$. Since $\mathcal{A}_n\mathcal{B}_n^{-1} \Longrightarrow f$, it follows that

$$\frac{\mathcal{A}_n + \nu_1\mathcal{B}_n^{\#}f_{n+1}}{\mathcal{B}_n} \Longrightarrow f \qquad (3.13)$$

and then, by (3.12) and (3.13), we deduce

$$\prod_{k=1}^{n}(1 - |\gamma_k|^2)^{1/2}\frac{1}{\mathcal{B}_n} \Longrightarrow F_f.$$

Then we prove (3.8). Using (4.2.10), we deduce that

$$(\mathcal{B}_n^{\#})^{(k)}(0) = k(\mathcal{B}_{n-1}^{\#})^{(k)}(0) + \gamma_n^*\mathcal{A}_{n-1}^{(k)}(0)$$

$((\mathcal{B}_n^{\#})^{(k)}(0)$ denotes the k-th derivative of $\mathcal{B}_n^{\#}$ at 0). Since $\gamma_n \to 0$ and $\{\mathcal{A}_n^{(k)}(0)\}_{n\ge 0}$ is a convergent sequence, it follows by induction on k that $(\mathcal{B}_n^{\#})^{(k)}(0)$ $\to 0$ when $n \to \infty$. On the other hand, $|\mathcal{B}_n^{\#}(z)| \le |\mathcal{B}_n(z)| \le M$ for $|z| \le r < 1$ by (3.7), consequently, $\mathcal{B}_n^{\#} \Longrightarrow 0$. Finally, we deduce from (4.2.8) that

$$(\mathcal{B}_n^{\#} - \mathcal{A}_n^{\#}f)(\mathcal{B}_n + \nu_1\mathcal{A}_n^{\#}f_{n+1}) = \prod_{k=1}^{n}(1 - |\gamma_k|^2)\nu_n.$$

Together with (3.11), this shows that $\mathcal{B}_n^{\#} - \mathcal{A}_n^{\#}f \Longrightarrow 0$, hence $\mathcal{A}_n^{\#} \Longrightarrow 0$. $\square$

Now, we investigate the effect of the same approximation procedure as in Theorem 3.2 when applied to Szegö polynomials. The result will be what is called the Szegö theory for the Szegö class of positive measures on the unit circle.

3.3 Theorem　*Let μ be a positive measure in the Szegö class and let $\{\varphi_n\}_{n\geq 0}$ be the Szegö polynomials of μ. If F_μ is the spectral factor of μ, then:*

(a)　$\varphi_n^\# \Longrightarrow F_\mu^{-1}$;

(b)　$\varphi_n \Longrightarrow 0$.

Proof　Let μ be a probability measure in the Szegö class and let f be the Schur function associated to μ by the formulae (3.3) and (3.4). Let $\{A_n\}_{n\geq 0}$ and $\{B_n\}_{n\geq 0}$ be the Schur polynomials of f. By Proposition 4.2.8,

$$\varphi_n^\# = \prod_{k=1}^n (1 - |\gamma_k|^2)^{-1/2}(B_n - A_n) = \prod_{k=1}^n (1 - |\gamma_k|^2)^{-1/2} B_n \left(1 - \frac{A_n}{B_n}\right).$$

By Theorem 4.2.6 and Theorem 3.2, we deduce that

$$\varphi_n^\# \Longrightarrow \frac{1}{F_f}(1 - f) = \frac{1}{F_\mu}$$

which is exactly (a). Based on Proposition 4.2.8, (b) is merely a restatement of (3.8). □

We can render explicit the process of approximating objects corresponding to a measure with Schur parameters $\{\Gamma_n\}_{n\geq 0}$ by similar objects corresponding to the sequence of measures with Schur parameters $\{\Gamma_0, \Gamma_1, \ldots, \Gamma_n, 0, 0, \ldots\}$.

3.4 Lemma　*A positive measure μ on the unit circle is absolutely continuous with respect to Lebesgue measure and $\mu' = 1/|p|^2$ with p a polynomial of degree m and zeros in the unit disc if and only if its Schur parameters $\{\Gamma_n\}_{n\geq 0}$ satisfy $|\Gamma_n| < 1$ for all $n \geq 0$ and $\Gamma_n = 0$ for $n > m$.*

Proof　It is readily checked that if μ is absolutely continuous with respect to Lebesgue measure and $\mu' = 1/|p|^2$ with p a polynomials of degree m and zeros in the unit disc, then the orthogonal polynomials of μ are $\varphi_n = \nu_{n-m}p$ for $n \geq m$. Therefore, $\Gamma_n = 0$ for $n > m$. Conversely, suppose $\Gamma_n = 0$ for $n > m$. Then, by Theorem 1.6.7, $\varphi_n^\# = \varphi_m^\#$ for $n \geq m$ and $F_\mu = (\varphi_m^\#)^{-1}$. On the other hand, by Proposition 4.2.7(b), the Schur function associated to μ by formula (3.4) is $f = A_m B_m^{-1}$ and the function g defined by (3.3) is $g = \frac{1}{2}(B_m + A_m)(B_m - A_m)^{-1}$. Therefore, g is bounded on the unit disc and μ is absolutely continuous with respect to Lebesgue measure. Finally, we choose $p = \varphi_m$ and μ has the required form. □

3.5 Proposition　*Let μ be a positive measure on the unit circle and let $\{\varphi_n\}_{n\geq 0}$ be the Szegö polynomials of μ. Then, for any continuous function u on the unit circle,*

$$\lim_{n\to\infty} \frac{1}{2\pi} \int_0^{2\pi} \frac{u(\zeta)}{|\varphi_n(\zeta)|^2} dt = \frac{1}{2\pi} \int_0^{2\pi} u(\zeta) d\mu(t), \qquad \zeta = e^{it}.$$

Proof Let $\{\Gamma_n\}_{n\geq 0}$ be the Schur parameters of μ and consider the measures μ_n with Schur parameters $\{\Gamma_0, \Gamma_1, \ldots, \Gamma_n, 0, 0, \ldots\}$. The first $n+1$ Fourier coefficients of μ and μ_n coincide and it follows from Lemma 3.4 that

$$\frac{1}{2\pi}\int_0^{2\pi}\frac{u(\zeta)}{|\varphi_n(\zeta)|^2}dt = \frac{1}{2\pi}\int_0^{2\pi}u(\zeta)d\mu(t), \qquad \zeta = e^{it},$$

for every trigonometric polynomial of degree less or equal to $n+1$. We can conclude the proof using the uniform density of the trigonometric polynomials in the set of continuous functions on the unit circle. □

In other words, Proposition 3.5 asserts the weak convergence of the measures with Schur parameters $\{\Gamma_0, \Gamma_1, \ldots, \Gamma_n, 0, 0, \ldots\}$ to the measure with Schur parameters $\{\Gamma_n\}_{n\geq 0}$. After this brief discussion it is apparent the critical role played by the condition

$$\lim_{n\to\infty}\Gamma_n = 0 \tag{3.14}$$

and it would be of interest to translate this property directly in terms of the measure with Schur parameters $\{\Gamma_n\}_{n\geq 0}$. A remarkable class of measures, which is larger than Szegö class and whose elements obey (3.14), is the *Erdös-Turán class* of the positive measures μ on the unit circle with the property that $\mu' > 0$ a.e.. To prove this result we first need a preliminary fact that reduces the discussion to absolutely continuous measures.

3.6 Lemma *Let τ be a finite positive measure on $[0, 2\pi)$ that is singular with respect to Lebesgue measure. Then there exists a sequence $\{h_n\}_{n\geq 1}$ of continuous $2\pi-$periodic functions such that $0 \leq h_n \leq 1$, $\lim_{n\to\infty}h_n(t) = 1$ a.e. and $\lim_{n\to\infty}\int_0^{2\pi}h_n(t)d\tau(t) = 0$.*

Proof Let $E \subset [0, 2\pi)$ be a Borel set of zero Lebesgue measure and $\tau([0, 2\pi) - E) = 0$. Let $\{E_n\}_{n\geq 1}$ be a decreasing sequence of open sets such that $E \subset E_n \subset [0, 2\pi)$, $\cap_{n=1}^{\infty}E_n = E$ and the Lebesgue measure of E_n is bounded by 2. For $t \in [0, 2\pi)$, define the functions

$$s_n(t) = \inf\{|t - s|/\, s \in [0, 2\pi) - E_n\}.$$

These are continuous functions on $[0, 2\pi)$, $s_n(t) > 0$ for $t \in E_n$ and $s_n(t) = 0$ for $t \in [0, 2\pi) - E_n$. Moreover, $s_n(t) < 1$ for all $t \in [0, 2\pi)$ since the Lebesgue measure of E_n is less than 2. Then, define the functions $h_{n,k}(t) = (1 - s_n(t))^k$ and remark that $\lim_{k\to\infty}h_{n,k}(t) = 0$ for all $t \in E$, and $0 \leq h_{n,k} \leq 1$. By Lebesgue's dominated convergence theorem, there exists an integer k_n such that

$$0 \leq \int_0^{2\pi}h_{n,k_n}(t)d\tau(t) \leq \frac{1}{n}.$$

Hence, we define $h_n = h_{n,k_n}$ and remark that these functions have the required properties. Thus, $h_n(0) = 1 = \lim_{t\to 2\pi}h_n(t)$. The next two properties, $0 \leq h_n(t) \leq 1$ for all $t \in [0, 2\pi)$ and $\lim_{n\to\infty}\int_0^{2\pi}h_n(t)d\tau(t) = 0$, are obvious. It remains to notice that since $h_n(t) = 1$ for $t \notin E_n$, it follows $\lim_{n\to\infty}h_n(t) = 1$ for

$t \notin \cap_{n=1}^{\infty} E_n = E$, which concludes the proof since the Lebesgue measure of E is zero. $\square$

Now, we can prove a first significant asymptotic result for the Erdös-Turán class.

3.7 Theorem *Let μ be a positive measure in the Erdös-Turán class and let $\{\varphi_n\}_{n \geq 0}$ be its Szegö polynomials. Then*

$$\lim_{n \to \infty} \int_0^{2\pi} |\frac{|\varphi_n(\zeta)|^2}{|\varphi_{n+1}(\zeta)|^2} - 1| dt = 0, \qquad \zeta = e^{it}.$$

Proof Using Schwartz's inequality, it follows that for $\zeta = e^{it}$,

$$(\frac{1}{2\pi} \int_0^{2\pi} |\frac{|\varphi_n(\zeta)|^2}{|\varphi_{n+1}(\zeta)|^2} - 1| dt)^2 \leq$$

$$\leq (\frac{1}{2\pi} \int_0^{2\pi} (\frac{|\varphi_n(\zeta)|}{|\varphi_{n+1}(\zeta)|} + 1)^2 dt)(\frac{1}{2\pi} \int_0^{2\pi} (\frac{|\varphi_n(\zeta)|}{|\varphi_{n+1}(\zeta)|} - 1)^2 dt) \leq$$

$$\leq 2(\frac{1}{2\pi} \int_0^{2\pi} \frac{|\varphi_n(\zeta)|^2}{|\varphi_{n+1}(\zeta)|^2} dt + 1)(\frac{1}{2\pi} \int_0^{2\pi} (\frac{|\varphi_n(\zeta)|}{|\varphi_{n+1}(\zeta)|} - 1)^2 dt).$$

It is a consequence of Proposition 3.5 that

$$\frac{1}{2\pi} \int_0^{2\pi} \frac{|\varphi_n(\zeta)|^2}{|\varphi_{n+1}(\zeta)|^2} dt = \frac{1}{2\pi} \int_0^{2\pi} |\varphi_n(\zeta)|^2 d\mu(t) = 1,$$

hence

$$0 \leq \frac{1}{2\pi} \int_0^{2\pi} (\frac{|\varphi_n(\zeta)|}{|\varphi_{n+1}(\zeta)|} - 1)^2 dt = 2 - \frac{1}{2\pi} \int_0^{2\pi} \frac{|\varphi_n(\zeta)|}{|\varphi_{n+1}(\zeta)|} dt.$$

Therefore, it is sufficient to prove that

$$\liminf_{n \to \infty} \frac{1}{2\pi} \int_0^{2\pi} \frac{|\varphi_n(\zeta)|}{|\varphi_{n+1}(\zeta)|} dt \geq 1. \tag{3.15}$$

For this purpose, let u be a positive continuous function on the unit circle and note that, by Schwartz's inequality and Proposition 3.5,

$$(\frac{1}{2\pi} \int_0^{2\pi} (\mu'(t)u(\zeta))^{1/4} dt)^4 \leq (\frac{1}{2\pi} \int_0^{2\pi} \frac{|\varphi_n(\zeta)|}{|\varphi_{n+1}(\zeta)|} dt)^2$$

$$\times \ (\frac{1}{2\pi} \int_0^{2\pi} |\varphi_{n+1}(\zeta)|^2 \mu'(t) dt)(\frac{1}{2\pi} \int_0^{2\pi} \frac{u(\zeta)}{|\varphi_{n+1}(\zeta)|^2} dt)$$

$$\leq \ (\liminf_{n \to \infty} \frac{1}{2\pi} \int_0^{2\pi} \frac{|\varphi_n(\zeta)|}{|\varphi_{n+1}(\zeta)|} dt)^2 (\frac{1}{2\pi} \int_0^{2\pi} u(\zeta) d\mu(t))$$

for $\zeta = e^{it}$. Now, let $\{h_m\}_{m \geq 1}$ be a family of continuous functions satisfying the conditions in Lemma 3.6 with respect to the singular part of μ. For each $\varepsilon > 0$, we obtain a family $\{g_k\}_{k\geq 1}$ of continuous functions on $[0, 2\pi]$ such that $g_k(0) = g_k(2\pi)$, $0 < g_k(t) < \varepsilon^{-1}$ and $\lim_{k\to\infty} g_k(t) = (\mu'(t) + \varepsilon)^{-1}$ a.e. By hypothesis, $\mu'(t) > 0$ a.e., hence $\lim_{\varepsilon\to 0} \mu'(t)(\mu'(t) + \varepsilon)^{-1} = 1$ a.e. Consequently,

$$\left(\frac{1}{2\pi}\int_0^{2\pi} (\mu'(t)h_m(t)g_k(t))^{1/4}dt\right)^4 \leq$$

$$\leq (\liminf_{n\to\infty} \frac{1}{2\pi}\int_0^{2\pi} \frac{|\varphi_n(\zeta)|}{|\varphi_{n+1}(\zeta)|}dt)^2(\frac{1}{2\pi}\int_0^{2\pi} h_m(t)g_k(t)d\mu(t)). \tag{3.16}$$

and letting successively $m \to \infty$, $k \to \infty$ and $\varepsilon \to 0$, we obtain (3.15). $\qquad\square$

Finally, we can prove that the elements of the Erdös-Turán class obey (3.14).

3.8 Theorem *Let μ be a positive measure in the Erdös-Turán class and let $\{\Gamma_n\}_{n\geq 0}$ be its Schur parameters. Then $\lim_{n\to\infty} \Gamma_n = 0$.*

Proof It follows from (1.6.19) that $|\psi_{n+1}(0)| = |\frac{\psi_{n+1}^\#(\zeta)}{\psi_n^\#(\zeta)} - 1|$ for $\zeta = e^{it}$. Since $|\psi_{n+1}(0)| < 1$, we deduce that $|\psi_{n+1}^\#(\zeta)| \leq 2|\psi_n^\#(\zeta)|$, hence

$$|\psi_{n+1}(0)| \leq 2|\frac{\psi_n^\#(\zeta)}{\psi_{n+1}^\#(\zeta)} - 1|. \tag{3.17}$$

As another consequence of the relations (1.6.19), one obtains that

$$\frac{\psi_n^\#(z)}{\psi_{n+1}^\#(z)} - 1 = -z\psi_{n+1}^*(0)\frac{\psi_n(z)}{\psi_{n+1}^\#(z)}, \qquad |z| \leq 1.$$

Consequently, for $\zeta = e^{it}$,

$$\mathrm{Re}(\frac{\psi_n^\#(\zeta)}{\psi_{n+1}^\#(\zeta)} - 1) = \frac{1}{2}(\frac{|\varphi_n(\zeta)|^2}{|\varphi_{n+1}(\zeta)|^2} - 1). \tag{3.18}$$

Consider the function $g = \psi_n^\#(\psi_{n+1}^\#)^{-1} - 1$ which is analytic in the closed unit disc and vanishes at $z = 0$. The function $u = \mathrm{Re}\,g$ is harmonic and by Kolmogorov's inequality and with $\tilde{u}$ denoting the complex conjugate function, we obtain

$$\left(\int_0^{2\pi} |g(\zeta)|^{1/2}dt\right)^2 = \left(\int_0^{2\pi} |u(\zeta) + \tilde{u}(\zeta)|^{1/2}dt\right)^2 \leq \left(\int_0^{2\pi} (|u(\zeta)| + |\tilde{u}(\zeta)|)^{1/2}dt\right)^2$$

$$\leq 2\left(\left(\int_0^{2\pi} |u(\zeta)|^{1/2}dt\right)^2 + \left(\int_0^{2\pi} |\tilde{u}(\zeta)|^{1/2}dt\right)^2\right)$$

$$\leq 2(2\pi + B)\int_0^{2\pi} |u(\zeta)|dt.$$

It follows from (3.16) that

$$|\psi_{n+1}(0)| \leq C\int_0^{2\pi} |\frac{|\varphi_n(\zeta)|^2}{|\varphi_{n+1}(\zeta)|^2} - 1|dt$$

and by Theorem 3.7, we can conclude the proof. □

Based on formula (3.8), Theorem 3.8 has the following translation to the Schur class.

3.9 Corollary Let f be a Schur function with the property that $|f(e^{it})| < 1$ a.e. and let $\{\gamma_n\}_{n \geq 0}$ be the Schur parameters of f. Then $\lim_{n \to \infty} \gamma_n = 0$. □

We can mention that the converse of Theorem 3.8 is not true.

5.4 Maximum Entropy

In this section we address a so-called maximum entropy problem. Its motivation comes from the introduction of a basic information theoretic quantity which is

$$D(f\|g) = \int f(\mathbf{x}) \log(f(\mathbf{x})/g(\mathbf{x}))d\mathbf{x},$$

where f and g are probability densities and $\mathbf{x} = (x_1, x_2, \ldots, x_n)$. The number $D(f\|g)$ is called the *relative entropy*, or the Kulback-Leibler *information number*, *I-divergence* and *information distance*. When $f = 1$, then $D(1\|g) = - \int \log g(\mathbf{x})d\mathbf{x}$ and it appears to be of interest to find the minimum of $D(1\|g)$ when g belongs to a certain specified set. Next, remark that if g is a function on $[0, 2\pi)$ such that $0 \leq g \leq 1$, then

$$D(1\|g) = - \int_0^{2\pi} \log g(t)dt = -4\pi \log F(0),$$

where, according to (2.7), F is the spectral factor of $g^{1/2}$. Based on this connection, we formulate the following problem. Denote by $\mathcal{A}$ the set of all upper triangular strict contractions S that satisfy (4.4.3) and denote by $D(S)$ the diagonal of S. Moreover, denote by F_S the upper triangular spectral factor of the positive operator $I - S^*S$. The *maximum entropy* problem is to solve the following optimization criterion:

$$\max_{S \in \mathcal{A}}\{D^*(F_S)D(F_S)\}. \tag{4.1}$$

The solution of this problem is based on the following result.

4.1 Lemma Let S be an element of $\mathcal{A}$ and let K be the corresponding strict contraction in formula (4.4.16). Then the spectral factor F_S can be chosen according to the following formula:

$$F_S = F_K(\mathbf{T}_{21}K + \mathbf{T}_{22})^{-1}.$$

Proof It follows from the $\mathbf{J}$−innerness of $\mathbf{T}$ and (4.4.16) that

$$I - S^*S = (K^*\mathbf{T}_{21}^* + \mathbf{T}_{22}^*)^{-1}(I - K^*K)(\mathbf{T}_{21}K + \mathbf{T}_{22})^{-1}. \tag{4.2}$$

Let F_K in $\mathcal{L}(\oplus_{t\in\mathbb{Z}}\mathcal{V}(t), \oplus_{t\in\mathbb{Z}}\mathcal{V}'(t))$ be the spectral factor of K and define the operator $F = F_K(\mathbf{T}_{21}K + \mathbf{T}_{22})^{-1}$. It is seen that F is an upper triangular operator which obeys the condition that the space $F(\oplus_{j\le t}\mathcal{V}(j))$ is dense in $\oplus_{j\le t}\mathcal{V}'(j)$ for all $t \in \mathbb{Z}$. Since $F_K^*F_K \le I - K^*K$, it follows from (4.2) that $F^*F \le I - S^*S$. Now we consider any other upper triangular contraction Z in $\mathcal{L}(\oplus_{t\in\mathbb{Z}}\mathcal{V}(t), \oplus_{t\in\mathbb{Z}}\mathcal{V}''(t))$ such that $Z^*Z \le I - S^*S$. It follows from (4.2) that

$$(K^*\mathbf{T}_{21}^* + \mathbf{T}_{22}^*)Z^*Z(\mathbf{T}_{21}K + \mathbf{T}_{22}) \le I - K^*K,$$

and using the properties of the spectral factors, we must have

$$(K^*\mathbf{T}_{21}^* + \mathbf{T}_{22}^*)Z^*Z(\mathbf{T}_{21}K + \mathbf{T}_{22}) \le F_K^*F_K.$$

This implies that $Z^*Z \le F^*F$ and, consequently, $F = F_K(\mathbf{T}_{21}K + \mathbf{T}_{22})^{-1}$ can be chosen as the spectral factor of S. $\square$

We are now in a position to obtain the solution of (4.1). For this purpose, and for notational convenience, we define the upper triangular operators $\chi = \mathbf{T}_{22}^{-1}\mathbf{T}_{21}$ and $\varphi = \mathbf{T}_{22}^{-1}$.

4.2 Theorem   *Consider the setting of Theorem 4.4.4. Then*

$$\max_{S\in\mathcal{A}}\{D^*(F_S)D(F_S)\} = [D(\mathbf{T}_{22})D^*(\mathbf{T}_{22}) - D(\mathbf{T}_{21})D^*(\mathbf{T}_{21})]^{-1}$$

$$= D^*(\varphi)[I - D(\chi)D^*(\chi)]^{-1}D(\varphi).$$

Furthermore, the maximum is attained if and only if $S = S_0 = \mathbf{T}[D^*(\chi)]$. In particular, if $D(\chi) = 0$ (equivalently, $D(\mathbf{T}_{21}) = 0$), then $\max_{S\in\mathcal{A}}\{D^*(F_S)D(F_S)\} = [D^*(\mathbf{T}_{22})]^{-1}[D(\mathbf{T}_{22})]^{-1}$.

Proof   Assume first that $D(\chi) = 0$. Then, by Lemma 4.1,

$$D(F_S) = D(F_K)(D(\mathbf{T}_{21})D(K) + D(\mathbf{T}_{22}))^{-1}$$

$$= D(F_K)(I + D(\chi)D(K))^{-1}D(\varphi) = D(F_K)D(\varphi).$$

But $D^*(F_K)D(F_K) \le I$, and since $D^*(F_K)D(F_K) \le F_K^*F_K \le I - K^*K$, it follows that the equality $D^*(F_K)D(F_K) = I$ holds if and only if $K = 0$. Therefore,

$$\max_{S\in\mathcal{A}}\{D^*(F_S)D(F_S)\} = D^*(\varphi)D(\varphi) = [D^*(\mathbf{T}_{22})]^{-1}[D(\mathbf{T}_{22})]^{-1},$$

and the maximum is attained for $S = S_0 = \mathbf{T}[0] = -\mathbf{T}_{12}\mathbf{T}_{22}^{-1}$.

Next assume $D(\chi) \ne 0$. Since $\|\chi\| < 1$, it follows that $\|D(\chi)\| < 1$ and $\mathbf{T}^0 = S(D(\chi)) = [\mathbf{T}_{ij}^0]_{i,j=1}^2$, defined according to (4.1.12), is a $\mathbf{J}$-inner operator. Consequently, if K is an upper triangular strict contraction in $\mathcal{L}(\oplus_{t\in\mathbb{Z}}\mathcal{U}(t), \oplus_{t\in\mathbb{Z}}\mathcal{V}(t))$, then the operator defined by the formula

$$K_1 = -(\mathbf{T}_{11}^0K + \mathbf{T}_{12}^0)(\mathbf{T}_{21}^0K + \mathbf{T}_{22}^0)^{-1}, \tag{4.3}$$

is an upper triangular operator in $\mathcal{L}(\oplus_{t\in\mathbb{Z}}\mathcal{U}(t), \oplus_{t\in\mathbb{Z}}\mathcal{V}(t))$ with $\|K_1\| < 1$. Moreover, $K_1 = 0$ if and only if $K = D^*(\chi)$. We may also remark that for any upper triangular operator K in $\mathcal{L}(\oplus_{t\in\mathbb{Z}}\mathcal{U}(t), \oplus_{t\in\mathbb{Z}}\mathcal{V}(t))$ with $\|K\| < 1$, there exists an upper triangular operator K_1 in $\mathcal{L}(\oplus_{t\in\mathbb{Z}}\mathcal{U}(t), \oplus_{t\in\mathbb{Z}}\mathcal{V}(t))$ with $\|K_1\| < 1$ and such that (4.3) holds. Now,

$$F_{K_1} = F_K(\mathbf{T}_{21}^0 K + \mathbf{T}_{22}^0)^{-1} = F_K(I - D(\chi)K)^{-1}(I - D(\chi)D^*(\chi))^{1/2}.$$

Since for $S = \mathbf{T}[-K]$ we have $D(F_S) = D(F_K)(I - D(\chi)D(K))^{-1}D(\varphi)$, it follows that

$$D(F_S) = D(F_{K_1})(I - D(\chi)D^*(\chi)^*)^{-1/2}D(\varphi).$$

But $D^*(F_{K_1})D(F_{K_1}) \leq I$, and $D^*(F_{K_1})D(F_{K_1}) = I$ if and only if $K_1 = 0$, or, equivalently, if and only if $K = D^*(\chi)$. Therefore,

$$\max_{S\in\mathcal{A}}\{D^*(F_S)D(F_S)\} = [D(\mathbf{T}_{22})D^*(\mathbf{T}_{22}) - D(\mathbf{T}_{21})D^*(\mathbf{T}_{21})]^{-1},$$

and the maximum is attained if and only if $S = S_0 = \mathbf{T}[D^*(\chi)]$.   □

The unique $S_0 = \mathbf{T}[D^*(\chi)]$ is called the *maximum entropy* solution of (4.4.3). We have seen in Theorem 4.4.4 that the description of $\mathcal{A}$ depends on the transfer map $\mathbf{T}$ of the system (4.4.4). We show now that we can choose a system (4.4.4) such that the maximum entropy solution of (4.4.3) coincides with $\mathbf{T}[0]$. More precisely, we have the following result.

Theorem 4.3 *Suppose (H_3)-(H_6) hold and the displacement equation (4.1.4) has a Pick solution that satisfies (H). Then we can always choose uniformly bounded families $\{h_i(t)\}_{t\in\mathbb{Z}}$, $\{k_i(t)\}_{t\in\mathbb{Z}}$ such that the operator $\mathbf{T} = \mathbf{T}_0\mathbf{T}_1\ldots\mathbf{T}_{n-1}$, where $\mathbf{T}_i$ are the transfer maps of the systems (4.4.15), has the property that $S_0 = \mathbf{T}[0]$.*

Proof It follows from Theorem 4.2 that $S_0 = \mathbf{T}[0]$ if and only if $D(\mathbf{T}_{21}) = 0$. We show how to choose uniformly bounded families $\{h_i(t)\}_{t\in\mathbb{Z}}$, $\{k_i(t)\}_{t\in\mathbb{Z}}$ such that $D(\mathbf{T}_{21}^{(i)}) = 0$ for each $i = 0, 1, \ldots n - 1$, where $\mathbf{T}_i$ are the transfer maps of the systems (4.4.15) associated with $\{f_i(t), g_i(t), h_i(t), k_i(t)\}$. By Theorem 4.1.5(b), we can find uniformly bounded families $\{\bar{h}_i(t)\}_{t\in\mathbb{Z}}$, $\{\bar{k}_i(t)\}_{t\in\mathbb{Z}}$ such that (4.1.6) holds. Let $\bar{\mathbf{T}}_i = [\bar{\mathbf{T}}_{kl}^{(i)}]_{k,l=1}^2$ denote the transfer map of the system (4.4.15) associated with $\{f_i(t), g_i(t), \bar{h}_i(t), \bar{k}_i(t)\}$. By (4.1.6), we have that $\bar{h}_i(t)d_i(t-1)\bar{h}_i^*(t) + \bar{k}_i(t)J(t)\bar{k}_i^*(t) = J(t)$, therefore $J(t) - \bar{k}_i(t)J(t)\bar{k}_i^*(t) = \bar{h}_i(t)d_i(t-1)\bar{h}_i^*(t) \geq 0$. Since $\dim\mathcal{U}(t) < \infty$, $\dim\mathcal{V}(t) < \infty$ for all $t \in \mathbb{Z}$, it also follows that $J(t) - \bar{k}_i^*(t)J(t)\bar{k}_i(t) \geq 0$ for all $t \in \mathbb{Z}$. If we partition $\bar{k}_i(t)$ accordingly with $J(t)$, $\bar{k}_i(t) = [\bar{k}_i^{(pl)}]_{p,l=1}^2$, then $\bar{k}_i^{(22)*}(t)\bar{k}_i^{(22)}(t) \geq I + \bar{k}_i^{(12)*}(t)\bar{k}_i^{(12)}(t)$. Therefore, $\bar{k}_i^{(22)}(t)$ is invertible and $\|(\bar{k}_i^{(22)}(t))^{-1}\| \leq 1$. We also know that $\|\bar{k}_i^{(22)}(t)\| \leq M$ for certain $M > 0$. It follows that $\bar{\rho}_i(t) = -\bar{k}_i^{(12)}(t)(\bar{k}_i^{(22)}(t))^{-1}$ is a contraction in $\mathcal{L}(\mathcal{V}(t), \mathcal{U}(t))$ and that

$$I - \bar{\rho}_i^*(t)\bar{\rho}_i(t) \geq \left(\bar{k}_i^{(22)*}(t)\right)^{-1}\left(\bar{k}_i^{(22)}(t)\right)^{-1} \geq \frac{1}{M^2}.$$

Hence $(I - \bar{\rho}_i^*(t)\bar{\rho}_i(t))^{-1} \le M$. Moreover, from the identity

$$(I - \bar{\rho}_i(t)\bar{\rho}_i^*(t))^{-1} = I + \bar{\rho}_i(t)(I - \bar{\rho}_i^*(t)\bar{\rho}_i(t))^{-1}\bar{\rho}_i^*(t),$$

we also obtain that $I - \bar{\rho}_i(t)\bar{\rho}_i^*(t)) \le 1 + M^2$. We conclude that the choices

$$h_i(t) = S^{-1}(\bar{\rho}_i(t))\bar{h}_i(t), \quad k_i(t) = S^{-1}(\bar{\rho}_i(t))\bar{k}_i(t) \tag{4.4}$$

satisfy the embedding relation (4.1.6), are uniformly bounded over t and result in $D(\mathbf{T}_{21}^{(i)}) = 0$, where $\mathbf{T}_i$ is the transfer map of the system (4.4.15) associated to $\{f_i(t), g_i(t), h_i(t), k_i(t)\}$, with $\{h_i(t)\}_{t \in \mathbb{Z}}$ and $\{k_i(t)\}_{t \in \mathbb{Z}}$ given by (4.4). $\qquad \square$

To illustrate the main steps involved in the above construction we consider several examples. Let us first concentrate on the case of strictly lower triangular $F(t)$, when $f_i(t) = 0$ for all $t \in \mathbb{Z}$. We begin with an additional assumption that $\dim \mathcal{R}_i(t) = \dim \mathcal{U}(t)$ for all $t \in \mathbb{Z}$ and $i = 0, 1, \dots, n-1$. Let $g_i(t) = [\, u_i(t) \quad v_i(t) \,]$ denote the top (block) row of $G_i(t)$, and since $g_i(t)J(t)g_i(t) = d_i(t) > 0$, it follows the there exist uniquely determined matrices $\gamma_i(t)$, $\|\gamma_i(t)\| < 1$, such that $v_i(t) = u_i(t)\gamma_i(t)$. We remark that

$$g_i(t)S(\gamma_i(t)) = [\, * \quad 0_{\mathcal{V}(t)} \,]. \tag{4.5}$$

We refer to the $\gamma_i(t)$ as the *Schur parameters* associated to the displacement equation, when $F(t)$ is strictly lower triangular. Consider further

$$\bar{k}_i(t) = I - J(t)g_i^*(t)d_i^{-1}(t)g_i(t). \tag{4.6}$$

$$\bar{h}_i(t) = J(t)g_i^*(t)d_i^{-1/2}(t)\tau_i(t)d_i^{-1/2}(t-1), \tag{4.7}$$

where $\tau_i(t)$ is a unitary matrix. We partition $\bar{k}_i(t)$ accordingly to $J(t)$, and introduce the *generalized reflection coefficients*

$$\bar{\rho}_i(t) = -\bar{k}_i^{(12)}(t)(\bar{k}_i^{(22)}(t))^{-1}. \tag{4.8}$$

4.4 Proposition *Consider the setting of Theorem 4.3. Assume further that $F(t)$ is strictly lower triangular and $\dim \mathcal{R}_i(t) = \dim \mathcal{U}(t)$ for all $t \in \mathbb{Z}$ and $i = 0, 1, \dots, n-1$. Then the Schur parameters $\{\gamma_i(t)\}$ and the generalized reflection coefficients $\{\bar{\rho}_i(t)\}$ coincide,*

$$\bar{\rho}_i(t) = \gamma_i(t), \quad t \in \mathbb{Z}, \, i = 0, 1, \dots, n-1.$$

Proof Since $\dim \mathcal{R}_i(t) = \dim \mathcal{U}(t)$ for all $t \in \mathbb{Z}$, $i = 0, 1, \dots, n-1$ and $u_i(t)u_i^*(t) \ge \varepsilon + v_i(t)v_i^*(t)$ for a certain $\varepsilon > 0$, we get that $u_i(t)$ are invertible matrices. Consequently,

$$\begin{aligned}
\bar{\rho}_i(t) &= u_i^*(t)d_i^{-1}(t)v_i(t)(I + v_i^*(t)d_i^{-1}(t)v_i(t))^{-1} \\
&= u_i^*(t)(u_i(t)(I - \gamma_i(t)\gamma_i^*(t))u_i^*(t))^{-1}u_i(t)\gamma_i(t) \\
&\quad \times (I + \gamma_i^*(t)u_i^*(t)(u_i(t)(I - \gamma_i(t)\gamma_i^*(t))u_i^*(t))^{-1}u_i(t)\gamma_i(t))^{-1} \\
&= (I - \gamma_i(t)\gamma_i^*(t))^{-1}\gamma_i(t)(I + \gamma_i^*(t)(I - \gamma_i(t)\gamma_i^*(t))^{-1}\gamma_i(t))^{-1} \\
&= (I - \gamma_i(t)\gamma_i^*(t))^{-1}\gamma_i(t)(I - \gamma_i^*(t)\gamma_i(t)) = \gamma_i(t). \quad \square
\end{aligned}$$

We can explain this result by taking into account a simplification in the generator recursion. Thus, we can choose

$$k_i(t) = S^{-1}(\gamma_i(t))\bar{k}_i(t), \quad h_i(t) = S^{-1}(\gamma_i(t))\bar{h}_i(t),$$

where $\bar{h}_i(t)$ and $\bar{k}_i(t)$ are given by (4.6) and (4.7). We readily conclude that

$$J(t)k_i^*(t)J(t) = S(\gamma_i(t))\begin{bmatrix} 0 & 0 \\ 0 & I \end{bmatrix}, \quad h_i^*(t)J(t) = d_i^{-1/2}(t-1)\tau_i^*(t)d_i^{-1/2}(t)[\delta_i(t) \quad 0].$$

We also remark that since $\dim \mathcal{R}_i(t) = \dim \mathcal{U}(t)$ for all $t \in \mathbb{Z}$, $i = 0, 1, \ldots n-1$, and $\delta_i(t)\delta_i^*(t) = g_i^*(t)J(t)g_i(t) = d_i(t)$, it follows by a simple Schur complement argument that $I - \delta_i^*(t)d_i^{-1}(t)\delta_i(t) = 0$. The recursion in Corollary 4.2.3 becomes:

$$\begin{bmatrix} & 0 \\ l_i(t) & \\ & G_{i+1}(t) \end{bmatrix} = [F_i(t)l_i(t-1) \quad G_i(t)] \begin{bmatrix} & 0 & \beta_i(t) \\ S(\gamma_i(t))\begin{bmatrix} \delta_i^*(t) \\ 0 \end{bmatrix} & S(\gamma_i(t))\begin{bmatrix} 0 & 0 \\ 0 & I \end{bmatrix} \end{bmatrix},$$
$$\tag{4.9}$$

where $\beta_i(t) = d_i^{-1/2}(t-1)\tau_i^*(t)d_i^{-1/2}(t)[\delta_i(t) \quad 0]$. We remark that since $\delta_i^*(t-1)d_i^{-1}(t-1)\delta_i(t-1) = \delta_i^*(t)d_i^{-1}(t)\delta_i(t)$, we can choose the unitary matrix $\tau_i(t)$ so as to satisfy the relation $\delta_i^*(t-1)d_i^{-1/2}(t-1)\tau_i^*(t) = \delta_i^*(t)d_i^{-1/2}(t)$. From (4.9), we obtain the following formula

$$\begin{bmatrix} 0 \\ G_{i+1}(t) \end{bmatrix} = F_i(t)G_i(t-1)S(\gamma_i(t-1))\begin{bmatrix} I & 0 \\ 0 & 0 \end{bmatrix} + G_i(t)S(\gamma_i(t))\begin{bmatrix} 0 & 0 \\ 0 & I \end{bmatrix}, \tag{4.10}$$

which is a generalization of the array form of the Schur algorithm (see Theorem 4.2.5).

We can also obtain the following result.

4.5 Theorem *Consider the setting of Proposition 4.4 and let $\Delta = [\Delta_{tt}]_{t\in\mathbb{Z}}$ denote the optimal diagonal operator, $\Delta = \max_{S\in\mathcal{A}}\{D^*(F_S)D(F_S)\}$. Then*

$$\Delta_{tt} = (I - \gamma_0^*(t)\gamma_0(t))^{1/2}(I - \gamma_1^*(t)\gamma_1(t))^{1/2}\ldots(I - \gamma_{n-1}^*(t)\gamma_{n-1}(t))$$

$$\ldots(I - \gamma_1^*(t)\gamma_1(t))^{1/2}(I - \gamma_0^*(t)\gamma_0(t))^{1/2}.$$

Proof We already know that the solution $\mathbf{T}[0]$ corresponding to the cascade

$$\mathbf{T} = \mathbf{T}_0\mathbf{T}_1\ldots\mathbf{T}_{n-1},$$

coincides with the maximum entropy solution and, consequently,

$$\Delta = [D(\mathbf{T}_{22}^*)]^{-1}[D(\mathbf{T}_{22}]^{-1}.$$

But, for each section i, $D(\mathbf{T}_{22}^{(i)})_{tt} = (I - \gamma_i^*(t)\gamma_i(t))^{-1/2}$ and the required result now follows. □

We must remark that the previous discussion can be extended by dropping the assumption that $\dim \mathcal{R}_i(t) = \dim \mathcal{U}(t)$ for all $t \in \mathbb{Z}$ and $i = 0, 1, \ldots, n-1$, but the details can be omited.

We now discuss the classical Nevanlinna recursion, which maps scalar Schur functions s_i, $\|s_i\|_\infty < 1$, to Schur functions s_{i+1} as follows:

$$s_{i+1}(z) = \frac{1 - z_i^* z}{z - z_i} \frac{s_i(z) - \gamma_i}{1 - \gamma_i^* s_i(z)} \quad , \quad \gamma_i = s_i(z_i) \quad , \quad s_0(z) = s(z) \quad , \quad i \geq 0. \quad (4.11)$$

We show the connection of the algorithm (4.11) with the displacement equations. We remark that (4.11) is a nonlinear recursion in $s_i(z)$, and it can be linearized by expressing $s_i(z)$ as the ratio of two power series, $s_i(z) = v_i(z)/u_i(z)$. It follows from (4.11) that we can also write

$$(z - z_i) \begin{bmatrix} u_{i+1}(z) & v_{i+1}(z) \end{bmatrix} = \begin{bmatrix} u_i(z) & v_i(z) \end{bmatrix} S(\gamma_i) \begin{bmatrix} B_i(z) & 0 \\ 0 & 1 \end{bmatrix}, \quad (4.12)$$

where $\gamma_i = \lim_{z \to z_i} v_i(z)/u_i(z)$ and $B_i(z) = \dfrac{z - z_i}{1 - z_i^* z}$. We see that each step of (4.12) gives rise to a first-order J−lossless section

$$T_i(z) = S(\gamma_i) \begin{bmatrix} B_i(z) & 0 \\ 0 & 1 \end{bmatrix}, \quad (4.13)$$

Next, we invoke the Newton power-series expansions of $u_i(z)$ and $v_i(z)$ and compare terms on both sides of (4.12): let $P_i(z)$ denote the Newton-series basis associated with the points $\{z_i, z_{i+1}, \ldots\}$, viz.,

$$P_i(z) = \begin{bmatrix} 1 & (z - z_i) & (z - z_i)(z - z_{i+1}) & (z - z_i)(z - z_{i+1})(z - z_{i+2}) & \ldots \end{bmatrix}$$

and assume we expand $u_i(z)$ and $v_i(z)$ with respect to this Newton-series basis,

$$u_i(z) = u_{ii} + u_{i+1,i}(z - z_i) + u_{i+2,i}(z - z_i)(z - z_{i+1}) + \ldots \quad (4.14)$$

$$v_i(z) = v_{ii} + v_{i+1,i}(z - z_i) + v_{i+2,i}(z - z_i)(z - z_{i+1}) + \ldots \quad (4.15)$$

[We remark that if we are given a function $h(z)$ then the coefficients of its Newton series expansion with respect to given points $\{z_0, z_1, z_2, \ldots\}$,

$$h(z) = h_0 + h_1(z - z_0) + h_2(z - z_0)(z - z_1) + \ldots,$$

can be computed recursively via the so-called *divided difference recursion* as follows: start with $h_0(z) = h(z)$ and then use

$$h_i(z) = \frac{h_{i-1}(z) - h_{i-1}}{z - z_{i-1}} \quad , \quad h_i = h_i(z_i) \ ].$$

Returning to (4.12), if we now introduce the two-column (semi-infinite) array G_i composed of the power series coefficients of u_i and v_i,

$$G_i = \begin{bmatrix} u_{ii} & v_{ii} \\ u_{i+1,i} & v_{i+1,i} \\ u_{i+2,i} & v_{i+2,i} \\ \vdots & \vdots \end{bmatrix} = \begin{bmatrix} \mathbf{u}_i & \mathbf{v}_i \end{bmatrix}, \quad (4.16)$$

then $[u_i(z) \quad v_i(z)] = P_i(z)G_i$ and we can rewrite (4.11) in the equivalent array form

$$\begin{bmatrix} 0 & 0 \\ G_{i+1} \end{bmatrix} = \Phi_i G_i S(\gamma_i) \begin{bmatrix} 1 & 0 \\ 0 & 0 \end{bmatrix} + G_i S(\gamma_i) \begin{bmatrix} 0 & 0 \\ 0 & 1 \end{bmatrix}, \qquad (4.17)$$

where Φ_i is the (semi-infinite) "Blaschke" matrix given by $\Phi_i = (F_i - z_i I)(I - z_i^* F_i)^{-1}$, and F_i is the submatrix obtained after deleting the first i columns and rows of the following bidiagonal matrix

$$F = \begin{bmatrix} z_0 \\ 1 & z_1 \\ & 1 & z_2 \\ & & \ddots & \ddots \end{bmatrix}. \qquad (4.18)$$

Therefore, the Nevanlinna algorithm reduces to the algorithm described in Theorem 4.2.2 for the displacement equation with generators F and $G = G_0$ given by (4.18) and (4.16) respectively, and $J = (1 \oplus -1)$.

The resulting cascade $T(z)$ that can be associated with n steps of the above recursion is given by

$$T(z) = T_0(z)T_1(z)\dots T_{n-1}(z), \quad T_i(z) = S(\gamma_i) \begin{bmatrix} B_i(z) & 0 \\ 0 & 1 \end{bmatrix}, \qquad (4.19)$$

and can be used to solve the Nevanlinna-Pick problem for $\{z_i\}_{i=0}^{n-1}$ and $\{w_i\}_{i=0}^{n-1}$. The only thing we have to do is to choose the initial column vectors $\mathbf{u}_0$ and $\mathbf{v}_0$ such that $P_0(z_i)\mathbf{v}_0/P_0(z_i)\mathbf{u}_0 = w_i$ for $i = 0, 1, \dots, n-1$. It is easily seen that this cascade has not the property in Theorem 4.3 for all data $\{z_i\}_{i=0}^{n-1}$ and $\{w_i\}_{i=0}^{n-1}$.

We now show how to use the construction in Theorem 4.3 in order to modify the Nevanlinna recursion and obtain an algorithm that leads to a cascade whose central solution coincides with the maximum-entropy solution.

To clarify this, we first elaborate on the connection of the Schur parameters $\{\gamma_i\}$ and the generalized reflection coefficients $\{\bar{\rho}_i\}$. Indeed, we choose $\Theta_i = I$ and τ_i such that

$$\bar{k}_i = I - Jg_i^* d_i^{-1}(1 - \frac{1+z_i}{1+z_i^*} z_i^*)^{-1} g_i = \begin{bmatrix} 1 - \dfrac{1+z_i^*}{1-|\gamma_i|^2} & -\dfrac{1+z_i^*}{1-|\gamma_i|^2}\gamma_i \\ \dfrac{1+z_i^*}{1-|\gamma_i|^2}\gamma_i^* & 1 + \dfrac{1+z_i^*}{1-|\gamma_i|^2}|\gamma_i|^2 \end{bmatrix}.$$

The generalized reflection coefficient is then related to the Schur parameter γ_i via

$$\bar{\rho}_i = \frac{1+z_i^*}{1+z_i^*|\gamma_i|^2}\, \gamma_i.$$

This leads to the choices

$$h_i = S(\bar{\rho}_i)^{-1} d_i^{-1} Jg_i^*, \quad k_i = S(\bar{\rho}_i)^{-1} \bar{k}_i,$$

and to the first-order sections

$$T_{\rho,i}(z) = \left\{ I + [B_i(z) - 1]\frac{Jg_i^* g_i}{g_i Jg_i^*} \right\} S(\bar{\rho}_i).$$

These sections are related to the earlier $T_i(z)$ via

$$T_{\rho,i}(z) = T_i(z)S(\gamma_i)^{-1}S(\bar{\rho}_i) = S(\gamma_i)\begin{bmatrix} B_i(z) & 0 \\ 0 & 1 \end{bmatrix} S(\gamma_i)^{-1}S(\bar{\rho}_i).$$

The corresponding generator recursion is given by

$$\begin{bmatrix} 0 \\ G_{i+1} \end{bmatrix} = \begin{bmatrix} G_i + (\Phi_i - I)G_i \dfrac{Jg_i^* g_i}{g_i Jg_i^*} \end{bmatrix} S(\bar{\rho}_i). \tag{4.20}$$

A simple computation shows that

$$S(\gamma_i)^{-1}S(\bar{\rho}_i) = \frac{|1 + z_i^*|\gamma_i|^2|}{(1 - |z_i|^2|\gamma_i|^2)^{\frac{1}{2}}} \begin{bmatrix} \dfrac{1}{1 + z_i|\gamma_i|^2} & -\dfrac{z_i^*\gamma_i}{1 + z_i^*|\gamma_i|^2} \\ -\dfrac{z_i\gamma_i^*}{1 + z_i|\gamma_i|^2} & \dfrac{1}{1 + z_i^*|\gamma_i|^2} \end{bmatrix}.$$

If we define

$$\zeta_i = \frac{|1 + z_i^*|\gamma_i|^2|}{1 + z_i|\gamma_i|^2}, \quad c_i = z_i^*\gamma_i,$$

then the generator recursion (4.20) leads to a modified Nevanlinna recursion of the type:

$$\frac{\zeta_i^* s_{i+1}(z) + c_i}{1 + c_i^*\zeta_i^* s_{i+1}(z)} = \frac{1 - z_i^* z}{z - z_i} \frac{s_i(z) - \gamma_i}{1 - \gamma_i^* s_i(z)}, \quad \gamma_i = s_i(z_i), \quad s_0 = s, i \geq 0. \tag{4.21}$$

The central solution of the cascade associated with this modified recursion now coincides with the maximum-entropy solution.

We conclude this section with an application of the maximum entropy principle. Let $c_0, c_1, \ldots, c_{n-1}$ be complex numbers such that

$$T = \begin{bmatrix} c_0 & 0 & \cdots & 0 \\ c_1 & c_0 & & \\ \vdots & & \ddots & \vdots \\ c_{n-1} & c_{n-2} & \cdots & c_0 \end{bmatrix}$$

is a strict contraction, $\|T\| = d_\infty < 1$. Moreover, denote by d_2 the Hilbert-Schmidt norm of T, $d_2^2 = \text{tr}(TT^*) = \sum_{k=0}^{n-1}|c_k|^2 = E_{n-1}^* TT^* E_{n-1}$. By Theorem 4.2.6, the set of solutions of the Schur problem for $\{c_k\}_{k=0}^{n-1}$ is given by the formula

$$f(z) = \frac{\mathcal{A}_{n-1}(z) + z\mathcal{B}_{n-1}^{\#}(z)f_n(z)}{\mathcal{B}_{n-1}(z) + z\mathcal{A}_{n-1}^{\#}(z)f_n(z)},$$

where f_n is an arbitrary Schur function. The same formula may be obtained using the methods in Chapter 3 and the central solution, denoted by f^0, corresponds to the choice $f_n = 0$. Therefore, $f^0 = \mathcal{A}_{n-1}/\mathcal{B}_{n-1}$. Let $\{\gamma_0, \gamma_1, \ldots, \gamma_{n-1}, 0, 0 \ldots\}$ be the set of the Schur parameters of f^0.

4.6 Theorem   *The central solution f^0 of the Schur problem for $\{c_k\}_{k=0}^{n-1}$ with $d_\infty < 1$, satisfies the following estimations: $\|f^0\|_\infty < 1$ and $\|f^0\|_2 \le d_2/\sqrt{1-d_\infty^2}$.*

Proof By Theorem 4.2.6, $\|f^0\|_\infty < 1$. By Theorem 4.5,

$$\frac{1}{2\pi} \int_0^{2\pi} \log(1 - |f^0(e^{it})|^2)dt = \log\Big(\prod_{k=0}^{n-1}(1 - |\gamma_k|^2)\Big).$$

It follows from Remark 2.2.2 that $\langle(I - TT^*)^{-1}E_{n-1}, E_{n-1}\rangle = \prod_{k=0}^{n-1}(1 - |\gamma_k|^2)^{-1}$. Therefore,

$$-\frac{1}{2\pi} \int_0^{2\pi} \log(1 - |f^0(e^{it})|^2)dt = \log\langle(I - TT^*)^{-1}E_{n-1}, E_{n-1}\rangle. \qquad (4.22)$$

The main remark is that

$$\begin{aligned} \langle(I - TT^*)^{-1}E_{n-1}, E_{n-1}\rangle &= E_{n-1}^*(I - TT^*)^{-1}E_{n-1} \\ &= 1 + E_{n-1}^*T(I - T^*T)^{-1}T^*E_{n-1} \\ &\le \exp(E_{n-1}^*T(I - T^*T)^{-1}T^*E_{n-1}), \end{aligned}$$

and it follows from (4.22) that

$$-\frac{1}{2\pi} \int_0^{2\pi} \log(1 - |f^0(e^{it})|^2)dt \le E_{n-1}^*T(I - T^*T)^{-1}T^*E_{n-1}.$$

Since $|x|^2 < -\log(1 - |x|^2)$ when $|x| < 1$, we get

$$\begin{aligned} \frac{1}{2\pi} \int_0^{2\pi} |f^0(e^{it})|^2 dt &\le -\frac{1}{2\pi} \int_0^{2\pi} \log(1 - |f^0(e^{it})|^2)dt \\ &\le E_{n-1}^*T(I - T^*T)^{-1}T^*E_{n-1} \le \frac{E_{n-1}^*TT^*E_{n-1}}{1 - d_\infty^2} = \frac{d_2^2}{1 - d_\infty^2}. \end{aligned}$$

The proof is concluded.   □

5.5   Notes

Detailed accounts about the history of the spectral factorization can be found in [BGK], [BS2], [Ga], [Ho], [IR], [RR1], [Sz.-NF2], [WM]. The applications to the prediction theory raised the question of extending the classical factorization results to matrix valued functions ([HL], [IR], [WM]) and the method of D. Lowdenslager (in [Lo]) of using the Wold decomposition proved to be useful in this respect—see also [Dev], [Do1-2]. Theorem 1.2 represents an extension of the results in [RR1], [Sz.-NF2], [SV]—we used [Co5] for its presentation. The particular cases considered in Section 4.2 are taken from [Sz.N-F2](Theorem 2.2), [SV](Theorem 2.1) and [RR1](Theorem 2.5). An interesting application of the last result is related to the so-called *Darlington synthesis* problem—see [Arov], [DH]. The Szegö theory (Theorem 3.3) discussed in Section 3 is presented in [Sz]. Proposition 3.1 is a result of Geronimus [Ge]. The proof of Theorem 3.2 is taken from [Bo]. Theorem 3.8 is proved in [Ra] and represents an important step towards the extension of the Szegö theory from the Szegö class to the Erdös-Túran class (introduced in [ET]). The proof of Theorem 3.8 and the presentation of the material follows [MNT], [Nev1-2]. The presentation of the material in Section 4 follows [CSK2]. For discussions about the maximum entropy principle, see [ADD1], [ArK1-2], [Bev], [Bur], [BLW], [BoL1-2], [Ch], [DeD2], [DG3], [DM], [EGL], [FFG], [GKW], [La], [LPK], [Nel]. Theorem 4.2 is a time-varying generalization of the main result in [ArK2]. Theorem 4.3 and Theorem 4.5 are proved in [CSK2]. The discussion of the Nevanlinna algorithm as a generalized Schur algorithm follows [SKLC]. Theorem 4.6 is a particular case of a result in [KLW]. For this proof we followed [FoF2], [FR]. There is a vast literature devoted the state-space computation of the spectral factor—see [BGK]. In connection with this chapter see [BS1-2], [BL], [BLM], [Cy], [Geo], [GeK], [GoK2], [KaR], [Vi], [YJN].

Chapter 6

Nonstationary Processes

The main purpose of this chapter is show how to use the Schur parameters in computations regarding the geometry of nonstationary processes. A typical example in this direction is the Kolmogorov-Wiener problem. The solution of this problem in terms of the Schur parameters is then compared with the classical solution of Kolmogorov and Wiener. Moreover, the relation with some models developed by so-called parametric methods in spectral analysis is indicated. Finally, the asymptotic behavior of the operator angles of a process is related to the asymptotic behavior (Szegö type phenomena) of the determinants of the finite sections of a positive definite kernel.

6.1 Modeling Nonstationary Processes

In this section we introduce two models of nonstationary processes and some problems concerning their geometry. Let $(\Omega, \mathcal{U}, P)$ be a probability space, where $\mathcal{U}$ is a $\sigma-$algebra of subsets of Ω and P is a probability measure on $\mathcal{U}$. A function $x : \Omega \rightarrow \mathbb{C}$ which is measurable with respect to the $\sigma-$algebra $\mathcal{U}$ is called a *stochastic variable*. A *stochastic process* is a family $\{x_n\}_{n \in \mathbb{Z}}$ of stochastic variables. Let $L^2(P)$ be the Hilbert space of $\mathcal{U}-$measurable functions on Ω which are square integrable with respect to P, equipped with the inner product defined by

$$\langle f, g \rangle = \int_\Omega f(\omega) g^*(\omega) dP(\omega).$$

We will consider only stochastic processes with variables in $L^2(P)$. The *mean-value variable* is defined by

$$m_n = \int_\Omega x_n(\omega) dP(\omega)$$

and it is convenient to suppose that $m_n = 0$ for all $n \in \mathbb{Z}$. The *correlation kernel* of the stochastic process $\{x_n\}_{n \in \mathbb{Z}}$ is given by

$$A(m,n)(= A_{mn}) = \int_\Omega x_n(\omega) x_m^*(\omega) dP(\omega) = \langle x_n, x_m \rangle \qquad (1.1)$$

for $m, n \in \mathbb{Z}$. It is easy to see that the correlation kernel of a stochastic process is a positive definite kernel, since

$$\sum_{i,j=m}^{n} A_{ij} \lambda_j \lambda_i^* = \sum_{i,j=m}^{n} \langle \lambda_j x_j, \lambda_i x_i \rangle = \sum_{i,j=m}^{n} \| \lambda_i x_i \|^2 \geq 0$$

for all integers m, n, $m \leq n$, and arbitrary complex numbers λ_k, $k = m, m + 1$, ..., n.

A central question in the theory of stochastic processes concerns the measurement of the interaction between various parts of the process. For instance, it is assumed that the linear span of the variables $x_n, n > 0$ is known (this space may be viewed as the "future" of the process), and it is required to predict the values of the process at some other moment m of time, $m \leq 0$. This is the well-known *Kolmogorov-Wiener prediction problem*. Convenient measures of this interaction are given by various numerical characteristics (like norm or trace) of the angle between two spaces. We set our discussion of the prediction problem on a geometrical ground. However, the interaction between parts of a given process may be measured on a statistical level by various degrees of independence. For instance, denote by $\mathcal{U}(1, \infty)$ and $\mathcal{U}(0)$ the smallest $\sigma-$ algebras of subsets of Ω with the property that all the variables x_n, $n > 0$, respectively, x_0, are measurable. One usual measure for the degree of independence between $\mathcal{U}(1, \infty)$ and $\mathcal{U}(0)$ is given by the number

$$k(\mathcal{U}(1, \infty), \mathcal{U}(0)) = \sup\{|P(AB) - P(A)P(B)| / A \in \mathcal{U}(1, \infty), B \in \mathcal{U}(0)\}.$$

Due to the well-known interplay between (statistical) independence and (geometrical) orthogonality for Gaussian processes, it is apparent that $k(\mathcal{U}(1, \infty), \mathcal{U}(0))$ is an analogue of the angle between future and present.

It is useful to describe here the classical approach to prediction. A stochastic process $\{x_n\}_{n \in \mathbb{Z}}$ is said to be *stationary (in wide sense)* if its correlation kernel is a Toeplitz kernel (*i.e.* $A(m, n) = A_{n-m}$ for all $m, n \in \mathbb{Z}$). In this situation, we use Theorem 1.3.2 in order to associate to the stationary stochastic process $\{x_n\}_{n \in \mathbb{Z}}$ the following objects: the Hilbert space $\mathcal{H}_A$, the unitary operator $S \in \mathcal{L}(\mathcal{H}_A)$ and the operator $Q \in \mathcal{L}(\mathbb{C}, \mathcal{H}_A)$ such that

$$A_n = Q^* S^n Q, \qquad n \in \mathbb{Z}.$$

The space $\mathcal{H}_A$ is obtained by renorming l^2 with the positive definite Toeplitz kernel A. In l^2 we have the standard basis $\{E_n\}_{n \in \mathbb{Z}}$, where E_n is the vector with the n-th entry 1 and all the other entries zero. Because we also suppose $A_0 = 1$, we can write

$$A_n = \langle S^n[E_0], [E_0] \rangle_A$$

where $[E_n]$ denotes the class of E_n in $\mathcal{H}_A$. Let E be the spectral measure of the unitary operator S^* and let us define the probability measure μ by the formula

$$\mu(f) = \langle E(f)[E_0], [E_0] \rangle_A,$$

for every continuous function f on the unit circle. It follows that

$$A_n = \frac{1}{2\pi} \int_0^{2\pi} \nu_{-n}(e^{it}) d\mu(t).$$

The measure μ is called the *spectral measure* of the stochastic process $\{x_n\}_{n \in \mathbb{Z}}$ and the isomorphism $x_n \longleftrightarrow \nu_{-n}$ between $\bigvee\{x_n \mid n \in \mathbb{Z}\}$ and $L^2(\mu)$ is called the *trigonometric* (or *Kolmogorov*) *isomorphism*. This isomorphism translates the Kolmogorov-Wiener prediction problem into an approximation problem in $L^2(\mu)$.

The geometric setting for the prediction problem may be extended in order to deal with the multivariate case as well. To that end, we remark that a variable of a stochastic process can be viewed as an operator from $\mathbb{C}$ to $L^2(P)$ by defining

$$\bar{x}_n : \mathbb{C} \longrightarrow L^2(P)$$

$$\bar{x}_n \lambda = \lambda x_n,$$

and the elements of the correlation kernel of the process can be computed according to the rule $A(m,n) = (\bar{x}_m)^* \bar{x}_n$. We also note that there are many stochastic process which may have the same correlation kernel. These remarks show that it is convenient to adopt the following terminology. The main object describing a *multivariate process* will be its correlation kernel, which is assumed to be a positive definite kernel A such that $A(m,n)(= A_{mn})$ belongs to $\mathcal{L}(\mathcal{H}_n, \mathcal{H}_m)$ for any $m, n \in \mathbb{Z}$, where $\mathbf{H} = \{\mathcal{H}_n\}_{n \in \mathbb{Z}}$ is a given family of Hilbert spaces. A pair $[\mathcal{K}, \mathbf{x}]$, where $\mathcal{K}$ is a Hilbert space and $\mathbf{x} = \{x_n\}_{n \in \mathbb{Z}}$ is a family of operators x_n in $\mathcal{L}(\mathcal{H}_n, \mathcal{K})$, is called a *geometrical model of the multivariate process with correlation kernel A*, if

$$A(m,n) = \mathbf{x}_m^* \mathbf{x}_n.$$

Theorem 1.3.1 shows that once a positive definite kernel A is given, there exists a geometrical model of a multivariate process with covariance kernel A. Given a geometrical model $[\mathcal{K}, \mathbf{x}]$ of a multivariate process with covariance kernel A, we denote by $\mathcal{H}_{\mathbf{x}}$ the subspace of $\mathcal{K}$ generated by this model, *i.e.* ,

$$\mathcal{H}_{\mathbf{x}} = \bigvee_{n \in \mathbb{Z}} \mathbf{x}_n \mathcal{H}_n. \tag{1.2}$$

If $[\mathcal{K}', \mathbf{x}']$ is another geometrical model of the same process, then we know by Theorem 1.3.1 that there exists a unitary operator $\Phi : \mathcal{H}_{\mathbf{x}} \longrightarrow \mathcal{H}_{\mathbf{x}'}$ such that $\Phi \mathbf{x}_n = \mathbf{x}_n'$ for all $n \in \mathbb{Z}$. This shows that the geometry of the process is essentially determined by the choice of a geometrical model such that

$$\mathcal{K} = \bigvee_{n \in \mathbb{Z}} \mathbf{x}_n \mathcal{H}_n. \tag{1.3}$$

We now introduce the prediction problem for multivariate processes. For two subspaces $\mathcal{H}_1$ and $\mathcal{H}_2$ of a Hilbert space $\mathcal{H}$, define

$$B(\mathcal{H}_1, \mathcal{H}_2) = P_{\mathcal{H}_1} P_{\mathcal{H}_2} P_{\mathcal{H}_1}$$

and
$$\Delta(\mathcal{H}_1, \mathcal{H}_2) = I - P_{\mathcal{H}_1} P_{\mathcal{H}_2} P_{\mathcal{H}_1}.$$

Let us consider a positive definite kernel A and let $\mathbf{x} = \{\mathbf{x}_n\}_{n \in \mathbb{Z}}$ be a geometrical model of the nonstationary process with covariance kernel A. We define the following subspaces of $\mathcal{K}$:

$$\mathcal{H}_{p,q}(\mathbf{x}) = \bigvee_{k=p}^{q} \mathbf{x}_k \mathcal{H}_k, \tag{1.4}$$

for p, q integers or $\pm\infty$, such that $p \leq q$. We can formulate the following.

1.1 Problem   *Given a geometrical model* $[\mathcal{K}, \mathbf{x}]$ *of the multivariate process with correlation kernel* A, *it is required to compute the prediction error operators*

$$\Delta(\mathcal{H}_{q,0}(\mathbf{x}), \mathcal{H}_{1,p}(\mathbf{x})),$$

for $q \leq 0 < p$.

We conclude this section with the presentation of two models of a given multivariate process. Their introduction is motivated by the construction of the Kolmogorov isomorphism in the case of stationary stochastic process. The first model is described as follows. Let A be a positive definite kernel and let $\Gamma = \{\Gamma_{ij} \mid i, j \in \mathbb{Z}, i \leq j\}$ be the set of its Schur parameters. Then,

$$\mathcal{K} = \mathcal{K}_0, \tag{1.5}$$

where $\mathcal{K}_0$ is defined according to (1.6.2), and

$$\mathbf{x}_n = \begin{cases} W_{-1}^* W_{-2}^* \ldots W_n^* / \mathcal{H}_n & \text{if } n < 0 \\ P_{\mathcal{H}_0}^{\mathcal{K}_0} / \mathcal{H}_0 & \text{if } n = 0 \\ W_0 W_1 \ldots W_{n-1} / \mathcal{H}_n & \text{if } n > 0, \end{cases} \tag{1.6}$$

where the operators W_n, $n \in \mathbb{Z}$, are defined according to (1.6.3). Theorem 1.6.1 shows that this model $[\mathcal{K}_0, \mathbf{x}]$ satisfies (1.3). Using this model, the prediction problem will be solved in terms of the Schur parameters of the correlation kernel of the process.

The second model $[\mathcal{L}, \mathbf{y}]$ is obtained only for factorable kernels. Thus, suppose the correlation kernel A is factorable and let G_A be its spectral factor obtained in Theorem 5.1.2. Define

$$\mathcal{L} = \oplus_{n \in \mathbb{Z}} \mathcal{L}_n, \tag{1.7}$$

where $\mathcal{L}_n$ are the spaces introduced in the proof of Theorem 5.1.2, and

$$\mathbf{y}_n = col_n G_A \in \mathcal{L}(\mathcal{H}_n, \mathcal{K}). \tag{1.8}$$

Since S is factorable, it follows directly from the definition that this is a geometrical model of the multivariate process with the covariance kernel A. Besides, since G_A is outer, it follows that the condition (1.3) is also fulfilled.

6.2 Kolmogorov-Wiener Prediction

We present two solutions of the Kolmogorov-Wiener prediction problem (*i.e.* Problem 1.1 for $q = 0$ and $p = \infty$) corresponding to the geometric models introduced in the previous section.

Let $\mathbf{H} = \{\mathcal{H}_n\}_{n \in \mathbb{Z}}$ be a family of Hilbert spaces and let A be a positive definite kernel such that $A_{ij} = A(i,j)$ belongs to $\mathcal{L}(\mathcal{H}_j, \mathcal{H}_i)$ for all $i, j \in \mathbb{Z}$, and $A_{nn} = I_{\mathcal{H}_n}$ for all $n \in \mathbb{Z}$. Let $\Gamma = \{\Gamma_{ij} \mid i, j \in \mathbb{Z}, i \leq j\}$ be the set of the Schur parameters of A. The geometrical model $[\mathcal{K}_0, \mathbf{x}]$ of the multivariate process with correlation kernel A, given by (1.5) and (1.6), is based on the model of the Kolmogorov decomposition of A described in Theorem 1.6.1. Thus, for $i \in \mathbb{Z}$, we introduce the row contractions

$$L_i : \oplus_{k=i+1}^{\infty} \mathcal{D}_{\Gamma_{i+1,k}} \longrightarrow \mathcal{H}_i \tag{2.1}$$

$$L_i = \quad \text{the row contraction associated to} \\ \text{the parameters} \quad \{\Gamma_{ik} \mid i < k\}.$$

We also define the spaces $\mathcal{D}_i = \oplus_{k=i+1}^{\infty} \mathcal{D}_{\Gamma_{ik}}$ and $\mathcal{D}_{i,*} = \mathrm{cl}\mathcal{R}(H_\infty(L_i))$, where $H_\infty(L_i)$ was defined by (1.4.10). Further, we define the spaces $\mathcal{K}_i^+ = \mathcal{H}_i \oplus \mathcal{D}_i$ and $\mathcal{K}_i = \oplus_{j=-\infty}^{i-1} \mathcal{D}_{j,*} \oplus \mathcal{K}_i^+$, and the unitary operators

$$W_i : \mathcal{K}_{i+1} \longrightarrow \mathcal{K}_i \tag{2.2}$$

$$W_i = I \oplus \begin{bmatrix} I & 0 \\ 0 & \alpha_{L_i} \end{bmatrix} R(L_i) \begin{bmatrix} 0 & I \\ \beta_{L_i}^* & 0 \end{bmatrix}$$

with respect to the direct sum decompositions $\mathcal{K}_{i+1} = (\oplus_{j=-\infty}^{i-1} \mathcal{D}_{j,*}) \oplus (\mathcal{D}_{i,*} \oplus \mathcal{K}_{i+1}^+)$ and, respectively, $\mathcal{K}_i = (\oplus_{j=-\infty}^{i-1} \mathcal{D}_{j,*}) \oplus \mathcal{K}_i^+$. The operator $R(L_i)$ is the elementary rotation of L_i, while α_{L_i} and β_{L_i} are the unitary operators defined by (1.4.11) and (1.4.12). It is convenient to introduce the unitary operators

$$\Upsilon_n : \mathcal{K}_n \longrightarrow \mathcal{K}_0 \tag{2.3}$$

$$\Upsilon_n = \begin{cases} W_{-1}^* W_{-2}^* \dots W_n^* & \text{if} \quad n < 0 \\ I_{\mathcal{K}_0} & \text{if} \quad n = 0 \\ W_0 W_1 \dots W_{n-1} & \text{if} \quad n > 0, \end{cases}$$

so that, the operators defined by (1.6) can be expressed in the following alternative form

$$\mathbf{x}_n = \Upsilon_n / \mathcal{H}_n. \tag{2.4}$$

We also mention that in connection with the solution of the Kolmogorov-Wiener problem, it is of interest to compute the so-called *best predictor* of $\mathbf{x}_0$, defined by

$$\hat{\mathbf{x}}_0 = P_{\mathcal{H}_{1,\infty}(\mathbf{x})} \mathbf{x}_0.$$

We obtain the following result.

2.1 Theorem *The best predictor $\hat{\mathbf{x}}_0$ of $\mathbf{x}_0$ is given by the formula*

$$\hat{\mathbf{x}}_0 = W_0 \begin{bmatrix} 0 & L_n^* \end{bmatrix}^\top,$$

and the prediction error operator can be expressed as

$$\Delta(\mathcal{H}_{0,0}(\mathbf{x}), \mathcal{H}_{1,\infty}(\mathbf{x})) = H_\infty^2(L_0).$$

Proof The proof is based on the explicit representation of the space

$$\mathcal{H}_1 \vee W_1 \mathcal{H}_2 \vee W_1 W_2 \mathcal{H}_3 \vee \dots$$

inside $\mathcal{K}_1$. More precisely, this space is exactly $\mathcal{K}_1^+$. Hence, $\mathcal{H}_{1,\infty}(\mathbf{x}) = \Upsilon_1 \mathcal{K}_1^+$ and then $P_{\mathcal{H}_{1,\infty}(\mathbf{x})} = \Upsilon_1 P_{\mathcal{K}_1^+} \Upsilon_1^*$. It follows that the best predictor $\hat{\mathbf{x}}_0$ of $\mathbf{x}_0$ is given by

$$
\begin{aligned}
\hat{\mathbf{x}}_0 &= P_{\mathcal{H}_{1,\infty}(\mathbf{x})} \mathbf{x}_0 = \Upsilon_1 P_{\mathcal{K}_1^+} \Upsilon_1^* \Upsilon_0 / \mathcal{H}_0 \\
&= \Upsilon_1 P_{\mathcal{K}_1^+} W_0^* / \mathcal{H}_0 = \Upsilon_1 \begin{bmatrix} 0 & L_0^* \end{bmatrix}^\top.
\end{aligned}
$$

Then, the prediction error can be computed as follows:

$$
\begin{aligned}
\Delta(\mathcal{H}_{0,0}(\mathbf{x}), \mathcal{H}_{1,\infty}(\mathbf{x})) &= I_{\mathcal{H}_0} - P_{\mathcal{H}_{0,0}(\mathbf{x})} P_{\mathcal{H}_{1,\infty}(\mathbf{x})} P_{\mathcal{H}_{0,0}(\mathbf{x})} \\
&= I_{\mathcal{H}_0} - P_{\mathcal{H}_0} \Upsilon_1 P_{\mathcal{K}_1^+} \Upsilon_1^* P_{\mathcal{H}_0} \\
&= I_{\mathcal{H}_0} - P_{\mathcal{H}_0} W_0 P_{\mathcal{K}_1^+} W_0^* P_{\mathcal{H}_0} \\
&= I - L_0 L_0^* = H_\infty^2(L_0).
\end{aligned}
$$

The proof is complete. $\square$

The Kolmogorov-Wiener problem is now solved using the geometrical model $[\mathcal{L}, \mathbf{y}]$ given by (1.7) and (1.8). Since the equality (1.3) holds for both geometrical models $[\mathcal{K}_0, \mathbf{x}]$ and $[\mathcal{L}, \mathbf{y}]$, and $A_{nn} = I_{\mathcal{H}_n}$ for all $n \in \mathbb{Z}$, it follows by Theorem 1.3.1 that there exists a unitary operator Φ from $\mathcal{K}_0$ to $\mathcal{L}$ such that $\Phi \mathbf{x}_n = \mathbf{y}_n$ for all $n \in \mathbb{Z}$ and $\Phi / \mathcal{H}_0 = I_{\mathcal{H}_0}$. Consequently,

$$\Delta(\mathcal{H}_{0,0}(\mathbf{x}), \mathcal{H}_{1,\infty}(\mathbf{x})) = \Delta(\mathcal{H}_{0,0}(\mathbf{y}), \mathcal{H}_{1,\infty}(\mathbf{y})).$$

We have the following result.

2.2 Theorem *Suppose the positive definite kernel A is factorable and let G_A denote its spectral factor. The best predictor $\hat{\mathbf{y}}_0$ of $\mathbf{y}_0$ is given by the formula*

$$\hat{\mathbf{y}}_0 = \begin{bmatrix} \dots & 0 & 0_{\mathcal{L}_0} & (G_A)_{01} & (G_A)_{02} & \dots \end{bmatrix}^\top,$$

and the prediction error operator can be expressed as

$$\Delta(\mathcal{H}_{0,0}(\mathbf{y}), \mathcal{H}_{1,\infty}(\mathbf{y})) = (G_A)_{00}^*(G_A)_{00}.$$

Proof By the definition of the spectral factor G_A, it follows that

$$(G_A)_{00}^*(G_A)_{00} = H_\infty^2(L_0),$$

while Theorem 2.1 shows that $\Delta(\mathcal{H}_{0,0}(\mathbf{y}), \mathcal{H}_{1,\infty}(\mathbf{y})) = (G_A)_{00}^*(G_A)_{00}$. Remark that the equalities

$$\Delta(\mathcal{H}_{0,0}(\mathbf{x}), \mathcal{H}_{1,\infty}(\mathbf{x})) = \Delta(\mathcal{H}_{0,0}(\mathbf{y}), \mathcal{H}_{1,\infty}(\mathbf{y}))$$
$$= H_\infty^2(L_0) = (G_A)_{00}^*(G_A)_{00}$$

hold without the assumption of A being factorable. Finally, the equality (5.1.1) implies that $\mathcal{H}_{1,\infty}(\mathbf{y}) = \oplus_{k \geq 1} \mathcal{L}_k$, hence

$$\hat{\mathbf{y}}_0 = P_{\mathcal{H}_{1,\infty}(\mathbf{y})} \mathbf{y}_0 = P_{\oplus_{k \geq 1} \mathcal{L}_k} col_0 G_A = [\ldots \quad 0 \quad 0_{\mathcal{L}_0} \quad (G_A)_{01} \quad (G_A)_{02} \quad \ldots]^\mathsf{T}.$$

The proof is complete. □

2.3 Remark Let us consider a stationary stochastic process $\{x_n\}_{n \in \mathbb{Z}}$. Each x_n is a stochastic variable on the probability space $(\Omega, \mathcal{U}, P)$. Let A be the correlation kernel of the process, hence A is a positive definite Toeplitz kernel. Let μ be the spectral measure of the process and let $\{\Gamma_n\}_{n \geq 0}$ be the set of the Schur parameters of A. It is seen that the prediction error operator reduces in this case to the quantity $\|x_0 - \hat{x}_0\|_{L^2(P)}^2$.

According to Theorem 2.1,

$$\|x_0 - \hat{x}_0\|_{L^2(P)}^2 = \prod_{n=0}^{\infty} (1 - |\Gamma_n|^2), \tag{2.5}$$

which is a classical formula of Verblunsky, and according to Theorem 2.2,

$$\|x_0 - \hat{x}_0\|_{L^2(P)}^2 = \exp\left(\frac{1}{2\pi} \int_0^{2\pi} \log \mu'(t) dt\right), \tag{2.6}$$

which is a classical formula of Szegö. □

2.4 Remark Based on the Kolmogorov isomorphism described in Section 1, we see that Theorem 2.1 and Theorem 2.2 give the answer to an extremal problem of Szegö which requires the computation of the numbers

$$\alpha = \inf\left\{\frac{1}{2\pi} \int_0^{2\pi} |p(e^{it})|^2 d\mu(t) \mid p \in \mathcal{P}, p(0) = 1\right\}$$

and

$$\alpha_n = \inf\left\{\frac{1}{2\pi} \int_0^{2\pi} |p(e^{it})|^2 d\mu(t) \mid p \in \mathcal{P}_n, p(0) = 1\right\}$$

(remember that $\mathcal{P}_n$ denotes the set of polynomials of degree at most n and $\mathcal{P} = \cup_{n \geq 0} \mathcal{P}_n$). Indeed, one obtains

$$\alpha = \prod_{n=0}^{\infty} (1 - |\Gamma_n|^2) = \exp\left(\frac{1}{2\pi} \int_0^{2\pi} \log \mu'(t) dt\right)$$

and

$$\alpha_n = \prod_{k=0}^{n}(1 - |\Gamma_k|^2) = |\varphi_n^{\#}(0)|^{-2},$$

where $\{\Gamma_n\}_{n\geq 0}$ is the set of the Schur parameters of μ and $\{\varphi_n\}_{n\geq 0}$ is the set of the orthogonal polynomials of μ. $\square$

The next two examples illustrate Theorem 2.1 and Theorem 2.2 for some stochastic processes which are widely used in engineering applications.

2.5 Example   Let $\{x_n\}_{n\in\mathbb{Z}}$ be a stochastic process which satisfies the stochastic difference equation

$$x_n + a_1(n)x_{n+1} + \ldots + a_M(n)x_{n+M} = b(n)v_n. \tag{2.7}$$

In this equation, $\{v_n\}_{n\in\mathbb{Z}}$ is a *white noise process*, *i.e.* the correlation kernel δ of $\{v_n\}_{n\in\mathbb{Z}}$ is given by:

$$\delta_{ij} = \begin{cases} 1 & \text{if } i = j \\ 0 & \text{if } i \neq j. \end{cases}$$

Moreover, $\{a_k(n) \mid k = 1, 2, \ldots, M, \, n \in \mathbb{Z}\}$ and $\{b(n)\}_{n\in\mathbb{Z}}$ are two sets of given complex numbers. We suppose, for a certain simplicity, that $x_n = 0$ for $n > 0$. A stochastic process $\{x_n\}_{n\in\mathbb{Z}}$ which satisfies (2.7) is called *autoregressive process of order M*. For $n \leq 0$, it follows that the linear space generated by the set of variables $\{x_{-k}\}_{k=0}^{n}$ coincides with the linear space generated by the set of variables $\{v_{-k}\}_{k=0}^{n}$. Hence, for $0 \leq k < n$,

$$\langle x_{-k}, v_{-n} \rangle = 0. \tag{2.8}$$

These equalities show that the best predictor $\hat{x}_n$ of x_n, $n \leq 0$, is given by the formula

$$\hat{x}_n = P_{\mathcal{H}_{n+1,\infty}(\{x_n\}_{n\in\mathbb{Z}})}x_n = -a_1(n)x_{n+1} - a_2(n)x_{n+2}\ldots - a_M(n)x_{n+M}.$$

Moreover, the prediction error is $\|\hat{x}_n - x_n\| = |b(n)|^2\|v_n\|^2 = |b(n)|^2$. It follows from (2.7) and (2.8) that the correlation kernel of the process $\{x_n\}_{n\in\mathbb{Z}}$ satisfies a set of linear equations, namely, for $n \leq 0$,

$$A^{(n-M,n)}\begin{bmatrix} 1 \\ a_1(n-M) \\ \vdots \\ a_M(n-M) \end{bmatrix} = \begin{bmatrix} |b(n-M)|^2 \\ 0 \\ \vdots \\ 0 \end{bmatrix}, \tag{2.9}$$

where we used the notation $A^{(n-M,n)} = [A_{ij} \mid n - M \leq i,j \leq n]$ introduced by (1.5.9). The equations (2.9) are known as the *Yule-Walker equation*. If Γ_{ij}, $i,j \in \mathbb{Z}$, $i \leq j$, are the Schur parameters of the correlation kernel A of the process

$\{x_n\}_{n \in \mathbb{Z}}$, then it can be shown that $\Gamma_{ij} = 0$ for $j - i > M$ (obviously, $\Gamma_{ij} = 0$ for all $i, j \geq 0$). To that end, we first remark that the equations (2.9) yield

$$A^{(n-M,n)} X^{(n-M,n)} = \begin{bmatrix} |b(n-M)|^2 & * & & * \\ 0 & |b(n-M+1)|^2 & \ddots & \\ & & \ddots & \ddots & * \\ 0 & & & & |b(n)|^2 \end{bmatrix},$$

where the k-th column of the matrix $X^{(n-M,n)}$ is given by

$$\begin{bmatrix} 1 & a_1(n-M+k) & \cdots & a_{M-k+1}(n-M+k) \end{bmatrix}^{\top}$$

and the entries marked by $*$ play no role here. Thus, $(X^{(n-M,n)})^* B^{(n-M,n)} X^{(n-M,n)}$ is a selfadjoint upper triangular matrix, hence diagonal. Consequently, the lower triangular Cholesky factor $G_{n-M,n}$ of $A^{(n-M,n)}$ is exactly the matrix

$$\begin{bmatrix} b(n-M) & 0 & 0 \\ 0 & b(n-M+1) & \ddots \\ & \ddots & \ddots \\ 0 & & b(n) \end{bmatrix} (X^{(n-M,n)})^{-1}.$$

Now define $a_m(k) = 0$ for $m > M$, $k \in \mathbb{Z}$, and consider the corresponding matrices $X^{(n-m,n)}$ for $m > M$. By similar computations, it follows that the matrix

$$\begin{bmatrix} b(n-m) & 0 & 0 \\ 0 & b(n-m+1) & \ddots \\ & \ddots & \ddots \\ 0 & & b(n) \end{bmatrix} (X^{(n-m,n)})^{-1}$$

coincides with the lower triangular Cholesky factor $G_{n-m,n}$ of $A^{(n-m,n)}$. Using the formula (1.6.15), one obtains $\Gamma_{ij} = 0$ for $j - i > M$, exactly as it was claimed. Finally, we mention that an application of Theorem 2.1 leads to the equality

$$|b(n)|^2 = A_{nn} \prod_{k=1}^{M} (1 - |\Gamma_{n,n+k}|^2). \qquad \square$$

2.6 Example   We consider a stationary stochastic process $\{x_n\}_{n \in \mathbb{Z}}$ that satisfies a stochastic difference equation

$$x_n + a_1 x_{n+1} + \ldots + a_M x_{n+M} = b v_n, \qquad (2.10)$$

where $\{v_n\}_{n \in \mathbb{Z}}$ is a white noise process and $\{a_k\}_{k=1}^{M}$ is a set of complex numbers. We also suppose that the polynomial $z^M + a_1 z^{M-1} + \ldots + a_M = 0$ has all its roots inside the unit disc. The process $\{x_n\}_{n \in \mathbb{Z}}$ is referred to as a *stationary autoregressive process of order M.* The difference equation (2.10) can be rewritten in matrix form:

$$X_n = TX_{n+1} + V_n, \qquad (2.11)$$

where, for $n \in \mathbb{Z}$,

$$X_n = [\, x_n \quad x_{n+1} \quad \cdots \quad x_{n+M} \,]^\top,$$

$$V_n = [\, bv_n \quad 0 \quad 0 \quad 0 \,]^\top$$

and

$$T = \begin{bmatrix} -a_1 & -a_2 & & -a_M & 0 \\ 1 & 0 & \cdots & 0 & 0 \\ & & \ddots & \ddots & \\ 0 & & & 1 & 0 \end{bmatrix}.$$

Since $\det(z - T) = z^M + a_1 z^{M-1} + \ldots + a_M$, it follows that T is a strict contraction. Consequently, X_n belongs to the closed space generated by $\{V_k\}_{k \geq n}$ and x_n belongs to the closed space generated by $\{v_k\}_{k \geq n}$. Hence,

$$\langle x_n, v_k \rangle = 0$$

for $k < n$ and the best predictor $\hat{x}_n$ is computed as follows:

$$\hat{x}_n = -a_1 x_{n+1} - a_2 x_{n+2} \ldots - a_M x_{n+M}. \qquad (2.12)$$

The prediction error is

$$\|x_0 - \hat{x}_0\|^2 = |b|^2 \qquad (2.13)$$

and the Yule-Walker equations have the form

$$\begin{bmatrix} A_0 & A_1^* & & A_M^* \\ A_1 & A_0 & \ddots & A_{M-1}^* \\ & \ddots & \ddots & \\ A_M & A_{M-1} & & A_0 \end{bmatrix} \begin{bmatrix} 1 \\ a_1 \\ a_2 \\ \vdots \\ a_M \end{bmatrix} = \begin{bmatrix} |b|^2 \\ 0 \\ 0 \\ \vdots \\ 0 \end{bmatrix}, \qquad (2.14)$$

where $A_{m-n} = A_{nm}$, $n, m \in \mathbb{Z}$ and $A = [A_{nm}]_{n,m \in \mathbb{Z}}$ is the correlation kernel of the process $\{x_n\}_{n \in \mathbb{Z}}$. Let $\{\Gamma_n\}_{n \geq 0}$ be the set of the Schur parameters of the correlation kernel A. It follows from the analysis in Example 2.5 that $\Gamma_n = 0$ for $n > M$ and by Theorem 2.1, the prediction error can be computed by the formula

$$\|x_0 - \hat{x}_0\|^2 = \prod_{k=0}^{M} (1 - |\Gamma_k|^2). \qquad (2.15)$$

If $\{\varphi_n\}_{n \geq 0}$ denotes the set of the orthogonal polynomials of the positive measure μ with Fourier coefficients $\{A_n\}_{n \in \mathbb{Z}}$, then $1 + a_1 z + \ldots a_M z^M = \varphi_M^\#(z)$ and, by Lemma 5.3.4, μ is absolutely continuous with respect to Lebesgue measure, $\mu' = |\varphi_M^\#|^{-2}$. Moreover, the spectral factor of μ is $F_\mu = (\varphi_M^\#)^{-1}$. In the framework of autoregressive processes, Procedure 4.3.2 is used to solve the Yule-Walker equations and it is known as the *Levinson-Durbin algorithm*. The Schur parameters $\{\Gamma_n\}_{n\geq 0}$ are known as the *partial correlation* (PARCOR) *coefficients* and the transmission line associated to μ is interpreted as the *lattice filter implementation* of the so-called *M-th order linear prediction error*, which is $e_n(M) = x_n - \hat{x}_n$. The stationary autoregressive processes are used to approximate general stationary stochastic processes, and the asymptotic property of the orthogonal polynomials expressed by Theorem 5.3.3(a) is one of the motivations of this method. We also see that assuming the sequence $\{x_n\}_{n\in\mathbb{Z}}$ is a stationary autoregressive process of order M is equivalent to the fact that $\{x_n\}_{n\in\mathbb{Z}}$ satisfies the maximum entropy criterion. This is the basis of the so-called *Burg technique* in spectral analysis of stationary time series. The discussion in Example 2.5 shows that the same idea extends naturally to nonstationary time series.

6.3 Other Prediction Problems

In this section we address Problem 1.1 in more details. The solution is described in terms of Schur parameters.

Let $\mathbf{H} = \{\mathcal{H}_n\}_{n\in\mathbb{Z}}$ be family of Hilbert spaces and let A be a positive definite kernel such that $A_{ij} = A(i,j)$ belongs to $\mathcal{L}(\mathcal{H}_j, \mathcal{H}_i)$ for all $i, j \in \mathbb{Z}$, and $A_{nn} = I_{\mathcal{H}_n}$ for all $n \in \mathbb{Z}$. Let $\Gamma = \{\Gamma_{ij} \mid i, j \in \mathbb{Z}, i \leq j\}$ be the set of the Schur parameters of A. We consider the geometrical model $[\mathcal{K}_0, \mathbf{x}]$ of the multivariate process with correlation kernel A, given by (1.5) and (1.6). The main remark in the proof of Theorem 2.1 was that

$$\mathcal{H}_n \vee W_n\mathcal{H}_{n+1} \vee W_nW_{n+1}\mathcal{H}_{n+2}\ldots = \mathcal{K}_n^+ = \mathcal{H}_n \oplus \oplus_{k=n+1}^{\infty}\mathcal{D}_{\Gamma_{nk}}.$$

Even more, it may be noted that for $0 \leq m < \infty$,

$$\mathcal{H}_n \vee W_n\mathcal{H}_{n+1} \vee \ldots \vee W_nW_{n+1}\ldots W_{n+m}\mathcal{H}_{n+m+1} = \mathcal{H}_n \oplus \oplus_{k=n+1}^{n+m+1}\mathcal{D}_{\Gamma_{nk}}.$$

For $0 \leq m < \infty$, we denote the spaces $\mathcal{H}_n \oplus \oplus_{k=n+1}^{n+m+1}\mathcal{D}_{\Gamma_{nk}}$ by $\mathcal{K}_n^{m+1}$ and we also use the notation $\mathcal{K}_n^0 = \mathcal{H}_n$. One obtains that

$$P_{\mathcal{H}_{n+1,\infty}}(\mathbf{x}) = \Upsilon_{n+1}P_{\mathcal{K}_{n+1}^+}^{\mathcal{K}_{n+1}^{+}}\Upsilon_{n+1}^* \tag{3.1}$$

and for $p > n$,

$$P_{\mathcal{H}_{n+1,p}}(\mathbf{x}) = \Upsilon_{n+1}P_{\mathcal{K}_{n+1}^{n+1-p}}^{\mathcal{K}_{n+1}}\Upsilon_{n+1}^*. \tag{3.2}$$

However, the spaces $\mathcal{H}_{q,n}(\mathbf{x})$, $q < n$, do not have a convenient representation inside $\mathcal{K}_n$. Our purpose will be to find another geometrical model $[\tilde{\mathcal{K}}_0, \tilde{\mathbf{x}}]$ of the multivariate process with correlation kernel A, with the property that (1.3) holds and the spaces $\mathcal{H}_{q,n}(\tilde{\mathbf{x}})$ may be explicitly represented inside $\tilde{\mathcal{K}}_0$. Due to a remark in Section 1, there will exist a unitary operator Φ in $\mathcal{L}(\mathcal{K}_0, \tilde{\mathcal{K}}_0)$ such that $\tilde{\mathbf{x}}_n = \Phi\mathbf{x}_n$ for all $n \in \mathbb{Z}$. The knowledge of the structure of this operator together with the explicit representation of the spaces $\mathcal{H}_{n+1,p}(\mathbf{x})$ inside $\mathcal{K}_0$ and of the spaces $\mathcal{H}_{q,n}(\tilde{\mathbf{x}})$ inside $\tilde{\mathcal{K}}_0$ will lead to a solution of Problem 1.1.

Consider the family $\widetilde{\mathbf{H}} = \{\widetilde{\mathcal{H}}_n\}_{n\in\mathbb{Z}}$ of Hilbert spaces defined by $\widetilde{\mathcal{H}}_n = \mathcal{H}_{-n}$ and the family $\{\widetilde{\Gamma}_{ij} \mid i, j \in \mathbb{Z}, \ i \leq j\}$ of contractions defined by $\widetilde{\Gamma}_{ij} = \Gamma^*_{-j,-i}$. We see that this set belongs to $\Pi(\widetilde{\mathbf{H}})$. Let $\widetilde{A} = [\widetilde{A}_{ij}]_{i,j\in\mathbb{Z}}$ be the positive definite kernel associated to the set $\{\widetilde{\Gamma}_{ij} \mid i, j \in \mathbb{Z}, \ i \leq j\}$ of Schur parameters, and let $\widetilde{V}$ be the Kolmogorov decomposition of $\widetilde{A}$. For any $n \in \mathbb{Z}$, $\widetilde{V}(n)$ belongs to $\mathcal{L}(\widetilde{\mathcal{H}}_n, \widetilde{\mathcal{K}}_0)$. We have the following result.

3.1 Lemma   The pair $[\widetilde{\mathcal{K}}_0, \widetilde{\mathbf{x}}]$, where $\widetilde{\mathbf{x}} = \{\widetilde{\mathbf{x}}_n\}_{n\in\mathbb{Z}}$ and $\widetilde{\mathbf{x}}_n = \widetilde{V}(-n)$, is a geometrical model of the multivariate process with correlation kernel A.

Proof Directly from the construction of $\widetilde{A}$, it follows that $\widetilde{A}_{ij} = A^*_{-j,-i} = A_{-i,-j}$ for $i, j \in \mathbb{Z}$. Hence, $A_{ij} = A^*_{ji} = \widetilde{A}_{-i,-j} = \widetilde{V}^*(-i)\widetilde{V}(-j) = \widetilde{\mathbf{x}}^*_i\widetilde{\mathbf{x}}_j$, which means that $\widetilde{\mathbf{x}} = \{\widetilde{\mathbf{x}}_n\}_{n\in\mathbb{Z}}$ is a geometrical model of the multivariate process with correlation kernel A.   □

By construction, $\widetilde{\mathcal{K}}_0 = \vee_{n\in\mathbb{Z}}\widetilde{\mathbf{x}}_n\mathcal{H}_n$. It follows, as a consequence of Theorem 1.3.1, that there exists a unitary operator Φ in $\mathcal{L}(\mathcal{K}_0, \widetilde{\mathcal{K}}_0)$ such that $\widetilde{\mathbf{x}}_n = \Phi\mathbf{x}_n$ for all $n \in \mathbb{Z}$. Moreover, for $q < 0$,

$$\begin{aligned}
\mathcal{H}_{q,0}(\mathbf{x}) &= \Phi^*(\mathbf{x}_0\mathcal{H}_0 \vee \mathbf{x}_{-1}\mathcal{H}_{-1} \vee \ldots \vee \mathbf{x}_q\mathcal{H}_q) \\
&= \Phi^*(\widetilde{\mathcal{H}}_0 \vee \widetilde{V}(1)\widetilde{\mathcal{H}}_1 \vee \ldots \vee \widetilde{V}(-q)\mathcal{H}_{-q}) \\
&= \Phi^*(\widetilde{\mathcal{H}}_0 \oplus \mathcal{D}_{\widetilde{\Gamma}_{01}} \oplus \mathcal{D}_{\widetilde{\Gamma}_{02}} \oplus \ldots \oplus \mathcal{D}_{\widetilde{\Gamma}_{0,-q}}) \\
&= \Phi^*(\mathcal{H}_0 \oplus \mathcal{D}_{\Gamma^*_{-1,0}} \oplus \mathcal{D}_{\Gamma^*_{-2,0}} \oplus \ldots \oplus \mathcal{D}_{\Gamma^*_{q0}}),
\end{aligned}$$

therefore, for $q < 0 < p$,

$$B(\mathcal{H}_{q,0}(\mathbf{x}), \mathcal{H}_{1,p}(\mathbf{x})) = \Phi^* P^{\widetilde{\mathcal{K}}_0}_{\oplus^q_{k=0}\mathcal{D}_{\Gamma^*_{k0}}}\Phi W_0 P^{\mathcal{K}_1}_{\oplus^p_{k=1}\mathcal{D}_{\Gamma_{1k}}} W^*_0 \Phi^* P^{\widetilde{\mathcal{K}}_0}_{\oplus^q_{k=0}\mathcal{D}_{\Gamma^*_{k0}}}\Phi. \qquad (3.3)$$

We recall that the notation $T(\{\Gamma_{ij} \mid 1 \leq i \leq m, 1 \leq j \leq n\})$ was introduced in Remark 2.2.5 in order to denote the block contraction $T = [T_{ij} \mid 1 \leq i \leq m, 1 \leq j \leq n]$ associated to the Schur parameters $\{\Gamma_{ij} \mid 1 \leq i \leq m, 1 \leq j \leq n\}$. It is also convenient to introduce, for $q \in \{-\infty\}\cup\{k \in \mathbb{Z} \mid k \leq 0\}$, $p \in \{k \in \mathbb{Z} \mid k > 0\}\cup\{\infty\}$ and $n \in \mathbb{Z}$, the sets $\alpha^n_{q,p} = \{(i, j) \mid q \leq i + n \leq 0, 1 \leq j + n \leq p\}$ and $\alpha_{q,p} = \alpha^0_{q,p}$. We also consider the contractions $T[\alpha^n_{q,p}] = T(\{\Gamma_{ij} \mid (i, j) \in \alpha^n_{q,p}\})$.

3.2 Lemma   The i-th row of $T[\alpha_{-\infty,\infty}]$, $i \geq 0$, is given by the formula

$$l_i D_\infty(l_{i-1})\ldots D_\infty(l_0),$$

where l_i is the row contraction of infinite length associated to the parameters $\{\Gamma_{-i,n}\}_{k\geq 1}$.

Proof The proof can be simply concluded by using the rules mentioned in Remark 2.2.5 and Theorem 2.2.1 in order to associate Schur parameters to $T[\alpha_{-\infty,\infty}]$ and then, employing the formula $(\alpha)_k$ in the proof of Theorem 1.5.3 and the formula (1.5.7).   □

3.3 Remark We must note a dual formulation of Lemma 3.2. Thus, denote by c_j, $j \geq 1$, the column contraction of infinite length associated by Remark 1.4.7 to the parameters $\{\Gamma_{nj}\}_{n \leq 0}$. Define $\tilde{D}_\infty(c_j) = D_\infty^*(c_j^*)$ and then the j-th column of $T[\alpha_{-\infty,\infty}]$, $j \geq 1$, is given by

$$\tilde{D}_\infty^*(c_1) \ldots \tilde{D}_\infty^*(c_{j-1})c_j.$$

Another remark is that similar formulae (with an obvious change of notation) hold for all $T[\alpha^n_{-\infty,\infty}]$, $n \in \mathbb{Z}$. □

The contractions $T[\alpha_{q,p}]$ play the main role in the solution of Problem 1.1. We have the following result.

3.4 Theorem *Let A be a positive definite kernel and let Γ be the set of the Schur parameters of A. Let $[\mathcal{K}_0, \mathbf{x}]$ be the geometrical model of the nonstationary process with correlation kernel A, given by (1.5) and (1.6). Then, for $q \in \{-\infty\} \cup \{k \in \mathbb{Z} \mid k \leq 0\}$ and $p \in \{k \in \mathbb{Z} \mid k > 0\} \cup \{\infty\}$,*

$$B(\mathcal{H}_{q,0}(\mathbf{x}), \mathcal{H}_{1,p}(\mathbf{x})) = U^*T[\alpha_{q,p}]T[\alpha_{q,p}]^*U,$$

where U is a certain unitary operator.

Proof Roughly speaking, the proof exploits the Schur algorithm (layer peeling idea) at the level of the Kolmogorov decomposition. Let us illustrate this statement for a very simple instance. Thus, consider a contraction $T = [T_1 \quad T_2 \quad \ldots]$ in $\mathcal{L}(\oplus_{n=1}^\infty \mathcal{H}_n, \mathcal{H}')$ and let $\{\Gamma_n\}_{n=1}^\infty$ be the parameters of T associated by Proposition 1.4.2. For $k \geq 1$, there are also considered the contractions $T_{(k)}$ in $\mathcal{L}(\oplus_{n \geq k}\mathcal{H}_n, \mathcal{D}_{\Gamma_{k-1}^*})$ associated by Proposition 1.4.2 to the family of parameters $\{\Gamma_n\}_{n \geq k}$ (hence $T = T_{(1)}$). Obviously,

$$H_\infty^2(T_{(k)}) = D_{\Gamma_k^*}H_\infty^2(T_{(k+1)})D_{\Gamma_k^*}$$

and by Lemma 1.4.1, there exist unitary operators

$$r_k : \text{cl}\mathcal{R}(H_\infty(T_{(k)})) \longrightarrow \text{cl}\mathcal{R}(H_\infty(T_{(k+1)})) \tag{3.4}$$

$$r_k H_\infty(T_{(k)}) = H_\infty(T_{(k+1)})D_{\Gamma_k^*}$$

for any $k \geq 1$. Then, for $n \geq 1$, the operators $D_{*,n+1}(T_{(1)})$ mentioned in Remark 1.4.6 are exactly

$$D_{*,n+1}(T_{(1)}) = H_\infty(T_{(n+1)})r_n r_{n-1} \ldots r_1.$$

Now, the main remark is that the following identity holds:

$$R_{T_{(1)}}(I_{\oplus_{n \geq 1}\mathcal{H}_n} \oplus r_1^*) = (\begin{bmatrix} D_{\Gamma_1^*} & \Gamma_1 \\ -\Gamma_1^* & D_{\Gamma_1} \end{bmatrix} \oplus I_{\oplus_{k \geq 2}\mathcal{D}_{\Gamma_k}}) \begin{bmatrix} H_\infty(T_{(2)}) & 0 & T_{(2)} \\ 0 & I_{\mathcal{H}_1} & 0 \\ K(T_{(2)}) & 0 & D_\infty(T_{(2)}) \end{bmatrix},$$

$$\tag{3.5}$$

where $R_{T_{(1)}}$, $R_{T_{(2)}}$ are defined by (1.4.13), $K(T_{(2)})$ is the operator defined in Remark 1.4.6 and $\tilde{D}_\infty(T_{(2)})$ is the operator defined by (1.4.9).

The next step of the proof uses the dual formulation of Lemma 3.2 mentioned in Remark 3.3, in order to write down the parameters of $T[\alpha_{-\infty,\infty}]$, when it is viewed as a row contraction of infinite length. Thus, define $\bar{c}_1 = c_1$ and remark that $\tilde{D}_\infty(c_1) = \alpha_{c_1^*} D_{c_1^*}$ by (1.4.11). This allows the introduction of the operator

$$\bar{c}_2 : \mathcal{D}_{\Gamma_{12}} \longrightarrow \mathcal{D}_{\bar{c}_1^*}$$

$$\bar{c}_2 = (\alpha_{\bar{c}_1^*})^* c_2,$$

which is obviously a contraction. Then, the second column of $T[\alpha_{-\infty,\infty}]$ may be written in the form $D_{\bar{c}_1^*} \bar{c}_2$, as required. Besides,

$$\delta_2 : \mathcal{D}_{\bar{c}_2^*} \longrightarrow \mathcal{D}_{c_2^*}$$

$$\delta_2 D_{\bar{c}_2^*} = D_{c_2^*} \alpha_{\bar{c}_1^*}$$

is a unitary operator. By induction, and using Remark 3.3, it follows that the j-th column of $T[\alpha_{-\infty,\infty}]$ can be written in the form $D_{\bar{c}_1^*} \ldots D_{\bar{c}_{j-1}^*} \bar{c}_j$, where $\bar{c}_j$ are contractions in $\mathcal{L}(\mathcal{D}_{\Gamma_{1j}}, \mathcal{D}_{\bar{c}_{j-1}^*})$ for all $j \geq 1$. Moreover, unitary operators δ_j in $\mathcal{L}(\mathcal{D}_{\bar{c}_j^*}, \mathcal{D}_{c_j^*})$ can be constructed such that $\delta_j D_{\bar{c}_j^*} = D_{c_j^*} \alpha_{\bar{c}_{j-1}^*}$ ($\alpha_{\bar{c}_{j-1}^*}$ are the operators introduced by (1.4.11)). Using this structure of $T[\alpha_{-\infty,\infty}]$ and the unitary operators of type (3.4), it follows that the defect space $\mathcal{D}_{T[\alpha_{-\infty,\infty}]^*}$ may be identified with the space $\oplus_{j \leq 0} \mathcal{D}_{j,*}$. Similarly, if $T[\alpha_{-\infty,\infty}]$ is written as a column contraction of infinite length, then the defect space $\mathcal{D}_{T[\alpha_{-\infty,\infty}]}$ may be identified with the space $\oplus_{i \leq -1} \tilde{\mathcal{D}}_{i,*}$, where $\tilde{\mathcal{D}}_{i,*} = \mathrm{cl}\mathcal{R}(H_\infty(c_i^*))$. Using these identifications, the elementary rotation of the operator $T[\alpha_{-\infty,\infty}]$ gives rise to a unitary operator $R[\alpha_{-\infty,\infty}]$ in $\mathcal{L}(\mathcal{K}_1, \tilde{\mathcal{K}}_0)$. Actually, each family $\{\Gamma_{ij} \mid i \leq r, j \geq s\}$, $r < s$, gives rise to such a unitary operator $R(\{\Gamma_{ij} \mid i \leq r, j \geq s\})$.

These constructions and the equality (3.5) imply that, with a slight and obvious abuse of notation, the following identity holds:

$$R(\{\Gamma_{jk} \mid j \leq i, k > i\}) = \tilde{W}_{-i}^* (I_{\mathcal{H}_i} \oplus R(\{\Gamma_{jk} \mid j \leq i, k > i+1\})). \qquad (3.6)$$

Here, the family of operators $\{\tilde{W}_n\}_{n \in \mathbb{Z}}$ consists of the unitary operators associated to the kernel $\tilde{A}$ by (1.6.3). A dual result can be also derived, and the following equality holds:

$$R(\{\Gamma_{jk} \mid j \leq i, k > i\}) = (I_{\mathcal{H}_i} \oplus R(\{\Gamma_{jk} \mid j < i, k > i\})) W_i. \qquad (3.7)$$

Using the identities (3.6) and (3.7), one shows by induction that, for $n \geq 0$,

$$(I_{\mathcal{H}_0} \oplus R(\{\Gamma_{jk} \mid j < 0, k \geq 1\})) W_0 \ldots W_{n-1} / \mathcal{H}_n = \tilde{W}_{-1}^* \ldots \tilde{W}_{-n}^* / \mathcal{H}_n. \qquad (3.8)$$

The relation (3.8) applied to the positive definite kernel $\tilde{A}$ shows that, for $n < 0$,

$$(I_{\mathcal{H}_0} \oplus R(\{\Gamma_{jk} \mid j < 0, k \geq 1\})) W_{-1}^* \ldots W_n^* / \mathcal{H}_n = \tilde{W}_0 \ldots \tilde{W}_{-n-1} / \mathcal{H}_n. \qquad (3.9)$$

It follows from (3.8) and (3.9) that the unitary operator $U = I_{\mathcal{H}_0} \oplus R(\{\Gamma_{jk} \mid j < 0, k \geq 1\})$ satisfies the relations $U\mathbf{x}_n = \tilde{\mathbf{x}}_n$ for all $n \in \mathbb{Z}$. Besides, $UW_0 = R[\alpha_{-\infty,\infty}]$ and then,

$$P^{\tilde{\mathcal{K}}_0}_{\oplus_{k=0}^q \mathcal{D}_{\Gamma^*_{k0}}} UW_0 P^{\mathcal{K}_1}_{\oplus_{k=1}^p \mathcal{D}_{\Gamma_{1k}}} = P^{\tilde{\mathcal{K}}_0}_{\oplus_{k=0}^q \mathcal{D}_{\Gamma^*_{k0}}} T[\alpha_{-\infty,\infty}] P^{\mathcal{K}_1}_{\oplus_{k=1}^p \mathcal{D}_{\Gamma_{1k}}} = T[\alpha_{q,p}].$$

By (3.3), $B(\mathcal{H}_{q,0}(\mathbf{x}), \mathcal{H}_{1,p}(\mathbf{x})) = U^* T[\alpha_{q,p}] T[\alpha_{q,p}]^* U$ and the proof is complete.

$\square$

3.5 Remark   Problem 1.1 may be seen as a generalization of the Kolmogorov-Wiener prediction problem and Theorem 3.4 appears as an extension of Theorem 2.1. As noted in Theorem 2.2, the spectral factor of the correlation kernel A is also connected with the prediction problems. As another example, we consider here a problem which is referred to as the *prediction n units of time ahead* and which requires the computation of the operator $B(\mathcal{H}_{0,0}(\mathbf{x}), \mathcal{H}_{n,\infty}(\mathbf{x}))$ for $n > 0$. In order to obtain a solution of this problem, we use the Wold-von Neumann decomposition of the family of isometries defined by (4.1.5). Thus, for $n > 0$, we have the equality

$$\mathcal{K}_0^+ = (\mathcal{L}_0 \oplus W_0\mathcal{L}_1 \oplus \ldots \oplus W_0W_1 \ldots W_{n-2}\mathcal{L}_{n-1}) \oplus W_0 \ldots W_{n-1}\mathcal{K}_n^+,$$

where $\mathcal{L}_n$, $n \in \mathbb{Z}$, are the spaces introduced in the proof of Theorem 4.1.2. It follows that

$$\begin{aligned}
\mathcal{H}_{n,\infty}(\mathbf{x}) &= \vee_{k=n}^\infty \mathbf{x}_k \mathcal{H}_k = W_0 \ldots W_{n-1}\mathcal{K}_n^+ \\
&= \mathcal{K}_0^+ \ominus (\mathcal{L}_0 \oplus W_0\mathcal{L}_1 \oplus \ldots \oplus W_0W_1 \ldots W_{n-2}\mathcal{L}_{n-1}),
\end{aligned}$$

and then

$$P_{\mathcal{H}_{n,\infty}(\mathbf{x})} = P_{\mathcal{K}_0^+} - (P_{\mathcal{L}_0} + W_0 P_{\mathcal{L}_1} W_0^* + \ldots + W_0W_1 \ldots W_{n-2} P_{\mathcal{L}_{n-1}} W_{n-2}^* \ldots W_1^* W_0^*).$$

In conclusion, one obtains

$$B(\mathcal{H}_{0,0}(\mathbf{x}), \mathcal{H}_{n,\infty}(\mathbf{x})) = I_{\mathcal{H}_0} - \sum_{k=0}^{n-1} (G_A)_{k0}^* (G_A)_{k0},$$

which is a solution of the considered problem of prediction n units of time ahead. We remark that if $\mathcal{R}_0$ is the residual space of the process, then

$$\mathcal{R}_0 = \cap_{p \geq 0} W_0 W_1 \ldots W_{p-1} \mathcal{K}_p^+ = \cap_{p \geq 0} \mathcal{H}_{p,\infty}(\mathbf{x}),$$

and the following relations hold

$$B(\mathcal{H}_{0,0}(\mathbf{x}), \mathcal{R}_0) = s\text{-} \lim_{n \to \infty} B(\mathcal{H}_{0,0}(\mathbf{x}), \mathcal{H}_{n,\infty}(\mathbf{x}))$$

$$= I_{\mathcal{H}_0} - \sum_{n=0}^\infty (G_A)_{n0}^* (G_A)_{n0}.$$

$\square$

6.4   Szegö's Limit Theorems

In this section we indicate some connections between the asymptotic properties of a multivariate process and the asymptotic properties of certain determinants associated to the correlation kernel of the process. We express the main results in terms of Schur parameters, but we also illustrate these phenomena in the classical setting of the stationary stochastic processes.

We start again with a family $\mathbf{H} = \{\mathcal{H}_n\}_{n \in \mathbb{Z}}$ of Hilbert spaces and a positive definite kernel A such that $A_{ij} = A(i,j)$ belongs to $\mathcal{L}(\mathcal{H}_j, \mathcal{H}_i)$ for all $i, j \in \mathbb{Z}$, and $A_{nn} = I_{\mathcal{H}_n}$ for all $n \in \mathbb{Z}$. Let $\Gamma = \{\Gamma_{ij} \mid i, j \in \mathbb{Z}, \ i \leq j\}$ be the set of the Schur parameters of A and $[\mathcal{K}_o, \mathbf{x}]$ is the geometrical model of the multivariate process with correlation kernel A, given by (1.5) and (1.6). That is, $\mathbf{x}_n = \Upsilon_n/\mathcal{H}_n$, where the unitary operators Υ_n are given by the formula (2.3). We also recall the notation introduced in Chapter 1 by the formula (1.5.9). Thus, for $m \leq n$, we defined $A^{(mn)} = [A_{ij} \mid m \leq i, j \leq n]$. The upper triangular Cholesky factor F_{mn} of $A^{(mn)}$ was introduced by the formula (1.5.6). Moreover, in Section 3 we introduced the notation $\mathcal{K}_n^{m+1}$ for $0 \leq m < \infty$ and $n \in \mathbb{Z}$, in order to denote the space

$$\mathcal{H}_n \vee W_n\mathcal{H}_{n+1} \vee \ldots W_nW_{n+1}\ldots W_{n+m}\mathcal{H}_{n+m+1} = \mathcal{H}_n \oplus \oplus_{k=n+1}^{n+m+1}\mathcal{D}_{\Gamma_{nk}}.$$

The main remark in this section is that the prediction error operator $\Delta(\mathcal{H}_{q,0}(\mathbf{x}), \mathcal{H}_{1,p}(\mathbf{x}))$, $q \leq 0, p \geq 1$, is connected with a certain Schur complement in the matrix $A^{(qp)}$. Thus, we can write

$$A^{(qp)} = \begin{bmatrix} A^{(q0)} & \widetilde{Q}_{qp} \\ \widetilde{Q}_{qp}^* & A^{(1p)} \end{bmatrix}$$

with respect to the decomposition $(\oplus_{i=q}^0 \mathcal{H}_i) \oplus (\oplus_{i=1}^p \mathcal{H}_i)$ of the space $\oplus_{i=q}^p \mathcal{H}_i$ and we define the block matrix $\widetilde{Q}_{qp} = [A_{ij} \mid q \leq i \leq 0, 1 \leq j \leq p]$. We also define the contraction $\widetilde{G}_{qp} = P_{\mathcal{K}_q^{-q-1}} W_q \ldots W_{-1}W_0 P_{\mathcal{K}_1^{p-2}}$ and we can state the following result.

4.1 Lemma   *Let A be a positive definite kernel and $q \leq 0, p \geq 1$. Then*

(a) $\widetilde{Q}_{qp} = F_{q0}^* \widetilde{G}_{qp} F_{1p}.$

(b) $A^{(qp)} = (F_{q0}^* \oplus F_{1p}^*) \begin{bmatrix} I & \widetilde{G}_{qp} \\ \widetilde{G}_{qp}^* & I \end{bmatrix} (F_{q0} \oplus F_{1p}).$

(c)   *If the Schur parameters Γ_{ij}, $i, j \in \mathbb{Z}$, $i \leq j$ are strict contractions, then F_{ij} are invertible operators and*

$$\Delta(\mathcal{H}_{q,0}(\mathbf{x}), \mathcal{H}_{1,p}(\mathbf{x})) = \Upsilon_q(I - (F_{q0}^*)^{-1}\widetilde{Q}_{qp}(A^{(1p)})^{-1}\widetilde{Q}_{qp}^*F_{q0}^{-1})\Upsilon_q^*.$$

Proof Using the formulae (1.6.6) and (1.6.10), it follows for $i \leq j$ that

$$F_{ij} = \begin{bmatrix} P_{\mathcal{K}_i^{j-i-1}}/\mathcal{H}_i & P_{\mathcal{K}_i^{j-i-1}}W_i/\mathcal{H}_{i+1} & \ldots & P_{\mathcal{K}_i^{j-i-1}}W_i\ldots W_j/\mathcal{H}_j \end{bmatrix}$$

and then (a) is a consequence of Theorem 1.6.1. The relation (b) is a consequence of (a). Finally, it is remarked that the following equality of spaces holds for $q < 0$:

$$W_q \ldots W_{-1} \mathcal{H}_{q,0}(\mathbf{x}) = \mathcal{H}_q \vee W_q \mathcal{H}_{q+1} \vee \ldots W_q \ldots W_{-2} W_{-1} \mathcal{H}_0 = \mathcal{K}_q^{-q-1}.$$

Therefore, $P_{\mathcal{H}_{q,0}(\mathbf{x})} = \Upsilon_q P_{\mathcal{K}_q^{-q-1}} \Upsilon_q^*$ and together with the formula (3.2), this shows that for $q \leq 0$ and $p \geq 1$,

$$\Delta(\mathcal{H}_{q,0}(\mathbf{x}), \mathcal{H}_{1,p}(\mathbf{x})) = \Upsilon_q D_{\tilde{G}_{qp}}^2 \Upsilon_q^*.$$

The proof of (c) can be concluded using (a). $\qquad\qquad\qquad\qquad\square$

Hereafter it is assumed that all the spaces $\mathcal{H}_n$, $n \in \mathbb{Z}$, are finite dimensional. At the level of the geometry of the nonstationary process with correlation kernel A, several asymptotic properties are quite obvious. Thus, for instance, it is clear that

$$s\text{-} \lim_{n \to \infty} \Delta(\mathcal{H}_{0,0}(\mathbf{x}), \mathcal{H}_{1,n}(\mathbf{x})) = \Delta(\mathcal{H}_{0,0}(\mathbf{x}), \mathcal{H}_{1,\infty}(\mathbf{x})). \qquad (4.1)$$

On the other hand, if it is supposed that the Schur parameters Γ_{ij}, $i, j \in \mathbb{Z}$, $i \leq j$, are strict contractions, then it follows, by using Lemma 4.1 and the Frobenius-Schur identity (1.2.4), that

$$
\begin{aligned}
\det \Delta(\mathcal{H}_{0,0}(\mathbf{x}), \mathcal{H}_{1,n}(\mathbf{x})) &= \det(I - (F_{00}^*)^{-1} \widetilde{Q}_{0n}(A^{(1n)})^{-1} \widetilde{Q}_{0n}^* F_{00}^{-1}) \\
&= \det(I - \widetilde{Q}_{0n}(A^{(1n)})^{-1} \widetilde{Q}_{0n}^*) = \frac{\det A^{(0n)}}{\det A^{(1n)}} .
\end{aligned}
$$

This shows that the equality (4.1) reflects a certain asymptotic property of the determinants of finite sections of the kernel A. More precisely, we have the following result.

4.2 Theorem *Let A be a positive definite kernel and suppose that its Schur parameters Γ_{ij}, $i, j \in \mathbb{Z}$, $i \leq j$, are strict contractions. Let G_A be the spectral factor of A. Then, for each $m \in \mathbb{Z}$,*

$$\lim_{n \to \infty} \frac{\det A^{(mn)}}{\det A^{(m+1,n)}} = \prod_{k=m+1}^{\infty} \det D_{\Gamma_{km}}^2 = |\det(G_A)_{mm}|^2.$$

Proof The first equality can be derived using Theorem 1.5.10, while the second one can be obtained using Theorem 2.1 and Theorem 2.2. $\qquad\qquad\qquad\square$

Let A be a positive definite Toeplitz kernel and suppose $A_{ij} = A_{j-i} \in \mathbb{C}$ for all $i, j \in \mathbb{Z}$. Let $\{\Gamma_n\}_{n \geq 0}$ be the set of the Schur parameters of A and let μ be the positive measure with Fourier coefficients $\{A_n\}_{n \in \mathbb{Z}}$. In this case, $A^{(1,n)} = A^{(0,n-1)}$ and one obtains as a consequence of Theorem 4.2, the classical *first Szegö limit theorem*.

4.3 Theorem   *If μ belongs to the Szegö class, then*

$$\lim_{n\to\infty} \frac{\det A^{(0n)}}{\det A^{(0,n-1)}} = \prod_{k=0}^{\infty}(1 - |\Gamma_k|^2) = \exp(\frac{1}{2\pi}\int_0^{2\pi} \log\mu'(t)dt). \ \square$$

It is interesting to note that the first Szegö limit theorem provides information about the asymptotic distribution of the eigenvalues of the Toeplitz matrices $A^{(0n)}$. Indeed, let $\lambda_0^{(n)}$, $\lambda_1^{(n)}$, ... $\lambda_n^{(n)}$ be the eigenvalues of $A^{(0n)}$. Then, we obtain as a consequence of Theorem 4.3 that for any positive measure μ in the Szegö class, the following identity holds:

$$\lim_{n\to\infty} \log(\det A^{(0n)})^{\frac{1}{n+1}} = \lim_{n\to\infty} \frac{\log\lambda_0^{(n)} + \log\lambda_1^{(n)} + \ldots + \log\lambda_n^{(n)}}{n+1}$$

$$= \frac{1}{2\pi}\int_0^{2\pi} \log\mu'(t)dt < \infty.$$

Using some standard arguments of the theory of equal distributions, the preceding formula leads to the following result.

4.4 Corollary   *If $f = \mu'$ is a bounded positive function, $0 \leq m \leq f \leq M$, then for any continuous function F on the interval $[m, M]$,*

$$\lim_{n\to\infty} \frac{F(\lambda_0^{(n)}) + F(\lambda_1^{(n)}) + \ldots + F(\lambda_n^{(n)})}{n+1} = \frac{1}{2\pi}\int_0^{2\pi} F(f(t))dt < \infty. \ \square$$

Remark that Theorem 4.2 can be improved using the angles of the spaces $\mathcal{H}_{q,0}(\mathbf{x})$ and $\mathcal{H}_{1,p}(\mathbf{x})$. Thus, we define for $m \in \mathbb{Z}$,

$$g_m = \prod_{k=m+1}^{\infty} \det D_{\Gamma_{mk}}^2$$

and the next result reflects the following asymptotic properties of the multivariate process with correlation kernel A :

$$s\text{-}\lim_{p\to\infty} \Delta(\mathcal{H}_{q,0}(\mathbf{x}), \mathcal{H}_{1,p}(\mathbf{x})) = \Delta(\mathcal{H}_{q,0}(\mathbf{x}), \mathcal{H}_{1,\infty}(\mathbf{x}))$$

and, respectively,

$$s\text{-}\lim_{q\to-\infty} \Delta(\mathcal{H}_{q,0}(\mathbf{x}), \mathcal{H}_{1,\infty}(\mathbf{x})) = \Delta(\mathcal{H}_{-\infty,0}(\mathbf{x}), \mathcal{H}_{1,\infty}(\mathbf{x})).$$

4.5 Theorem   Let A be a positive definite kernel and suppose that its Schur parameters Γ_{ij}, $i,j \in \mathbb{Z}$, $i \leq j$, are strict contractions. Also suppose that $g_m > 0$ for all $m \in \mathbb{Z}$. Then

(i)   for each $q \leq 0$,

$$\lim_{p \to \infty} \frac{\det A^{(qn)}}{\det A^{(1n)}} = \prod_{k=q}^{0} g_k.$$

(ii)

$$\lim_{q \to -\infty} \frac{\det A^{(q0)}}{\prod_{k=q}^{0} g_k} = \prod_{i=-\infty}^{0} \prod_{j=1}^{\infty} \det D_{\Gamma_{ij}}^{-2}.$$

Moreover, the following are equivalent:

(a)

$$\lim_{q \to -\infty} \frac{\det A^{(q0)}}{\prod_{k=q}^{0} g_k} < \infty.$$

(b)   The operator $B(\mathcal{H}_{-\infty,0}(\mathbf{x}), \mathcal{H}_{1,\infty}(\mathbf{x}))$ is a trace class strict contraction.

(c)   The operator $T[\alpha_{-\infty,\infty}]T[\alpha_{-\infty,\infty}]^*$ is a trace class strict contraction.

Proof By Lemma 4.1,

$$\det \Delta(\mathcal{H}_{q,0}(\mathbf{x}), \mathcal{H}_{1,p}(\mathbf{x})) = \frac{\det A^{(qp)}}{\det A^{(q0)} \det A^{(1p)}}$$

and the proof of (i) and (ii) can be concluded by an application of Theorem 1.5.10. Then, suppose that $B(\mathcal{H}_{-\infty,0}(\mathbf{x}), \mathcal{H}_{1,\infty}(\mathbf{x}))$ is a trace class strict contraction. It follows that the sequence $\{B(\mathcal{H}_{q,0}(\mathbf{x}), \mathcal{H}_{1,\infty}(\mathbf{x}))\}_{q \leq 0}$ converges to $B(\mathcal{H}_{-\infty,0}(\mathbf{x}), \mathcal{H}_{1,\infty}(\mathbf{x}))$ in the topology of the trace class operators and

$$\lim_{q \to -\infty} \det \Delta(\mathcal{H}_{q,0}(\mathbf{x}), \mathcal{H}_{1,\infty}(\mathbf{x})) = \det \Delta(\mathcal{H}_{-\infty,0}(\mathbf{x}), \mathcal{H}_{1,\infty}(\mathbf{x})).$$

Moreover, $\det \Delta(\mathcal{H}_{-\infty,0}(\mathbf{x}), \mathcal{H}_{1,\infty}(\mathbf{x})) > 0$. Now, it is a consequence of the part (ii) of this theorem that

$$\det \Delta(\mathcal{H}_{q,0}(\mathbf{x}), \mathcal{H}_{1,\infty}(\mathbf{x})) = \frac{\prod_{k=q}^{0} g_k}{\det A^{(q0)}} \leq \prod_{i=-\infty}^{0} \prod_{j=1}^{\infty} \det D_{\Gamma_{ij}}^{2}.$$

Therefore,

$$\prod_{i=-\infty}^{0} \prod_{j=1}^{\infty} \det D_{\Gamma_{ij}}^{2} = \det \Delta(\mathcal{H}_{-\infty,0}(\mathbf{x}), \mathcal{H}_{1,\infty}(\mathbf{x})) > 0$$

and (b) implies (a). Conversely, suppose $\prod_{i=-\infty}^{0} \prod_{j=1}^{\infty} \det D_{\Gamma_{ij}}^{2} > 0$. It follows that

$$\lim_{q \to -\infty} \det \Delta(\mathcal{H}_{q,0}(\mathbf{x}), \mathcal{H}_{1,\infty}(\mathbf{x})) = \gamma > 0.$$

For each $q \leq 0$, $B(\mathcal{H}_{q,0}(\mathbf{x}), \mathcal{H}_{1,\infty}(\mathbf{x}))$ is a finite rank operator and let $\{\lambda_1^{(q)}, \ldots, \lambda_{N_q}^{(q)}\}$ be its eigenvalues. Consequently,

$$\begin{aligned}
\det \Delta(\mathcal{H}_{q,0}(\mathbf{x}), \mathcal{H}_{1,\infty}(\mathbf{x})) &= \det(I - B(\mathcal{H}_{q,0}(\mathbf{x}), \mathcal{H}_{1,\infty}(\mathbf{x})) \\
&= \prod_{k=1}^{N_q}(1 - \lambda_k^{(q)}) > \gamma > 0,
\end{aligned}$$

which shows that $0 \leq \lambda_k^{(q)} < 1$ for $k = 1, 2, \ldots, N_q$. Since

$$\prod_{k=1}^{N_q} (1 - \lambda_k^{(q)}) \leq \exp(-\sum_{k=1}^{N_q} \lambda_k^{(q)}),$$

it follows that $\operatorname{tr} B(\mathcal{H}_{q,0}(\mathbf{x}), \mathcal{H}_{1,\infty}(\mathbf{x})) \leq -\log \gamma$, hence

$$\sup_{q \leq 0} \operatorname{tr} B(\mathcal{H}_{q,0}(\mathbf{x}), \mathcal{H}_{1,\infty}(\mathbf{x})) < \infty.$$

This shows that $B(\mathcal{H}_{-\infty,0}(\mathbf{x}), \mathcal{H}_{1,\infty}(\mathbf{x}))$ is a trace class operator and it is easily seen that it is also a strict contraction. Finally, the equivalence between (b) and (c) is a consequence of Theorem 3.4. $\square$

4.6 Remark Let A be a positive definite Toeplitz kernel and suppose $A_{ij} = A_{j-i} \in \mathbb{C}$ for all $i, j \in \mathbb{Z}$. Let $\{\Gamma_n\}_{n \geq 0}$ be the set of the Schur parameters of A and let μ be the positive measure with Fourier coefficients $\{A_n\}_{n \in \mathbb{Z}}$. In this case, $A^{(q,0)} = A^{(0,-q)}$ for $q \leq 0$, and $g_m = g = \prod_{k=0}^{\infty} (1 - |\Gamma_k|^2)$ for all $m \in \mathbb{Z}$. As a consequence of Theorem 4.5(ii), we obtain an asymptotic formula for the Toeplitz determinants,

$$\lim_{n \to \infty} \frac{\det A^{(0n)}}{g^{n+1}} = \prod_{k=0}^{\infty} (1 - |\Gamma_k|^2)^{-k}$$

and the limit is finite if and only if $\sum_{k=0}^{\infty} k \log(1 - |\Gamma_k|^2) > -\infty$. $\square$

We conclude this section with some remarks concerning the role of the spectral factor F_μ of the measure μ in the computation of $\lim_{n \to \infty} g^{-(n+1)} \det A^{(0n)}$. This is, actually, the classical formulation of the *second (strong) Szegö limit theorem*. We consider here only a particular case which contains some of the basic computations.

4.7 Theorem   Let μ be a probability measure associated to the set $\{\Gamma_k\}_{k \geq 0}$ of Schur parameters such that $\Gamma_k = 0$ for $k > n$. Then, for $m \geq n$,

$$\frac{\det A^{(0m)}}{g^{m+1}} = \exp(\frac{1}{\pi} \int\int_{|z| \leq 1} |\frac{F'_\mu(z)}{F_\mu(z)}|^2 d\sigma).$$

Proof According to Theorem 5.3.4, the measure μ is absolutely continuous with respect to Lebesgue measure and $\mu' = |p|^{-2}$, where p a polynomial of degree n and zeros in the unit disc. In addition, $F_\mu = (p^\#)^{-1}$, and define

$$q(z) = p^\#(z) = c(z - z_0)(z - z_1) \ldots (z - z_{n-1}),$$

where z_k, $k = 0, 1, \ldots n - 1$, are complex numbers of modulus strictly greater than one. Moreover, suppose that the points $z_0, z_1, \ldots z_{n-1}$ are distinct. The proof consists of direct computations of the quantities $\exp(\frac{1}{\pi} \int\int_{|z| \leq 1} |F'_\mu(z)/F_\mu(z)|^2 d\sigma)$,

g and $\det A^{(0,n-1)}$ in terms of the zeros $z_0, z_1, \ldots z_{n-1}$ of q. First, it is remarked that

$$\frac{F'_\mu(z)}{F_\mu(z)} = -\frac{q'(z)}{q(z)} = -\left(\frac{1}{z-z_0} + \frac{1}{z-z_1} + \ldots + \frac{1}{z-z_{n-1}}\right),$$

hence

$$\frac{1}{\pi}\iint_{|z|\le 1}\left|\frac{F'_\mu(z)}{F_\mu(z)}\right|^2 d\sigma = \sum_{j,k=0}^{n-1}\frac{1}{\pi}\iint_{|z|\le 1}\frac{d\sigma}{(z-z_j)(z^*-z_k^*)}.$$

Passing to polar coordinates and using Cauchy's formula,

$$\iint_{|z|\le 1}\frac{d\sigma}{(z-z_j)(z^*-z_k^*)} = \int_0^1\int_0^{2\pi}\frac{r\,dr\,dt}{(re^{it}-z_j)(re^{-it}-z_k^*)}$$

$$= \int_0^1 r\int_0^{2\pi}\frac{dt}{(re^{it}-z_j)(re^{-it}-z_k^*)}dr$$

$$= 2\pi\int_0^1\frac{r\,dr}{z_jz_k^*-r^2} = \pi\log\frac{z_jz_k^*}{z_jz_k^*-1}.$$

and then,

$$\exp\left(\frac{1}{\pi}\iint_{|z|\le 1}\left|\frac{F'_\mu(z)}{F_\mu(z)}\right|^2 d\sigma\right) = \prod_{j,k=0}^{n-1}\frac{z_jz_k^*}{z_jz_k^*-1} = \frac{|z_0\ldots z_{n-1}|^2}{\prod_{j,k=0}^{n-1}(z_jz_k^*-1)}.$$

It is easy to see that

$$g = |F_\mu(0)|^2 = |c|^{-2}|z_0\ldots z_{n-1}|^{-2}.$$

Finally, $\det A^{(0,n-1)}$ is computed using Lagrange interpolation. It follows from (3.1.6) that

$$z^p = \sum_{k=0}^{n-1}\frac{z^p}{q'(z_k)}\frac{q(z)}{z-z_k}$$

for $p = 0, 1, \ldots, n-1$. Therefore,

$$[1 \quad z \quad \ldots \quad z^{n-1}]^\top = L[\frac{q(z)}{z-z_0} \quad \frac{q(z)}{z-z_1} \quad \ldots \quad \frac{q(z)}{z-z_{n-1}}]^\top,$$

where $L = [z_k^p/q'(z_k)]_{p,k=0}^{n-1}$ is obviously invertible since z_k, $k = 0, 1, \ldots n-1$, are distinct points. A new matrix is introduced by the formula

$$R^{(0,n-1)} = L^{-1}A^{(0,n-1)}(L^{-1})^*,$$

and then $\det R^{(0,n-1)} = |\det L^{-1}|^2\det A^{(0,n-1)}$. It is easy to see that

$$|\det L|^2 = \left|\frac{\prod_{k\ne j}(z_k-z_j)}{q'(z_0)\ldots q'(z_{n-1})}\right|^2 = |c|^{-2n}\prod_{k\ne j}|z_k-z_j|^{-2}.$$

Now, the usefulness of the previous transformation of $A^{(0,n-1)}$ consists in the fact that it is easy to compute $\det R^{(0,n-1)}$. Indeed,

$$
\begin{aligned}
[R^{(0,n-1)}]_{kj} &= \frac{1}{2\pi} \int_0^{2\pi} \frac{q(e^{it})}{e^{it} - z_j} \times \frac{q(e^{it})^*}{e^{-it} - z_k^*} \frac{dt}{|q(e^{it})|^2} \\
&= \frac{1}{2\pi} \int_0^{2\pi} \frac{dt}{(e^{it} - z_j)(e^{-it} - z_k^*)} = \frac{1}{z_j z_k^* - 1},
\end{aligned}
$$

hence

$$
\det R^{(0,n-1)} = \det[\frac{1}{z_j z_k^* - 1}]_{j,k=0}^{n-1} = \frac{\prod_{k \neq j} |z_k - z_j|^2}{\prod_{j,k=0}^{n-1}(z_j z_k^* - 1)}.
$$

Consequently,

$$
\frac{\det A^{(0,n-1)}}{g^n} = \frac{|z_0 \cdots z_{n-1}|^2}{\prod_{j,k=0}^{n-1}(z_j z_k^* - 1)}
$$

and the assumption that the zeros of q are distinct can be dropped. Besides, one obtains the equality

$$
\frac{\det A^{(0,n-1)}}{g^n} = \exp(\frac{1}{\pi} \iint_{|z| \leq 1} |\frac{F_\mu'(z)}{F_\mu(z)}|^2 d\sigma).
$$

Using the formulae for $\det A^{(0,n-1)}$ and g^n in terms of the Schur parameters of μ, the equalities

$$
\frac{\det A^{(0,m-1)}}{g^m} = \frac{\det A^{(0,n-1)}}{g^n} = \exp(\frac{1}{\pi} \iint_{|z| \leq 1} |\frac{F_\mu'(z)}{F_\mu(z)}|^2 d\sigma)
$$

are obtained for $m \geq n-1$. The proof is concluded.   □

This result may be used to obtain that under suitable conditions on the measure μ, the limit

$$
\lim_{n \to \infty} \frac{\det A^{(0,n)}}{g^{n+1}}
$$

is finite and equal to $\exp(\frac{1}{\pi} \iint_{|z| \leq 1} |F_\mu'(z)/F_\mu(z)|^2 d\sigma)$. However, it is beyond our scope to pursue these investigations.

6.5 Notes

The theory of stochastic processes is a well developed field and our purpose was only to illustrate some aspects involving Schur parameters. For monographs containing classical material close to the questions discussed in Section 2 and Section 3, we mention here [DM], [GrS], [IR] and [Wol], as well as [He] and [Sz.-NF2]. The solution of the Kolmogorov-Wiener problem is a classical result in [Kol-2] and [Wiener]. For the presentation of the Verblunsky formula (2.5) see [Ge]. For the Szegö formula (2.6) see [GrS], [Sz]. Autoregressive processes of finite order constitute a well known class of processes, largely used as a tool in the study of stochastic processes. For extensions of the Szegö formula (2.6) (in general, of prediction theory) to matrix and operatorial case we mention here [HL], [IR], [Ma1-3], [RR1], [SV], [Sz.NF2], [WM]. The extension of the Verblunsky formula to the operatorial, nonstationary case, was noticed in [Co5]. The presentation of the material in Section 3 follows the paper [AC]. For details about the Szegö limit theorems, see [GrS]. These results have a number of important applications and we mention here the monographs [BS1], [BS2] for a detailed account on this subject. The connection with the Schur parameters was established by Baxter [Bax] for the stationary stochastic processes. Theorem 4.2 was proved in [Co5] and Theorem 4.5 was proved in [AC].

Chapter 7

Graphs and Completion Problems

Some of the completion problems considered so far referred to matrices with specified entries lying about the main diagonal. It turns out that the tentative to approach similar problems for matrices with sparse specified entries leads to connections with certain classes of graphs. In this chapter we show some of these connections, especially emphasizing the role played by the chordal graphs in the extension of some of the results concerning the band matrices. Partial matrices which admit positive or contractive completions are studied in more details. Besides, it is pointed out that the Schur parameters still play a certain role in this direction.

7.1 Preliminaries

In this section we introduce some definitions and notation concerning graphs, with special attention directed to the class of chordal graphs.

An *undirected graph* is a pair $G = (V, E)$ consisting of a finite set V of *vertices* and a symmetric irreflexive binary relation E on V. Usually, we take $V = \{1, 2, \ldots, n\}$ and $|V| = n$ is the *cardinality* of V. We call E the set of the *edges* of the graph G. For a vertex v in V, the *adjacency* set of v is introduced by

$$Adj_G(v) = \{w \in V \mid (v, w) \in E\}.$$

When no confusion about G is possible, we use the symbol $Adj(v)$ instead of $Adj_G(v)$. If (v, w) belongs to E, then we say that v and w are *adjacent*.

Two graphs $G = (V, E)$ and $G' = (V', E')$ are called *isomorphic* if there exists a bijection $f : V \longrightarrow V'$ such that (v, w) belongs to E if and only if $(f(x), f(y))$ belongs to E' for all v, w in V.

A *subgraph* of $G = (V, E)$ is any graph $H = (V', E')$ satisfying $V' \subseteq V$ and $E' \subseteq E$. For a subset α of V, we define the *subgraph induced* by α to be $G_\alpha = (\alpha, E_\alpha)$, where $E_\alpha = \{(v, w) \in E \mid v, w \in \alpha\}$.

We mention here several examples of graphs which will frequently occur in our considerations. A graph $G = (V, E)$ is a *complete graph* on n vertices, denoted by K_n, if for every v and w in V, the pair (v, w) belongs to E. A *path* from v_0 to v_r of *length* r is a sequence of vertices in the order $[v_0, v_1, v_2, \ldots, v_r]$ such that (v_{i-1}, v_i) are edges for $i = 1, 2, \ldots, r$. The *n-cycle* (or the *chordless cycle* on n vertices), denoted by O_n, is the graph with n vertices written in the order $[v_1, v_2, \ldots, v_n]$, such that (v_n, v_1) and (v_{i-1}, v_i), $i = 2, 3, \ldots, n$ are all the edges of O_n. Another special type of undirected graphs are the so-called *bipartite graphs*. The vertex set of such a graph is represented by the union of two disjoint sets Y and Z and any edge has one endpoint in Y and the other one in Z. We denote a bipartite graph by $G = (Y, Z, E)$, such that $V = Y \cup Z$ and E is the set of edges.

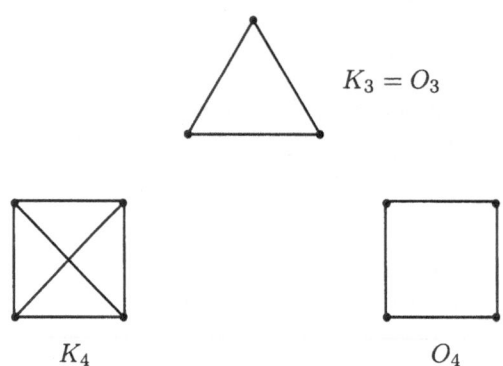

Figure 7.1: Examples of graphs

A graph G is called *chordal* if it does not contain induced subgraphs isomorphic to O_n, $n > 3$. Chordal graphs may be also found in literature under the name of *triangulated, rigid-circuit, monotone-transitive* or *perfect elimination* graphs. Several properties of the chordal graphs will be used in our developments and the necessary details are presented in the remainder of this section.

A graph $G = (V, E)$ is *connected* if between any two vertices there exists a path in G joining them. A subset S of V is a *vertex separator* of G if G_{V-S} is not connected. If two vertices a and b belong to two different connected components of G_{V-S}, then we call S an $a - b$ *separator*. S is called a *minimal vertex separator* if it is minimal (with respect to the inclusion of sets) $a - b$ separator for certain a, b in V. We obtain a first characterization of the chordal graphs.

1.1 Lemma   *G is a chordal graph if and only if every minimal vertex separator induces a complete subgraph of G.*

Proof Suppose every minimal vertex separator induces a complete subgraph of G and consider the $(k + 3)$-cycle $[a, x, b, y_1, \ldots, y_k]$ of G, $k \geq 1$. For the two nonadjacent vertices a and b there exists a minimal $a - b$ separator S. Consequently, x and one of the vertices y_i, $1 \leq i \leq k$, must belong to S, hence (x, y_i) is an edge of G. This is a contradiction with the fact that $[a, x, b, y_1, \ldots, y_k]$ is a $(k+3)$-cycle. In other words, G does not contain induced subgraphs isomorphic to O_n for $n > 3$, *i.e.* G is chordal.

Conversely, let S be a minimal $a - b$ separator. Moreover, let G_α and G_β be the connected components of G_{V-S} containing a and, respectively, b. We claim that if x belongs to S, then there exist a' in α and b' in β such that (x, a') and (x, b') are edges of G. For suppose $(x, a') \notin E$ for every a' in α. Define $S' = S - \{x\}$ and remark that the connected component in $G_{V-S'}$ containing a is still G_α, while the connected component in $G_{V-S'}$ containing b is either G_β or $G_{\beta \cup \{x\}}$. In any case, $S' \subset S$ is an $a - b$ separator, contradicting the minimality of S as an $a - b$ separator.

Finally, take x and y in S. By the above claim, there exist two paths of minimal length, $[x, a_1, \dots, a_r, y] \subset \alpha$ and $[y, b_1, \dots, b_t, x] \subset \beta$. Consequently, the graph G contains the cycle $[x, a_1, \dots, a_r, y, b_t, \dots, b_1]$ which is an n-cycle with $n > 3$, therefore (x, y) must be an edge of G. $\square$

1.2 Remark   It follows from the previous proof that if S is a minimal vertex separator of a chordal graph, then for all x, y in S there exists a vertex in each connected component of G_{V-S} which is adjacent to both x and y. Actually, we can prove that there exists a vertex in each connected component of G_{V-S} which is adjacent to all the vertices in S.

This statement was proved for every subset X of S with $|X| = 2$ and suppose it was also proved for every subset X of S with $|X| = k > 2$. Consider now a subset Y of S with $|Y| = k + 1$ and pick some z in Y. Let G_α be a connected component of G_{V-S}. By the induction hypothesis, there exist vertices $c_1, c_2, \dots, c_p$ in α adjacent to all the vertices in $Y - \{z\}$. Since S is minimal, there exists c in α such that (c, z) is an edge of G and assume $c \neq c_k$ for all $k = 1, 2, \dots, p$. Considering a path from c to c_k for each $k = 1, 2, \dots, p$, we get n-cycles of G with $n > 3$. It follows that at least one of the vertices $c_1, c_2, \dots, c_p$ should be adjacent to z, concluding the proof of the claim. $\square$

The next result plays an important role in the developments concerning chordal graphs.

1.3 Lemma   *Every chordal graph $G = (V, E)$ has a vertex v with the property that $G_{Adj(v)}$ is a complete graph. If $G \neq K_{|V|}$, then G has two nonadjacent vertices with this property.*

Proof If $G = K_{|V|}$, the result is obvious. Otherwise, let G be a chordal graph with two nonadjacent vertices a and b and suppose the lemma is true for all the graphs with fewer vertices than G. Let S be a minimal $a - b$ separator. Moreover, let G_α and G_β be the connected components of G_{V-S} containing a and, respectively, b. We distinguish two cases.

A. $G_{\alpha \cup S} = K_p$ for a certain $p < n$.

In this case, any vertex in α has the required property.

B. $G_{\alpha \cup S}$ has two nonadjacent vertices a' and b' whose adjacency sets induce complete graphs in $G_{\alpha \cup S}$.

Since G_S is a complete subgraph of G by Lemma 1.1, at least one of the vertices a' and b', which will be denoted by v, must be in α. But then, since

$Adj(v) \subset \alpha \cup S$ and $G_{Adj(v)}$ is a complete subgraph in $G_{\alpha \cup S}$, it follows that $G_{Adj(v)}$ is also a complete subgraph of G. Similarly, β will contain a vertex w such that $G_{Adj(w)}$ is a complete subgraph of G. The proof is complete.   □

A vertex v of G with the property that $G_{Adj(v)}$ is a complete subgraph of G is called a *simplicial vertex*. We introduced chordal graphs by means of a "forbidden subgraphs" characterization. The next result translates this characterization into a "positive" property. An ordering $\sigma = [v_1, v_2, \ldots, v_n]$ of the vertices of a graph G is called a *perfect vertex elimination scheme* (or *perfect scheme*) if each v_i, $i = 1, 2, \ldots, n-1$, is a simplicial vertex of $G_{\{v_i, v_{i+1}, \ldots, v_n\}}$ or, what is equivalent, if for each $i = 1, 2, \ldots, n-1$, the set

$$X_i = \{v_j \in Adj(v_i) \mid j > i\} \tag{1.1}$$

induces a complete subgraph of G.

1.4 Theorem   *G is a chordal graph if and only if it has a perfect scheme. Moreover, any simplicial vertex can start a perfect scheme.*

Proof Suppose G is a chordal graph. By Lemma 1.3, G has a simplicial vertex v_1. Since $G_{V-\{v_1\}}$ is also a chordal graph, this process may be continued in order to obtain a perfect scheme of G. Conversely, suppose $[x_1, x_2, \ldots, x_p], p > 3$, is a p-cycle in G and $\sigma = [v_1, v_2, \ldots, v_n]$ is a perfect scheme of G. Let x_r be the edge of the given p-cycle with lowest index in σ, say $x_r = v_k$. It follows that $x_{r+1} = v_m$ and $x_{r-1} = v_l$, where $m, l > k$. The vertex v_k is simplicial in $G_{\{v_k, \ldots, v_n\}}$, hence (x_{r-1}, x_{r+1}) is an edge of E, which is a contradiction showing that if G has a perfect scheme then G is a chordal graph.   □

Another result which is important for our purposes is the following.

1.5 Lemma   *Let $G = (V, E)$ and $G' = (V, E')$ be two chordal graphs with $E \subset E'$ and $|E'| \geq |E| + 2$. Then there exists a chordal graph $G'' = (V, E'')$ with $E \subset E'' \subset E'$.*

Proof If $|V| \leq 4$, the result is immediate. Then, suppose the result is true for every graph with at most $n - 1$ vertices. Consider $G = (V, E)$, $G' = (V, E')$ with $|V| = n$, $E \subset E'$ and $|E'| \geq |E| + 2$. Let $\sigma = [v_1, v_2, \ldots, v_n]$ be a perfect scheme of G and we distinguish two cases.

A. $G_{\{v_2, \ldots, v_n\}} \neq G'_{\{v_2, \ldots, v_n\}}$.

In this case, define $G_{\{v_2, \ldots, v_n\}} = (V - \{v_1\}, \hat{E})$ and $G'_{\{v_2, \ldots, v_n\}} = (V - \{v_1\}, \hat{E}')$. Then, by the induction hypothesis, there exists a chordal graph $\hat{G}'' = (V - \{v_1\}, \hat{E}'')$ with $\hat{E} \subset \hat{E}'' \subset \hat{E}'$. If $G'_{\{v_2, \ldots, v_n\}}$ has only one edge more than $G_{\{v_2, \ldots, v_n\}}$, we can choose $\hat{G}'' = G'_{\{v_2, \ldots, v_n\}}$. Then, the graph G'' is constructed by adding to $V - \{v_1\}$ the vertex v_1 and to $\hat{E}''$ all the edges in G having v_1 as an endpoint. If $[w_2, \ldots, w_n]$ is a perfect scheme of $\hat{G}''$, then $[v_1, w_2, \ldots, w_n]$ is a perfect scheme of G''. By Theorem 1.4, it is concluded that G'' is a chordal graph satisfying the required properties.

B. $G_{\{v_2,\dots,v_n\}} = G'_{\{v_2,\dots,v_n\}}.$

Hence, by Lemma 1.3, there exists another simplicial vertex of G starting a perfect scheme of G and which satisfies condition **A.** $\square$

As a consequence of Lemma 1.5, we obtain the following result which will play a key role in the study of the completion problem considered in the next section.

1.6 Theorem   *Let two chordal graphs $G = (V, E)$ and $G' = (V, E')$ with $E \subset E'$ be given. Then there exists a sequence of chordal graphs $G = G_0, G_1, \dots, G_s = G'$ such that G_j is obtained by adding exactly one edge to G_{j-1}, for all $j = 1, 2, \dots, s.$* $\square$

Another remarkable characterization of the chordal graphs will be used in the next chapter in connection with certain determinantal formulae. We need a few additional definitions. A *tree* is a connected graph without cycles. Let $\mathcal{F} = \{F_k\}_{k=1}^n$ be a family of nonempty sets. The *intersection graph* of $\mathcal{F}$ is defined by $G_{\mathcal{F}} = \{\mathcal{F}, \mathcal{E}\}$, where $(F_i, F_j) \in \mathcal{E}$ if and only if $i \neq j$ and $F_i \cap F_j \neq \emptyset$. A tree $T = (V, E(T))$ with $V = \{F_1, F_2, \dots F_m\}$ has the *intersection property* if $F_i \cap F_j \subseteq F_k$ whenever F_k lies on the (unique) path from F_i to F_j in T. A special role in connection with intersection graphs is played by the set $\mathcal{C} = \{\omega_k\}_{k=1}^p$ of the maximal cliques of a graph $G = (V, E)$. A subset $\omega \subseteq V$ is called a *clique* of G if G_ω is a complete subgraph and the maximality is understood with respect to the inclusion of sets.

1.7 Theorem   *Let $G = (V, E)$ be a connected graph. The following are equivalent:*

(a)   *G is a chordal graph.*
(b)   *G is the intersection graph of a family of subtrees of a tree.*
(c)   *There exists a tree $T = (\mathcal{C}, \mathcal{E})$ such that each of the induced subgraphs $T_{\mathcal{C}_v}$ ($v \in V$) is connected, where $\mathcal{C}_v$ denotes the set of those maximal cliques in G containing v.*
(d)   *There exists a tree $T = (\mathcal{C}, \mathcal{E})$ with the intersection property.*

Proof $(a) \Rightarrow (c)$ This implication is proven by induction on $|V|$. The result is obvious for small values of $|V|$. Consider $G = (V, E)$ a connected graph which is not complete and assume that the result is true for all the graphs having fewer vertices than G. Let $\sigma = [v_1, v_2, \dots, v_n]$ be a perfect scheme of G. Then $G_{\{v_2,\dots,v_n\}}$ is a chordal graph with fewer vertices than G and by the induction hypothesis, there exists a tree $T' = (\mathcal{C}', \mathcal{E}')$ with the property that each of the induced subgraphs $T'_{\mathcal{C}'_v}$ ($v \in \{v_2, \dots, v_n\}$) is connected, where $\mathcal{C}'$ is the set of the maximal cliques of $G_{\{v_2,\dots,v_n\}}$. We can distinguish two cases (remember that X_1 is defined by (1.1)).

A.   The clique X_1 is a maximal clique in $G_{\{v_2,\dots,v_n\}}$.

Then, one obtains the tree $T = (\mathcal{C}, \mathcal{E})$ by renaming X_1 with $\{v_1\} \cup X_1$ and taking $\mathcal{E} = \mathcal{E}'$. It is easy to see that

$$\mathcal{C}_v = \begin{cases} \mathcal{C}'_v & \text{if } v \neq v_1 \\ \{\{v_1\} \cup X_1\} & \text{if } v = v_1, \end{cases}$$

consequently the tree T has the required property.

B.   The clique X_1 is not a maximal clique in $G_{\{v_2,\ldots,v_n\}}$.

To the tree T' it is added a new vertex corresponding to $\{v_1\} \cup X_1$ and a new edge joining this vertex with the vertex of T' corresponding to the maximal clique of $G_{\{v_2,\ldots,v_n\}}$ containing X_1. This construction gives

$$
\mathcal{C}_v = \begin{cases}
\mathcal{C}'_v & \text{if } v \notin \{v_1\} \cup X_1 \\
\{\{v_1\} \cup X_1\} & \text{if } v = v_1 \\
\mathcal{C}'_v \cup \{\{v_1\} \cup X_1\} & \text{if } v \in X_1,
\end{cases}
$$

consequently the tree T has the required property.

$(c) \Rightarrow (b)$ Assume that there exists a tree $T = (\mathcal{C}, \mathcal{E})$ as in the statement (c). Consider the family of subtrees $\mathcal{F} = \{T_{\mathcal{C}_v} \mid v \in V\}$. Let (v, w) be an edge of G and let ω be a maximal clique in $\mathcal{C}$ containing v and w. Consequently, ω belongs to $\mathcal{C}_v \cap \mathcal{C}_w$ and $T_{\mathcal{C}_v} \cap T_{\mathcal{C}_w} \neq \emptyset$, so that G can be identified (by an isomorphism of graphs) with the intersection graph of $\mathcal{F}$.

$(b) \Rightarrow (a)$ Let $\{T_v\}_{v \in V}$ be a family of subtrees of a tree T such that (v, w) is an edge of G if and only if $T_v \cap T_w \neq \emptyset$. Consider a k-cycle $[v_1, \ldots, v_k]$ of G with $k > 3$ and choose α_i in $T_{v_i} \cap T_{v_{i+1}}$ for $i = 1, 2, \ldots, k-1$, and α_k in $T_{v_k} \cap T_{v_1}$. Moreover, consider the (unique) paths from α_i to α_{i+1} if $i = 1, 2, \ldots, k-1$ and the path from α_k to α_1. Eliminating possible common parts of these paths we get a k-cycle of T, a contradiction which shows that G is a chordal graph if it is the intersection graph of a family of subtrees of a tree.

$(d) \Rightarrow (c)$ Choose v in V and let us show that $T_{\mathcal{C}_v}$ is connected. Let ω_i, ω_j be two maximal cliques in $\mathcal{C}_v$ and suppose ω_k lies on the path of T from ω_i to ω_j. Consequently, $v \in \omega_i \cap \omega_j \subseteq \omega_k$. That is, the path in T from ω_i to ω_j is also a path in $T_{\mathcal{C}_v}$.

$(c) \Rightarrow (d)$ Let ω_i, ω_j be maximal cliques in $\mathcal{C}_v$ and suppose ω_k lies on the path of T from ω_i to ω_j. Pick some v in $\omega_i \cap \omega_j$, then ω_i and ω_j belong to $\mathcal{C}_v$. Since $T_{\mathcal{C}_v}$ is connected, it follows that ω_k belongs to $\mathcal{C}_v$, i.e. v belongs to ω_k. Therefore, $\omega_i \cap \omega_j \subseteq \omega_k$. $\square$

We mention now two interesting subclasses of chordal graphs. A graph G is called *proper interval graph* if it can be identified (by an isomorphism of graphs) with the intersection graph of a family of intervals on the real line such that no interval properly contains another. A graph G is called *interval graph* if it can be identified with the intersection graph of a family of intervals on the real line. It is easy to see that it does not matter whether we use open intervals or closed intervals, because the resulting class of (proper) interval graphs is the same. An interval graph, denoted by $K_{1,3}$, which is not a proper interval graph is depicted in Figure 7.2.

We note a characterization of the proper interval graphs which will allow us in the next section to conveniently rephrase the problem of completing block band matrices.

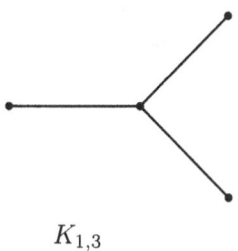

$K_{1,3}$

Figure 7.2: The graph $K_{1,3}$

1.8 Proposition   *A graph $G = (V, E)$ is a proper interval graph if and only if there exists an ordering $\sigma = [v_1, v_2, \ldots, v_n]$ of V such that $(v_i, v_j) \in E$ for $i < j$ implies that $(v_p, v_q) \in E$ for every p, q with $i \leq p < q \leq j$.*

Proof Suppose the proper interval graph G is the intersection graph of the family of intervals $\{I_k\}_{k=1}^{n}$, $I_k = (a_k, b_k)$, with $a_1 < a_2 < \ldots < a_n$. Since G is proper interval, we can consider $b_1 < b_2 < \ldots < b_n$. Suppose $I_i \cap I_j \neq \emptyset$ and $i < j$. Then $a_j < b_i$ and a_j belongs to I_p for every p with $i \leq p \leq j$, hence $I_p \cap I_q \neq \emptyset$ for any p, q, $i \leq p < q \leq j$. Conversely, the existence of the ordering σ implies that for each $k = 1, 2, \ldots, n$, there exist positive integers $\alpha(k)$ and $\beta(k)$ such that $1 \leq \alpha(k) \leq k \leq \beta(k) \leq n$ and $\{v_k\} \cup Adj(v_k) = \{v_{\alpha(k)}, v_{\alpha(k)+1}, \ldots v_{\beta(k)}\}$. Moreover, $1 = \alpha(1) \leq \alpha(2) \leq \ldots \leq \alpha(n) \leq n$ and $1 \leq \beta(1) \leq \beta(2) \leq \ldots \leq \beta(n) = n$. Choose $0 < \varepsilon_1 < \varepsilon_2 < \ldots < \varepsilon_n < \frac{1}{2}$, and define $I_k = (\alpha(k) - \varepsilon_{n-k+1}, \beta(k) + \varepsilon_k)$. It is readily checked that (v_i, v_j) is an edge of G if and only if $I_i \cap I_j \neq \emptyset$, hence G is a proper interval graph.   □

In Figure 7.3 there are depicted two graphs which are chordal but not proper interval. We also include here one characterization of the interval graphs.

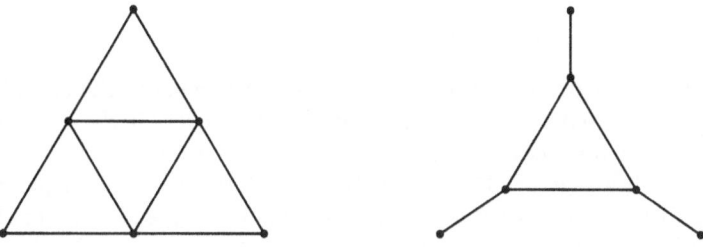

Figure 7.3: Two graphs which are chordal but not proper interval

1.9 Proposition   *A graph G is an interval graph if and only if there exists a linear ordering of its maximal cliques such that for every vertex v of G, the maximal cliques containing v occur consecutively.*

Proof Suppose that the maximal cliques $\omega_1 < \omega_2 < \ldots < \omega_s$ are identified with the integers $1, 2, \ldots, s$ on the real axis. For v in V, consider $I(v)$ the interval $(p - \frac{2}{3}, q + \frac{2}{3})$ with the property that $\{p, p + 1, \ldots, q - 1, q\}$ is the set of all the maximal cliques of G containing v. The fact that (v, w) belongs to E implies that there exists a maximal clique ω_k containing v and w, hence $I(v) \cap I(w) \neq \emptyset$. In other words, G is the intersection graph of the family of intervals $\{I(v)\}_{v \in V}$. Conversely, let G be the intersection graph of the family of intervals $\{I_k\}_{k=1}^n$, $I_k = (a_k, b_k)$, $k = 1, 2, \ldots, n$, with $a_1 < a_2 < \ldots < a_n$ (we can suppose, without loss of generality, that all the numbers contained in the set $\{a_k, b_k\}_{k=1}^n$ are distinct). A partition of the set $\{a_k\}_{k=1}^n$ is defined in the following way:

$$\alpha_1 = \{a_p \mid 1 \leq p \leq k_1 - 1, \, a_p < b_1\}$$

and for $r > 1$,

$$\alpha_r = \{a_p \mid k_{r-1} \leq p \leq k_r - 1, \, a_p < b_{k_{r-1}}\}.$$

Certainly, this procedure stops after a finite number s of steps and define for $1 \leq r \leq s$ the sets $\omega_r = \{(a_p, b_p) \mid a_p \in \alpha_r\}$. It is a simple exercise to check that $\{\omega_r\}_{r=1}^s$ is the set of the maximal cliques of G and that $\omega_1 < \omega_2 < \ldots < \omega_s$ is a linear ordering with the required properties. $\qquad \square$

7.2 Completing Positive Partial Matrices. I

The next three sections contain material devoted to certain generalizations of the completion problems studied in Chapter 3. Additional terminology is necessary for the presentation of the new problems.

By *partial matrix* we mean a rectangular array in which some entries are specified (*i.e.* they are known complex numbers, matrices or operators) and the others are unspecified. For instance

$$\begin{bmatrix} \circ & \circ & ? & \circ \\ \circ & \circ & \circ & ? \\ ? & \circ & \circ & \circ \\ \circ & ? & \circ & \circ \end{bmatrix} \quad \text{and} \quad \begin{bmatrix} ? & \circ & \circ \\ \circ & ? & \circ \end{bmatrix}$$

represent two such partial matrices, where by $\circ$ we denoted the specified entries and by ? we denoted the unspecified entries. Moreover, if the specified entries are operators or matrices we will sometimes use the term *partial block matrix* to refer to such a matrix. By *positive partial matrix* we mean a partial matrix $A_0 = [A_{ij} \mid 1 \leq i, j \leq n]$ with the properties: (1) its diagonal consists of specified positive elements; (2) if A_{ij} is specified then so is A_{ji} and $A_{ji} = A_{ij}^*$; (3) each of its specified principal submatrices $A_0[\alpha]$, $\alpha \subset \{1, 2, \ldots, n\}$, is positive. Throughout the text, $A_0[\alpha]$ denotes the partial matrix lying in the rows and columns indicated by α. The undirected graph $G(A_0)$ associated to a positive partial matrix $A_0 = [A_{ij} \mid 1 \leq i, j \leq n]$ has $\{1, 2, \ldots n\}$ as the set of vertices and an edge between i and j if A_{ij} is a specified entry of A_0. We will say that A_0 is *subjacent* to the graph G when $G(A_0) = G$. A *completion* of a partial matrix A_0 is a matrix A which coincides with A_0 on the specified entries of A_0.

Using Proposition 1.8, we can easily notice that any block band matrix as considered in Chapter 3 is subjacent to a proper interval graph. Consequently, we see that Theorem 3.4.9 concerning the band completion problem can be rephrased as follows: Every positive partial block matrix subjacent to a proper interval graph has positive completions.

This remark leads to the following generalization of the band completion problem.

2.1 Problem   *Characterize those graphs G with the property that every positive partial block matrix subjacent to G has positive completions.*

A first step towards a solution of Problem 2.1 is to exhibit a general class of graphs which do not have the required property. Namely, we show that the n-cycles with $n \geq 4$ are not solutions to Problem 2.1.

2.2 Lemma   *For any fixed complex number z with $|z| < 1$ and any $n \geq 4$, there exist complex numbers $a_2, a_3, \ldots, a_n$ with $|a_k| < 1$ for all $k \in \{2, 3, \ldots, n\}$, such that the strictly positive partial matrix*

$$
A_0 = \begin{bmatrix}
1 & z & ? & & a_n \\
z^* & 1 & a_2 & ? & \\
? & a_2^* & 1 & \ddots & \\
& ? & \ddots & \ddots & a_{n-1} \\
a_n^* & & & a_{n-1}^* & 1
\end{bmatrix}
$$

has no positive completion.

Proof   First, it is considered the case $n = 4$. Suppose A_0 has a positive completion

$$
A = \begin{bmatrix}
1 & z & b_{13} & a_4 \\
z^* & 1 & a_2 & b_{24} \\
b_{13}^* & a_2^* & 1 & a_3 \\
a_4^* & b_{24}^* & a_3^* & 1
\end{bmatrix}.
$$

Since the matrix

$$
\begin{bmatrix}
1 & z & b_{13} \\
z^* & 1 & a_2 \\
b_{13}^* & a_2^* & 1
\end{bmatrix}
$$

is positive, Theorem 1.5.3 shows that

$$
b_{13} = za_2 + (1 - |z|^2)^{1/2} g (1 - |a_2|^2)^{1/2}, \tag{2.1}
$$

where $|g| < 1$. Similarly, since the matrix

$$
\begin{bmatrix}
1 & b_{13} & a_4 \\
b_{13}^* & 1 & a_3 \\
a_4^* & a_3^* & 1
\end{bmatrix}
$$

is positive, it follows that

$$b_{13} = a_4 a_3^* + (1 - |a_4|^2)^{1/2} g'(1 - |a_3|^2)^{1/2}, \tag{2.2}$$

where $|g'| < 1$. Suppose $z \neq 0$, then take $a_2 = 0$ and $a_3 = a_4 = \delta > 0$, where $1 > \delta^2 > \frac{1}{2}(1 + (1 - |z|^2)^{1/2})$. With this setting, the attempt to express b_{13} both as in (2.1) and (2.2) leads to a contradiction. If $z = 0$, then the roles of z and a_2 are interchanged and the same contradiction as above is easily obtained.

Consider now $n > 4$ and suppose A_0 has a positive completion

$$A = \begin{bmatrix} 1 & z & b_{13} & & b_{1,n-1} & a_n \\ z^* & 1 & a_2 & & & b_{2,n} \\ b_{13}^* & a_2^* & 1 & \ddots & & \\ & & \ddots & \ddots & \ddots & \\ b_{1,n-1}^* & & & & 1 & a_{n-1} \\ a_n^* & b_{2,n}^* & & & a_{n-1}^* & 1 \end{bmatrix}.$$

Consequently, the following principal submatrices of A,

$$\begin{bmatrix} 1 & z & b_{13} \\ z^* & 1 & a_2 \\ b_{13}^* & a_2^* & 1 \end{bmatrix}, \quad \begin{bmatrix} 1 & b_{13} & a_n \\ b_{13}^* & 1 & b_{3,n} \\ a_n^* & b_{3,n}^* & 1 \end{bmatrix} \tag{2.3}$$

and

$$\begin{bmatrix} 1 & a_3 & b_{35} & & b_{3,n} \\ a_3^* & 1 & a_4 & & b_{4,n} \\ b_{35}^* & a_4^* & 1 & \ddots & \\ & & \ddots & \ddots & a_{n-1} \\ b_{3,n}^* & b_{4,n}^* & & a_{n-1}^* & 1 \end{bmatrix} \tag{2.4}$$

are positive matrices. Since the two matrices in (2.3) are positive, it follows that

$$b_{13} = z a_2 + (1 - |z|^2)^{1/2} g(1 - |a_2|^2)^{1/2}, \tag{2.5}$$

where $|g| < 1$ and, respectively,

$$b_{13} = a_n b_{3,n}^* + (1 - |a_n|^2)^{1/2} g'(1 - |b_{3,n}|^2)^{1/2}, \tag{2.6}$$

where $|g'| < 1$. It is convenient to treat separately the cases $z \neq 0$ and $z = 0$. If $z \neq 0$, then take $a_2 = 0$. Suppose $\varepsilon > 0$ and consider $a_3 = a_4 = \ldots = a_{n-1} = \delta > 0$. By Theorem 1.5.3 and the positivity of the matrix introduced in (2.4), it follows that $b_{3,n}$ belongs to a disc of radius $r \leq 1 - \delta^2$ and center of the form $c = \delta^{n-3} + (1 - \delta^2)k$, where $|k| \leq (n-3)^2 - 1$. Hence

$$|c| > \delta^{n-3} - (1 - \delta^2)|k| \tag{2.7}$$

and choosing $1 > \delta^{n-3} > 1 - \varepsilon((n-3)^2 - 1)^{-1}$, it follows by (2.7) that $|c| > 1 - 2\varepsilon$. Consequently, $|b_{3,n}| > 1 - 3\varepsilon$. By (2.6), $|b_{13}| > |a_n b_{3,n}| - (1 - |a_n|^2)^{1/2}(1 - |b_{3,n}|^2)^{1/2}$

and taking $a_n = 1 - 3\varepsilon$ gives $|b_{13}| > 2(1 - 3\varepsilon)^2 - 1$. The last step is to choose $\varepsilon > 0$ such that $(1 - |z|^2)^{1/2} < |2(1 - 3\varepsilon)^2 - 1|$. With this setting of $a_2, a_3, \ldots a_n$, it follows that b_{13} cannot simultaneously satisfy (2.5) and (2.6). Therefore, this partial matrix A_0 has no positive completion. If $z = 0$, then the roles of z and a_2 are interchanged and a similar contradiction may be obtained. The proof is complete.   □

The preceding result suggests that the chordal graphs may provide the solution to Problem 2.1. This is indeed the case, as is shown by the following.

2.3 Theorem   *Every positive partial block matrix subjacent to a graph G has positive completions if and only if G is chordal.*

Proof   Suppose G is not a chordal graph and let O_p be a p-cycle of G with $p \geq 4$. Construct a positive partial block matrix A_0 subjacent to G in the following way: choose the entries corresponding to the edges of O_p as in Lemma 2.2. Since no maximal clique of G contains more than one edge of O_p, it follows that a positive partial matrix A_0 subjacent to G may be constructed using Theorem 1.5.3. By Lemma 2.2, this positive partial matrix has no positive completions.

Conversely, suppose that G is a chordal graph. By Theorem 1.6, there exists a sequence of chordal graphs $G = G_0, G_1, \ldots, G_t = K_n$ such that, for $j = 1, 2, \ldots, t$, the graph G_j is obtained by adding exactly one edge (u_j, w_j) to G_{j-1}. Moreover, we claim that there exists exactly one maximal clique ϖ_j in G_j which is not a clique in G_{j-1}. For let ω', ω'' be two cliques in G_j containing both u_j and w_j. Choose w' in ω' and w'' in ω'' such that w', w'', u_j and w_j are distinct vertices. The graph G_{j-1} is chordal, hence (w', w'') belongs to E_j and $\omega' \cup \omega''$ is a clique of G_j. Now, let A_0 be a positive partial block matrix subjacent to G. Using the above claim, a positive completion of A_0 is obtained in the following way. After a reordering of ϖ_j, if necessary, $A_0[\varpi_j]$ has the following matrix representation:

$$A_0[\varpi_j] = \begin{bmatrix} B_j & H_j & ? \\ H_j^* & D_j & E_j \\ ? & E_j^* & F_j \end{bmatrix},$$

where

$$\begin{bmatrix} B_j & H_j \\ H_j^* & D_j \end{bmatrix} \quad \text{and} \quad \begin{bmatrix} D_j & E_j \\ E_j^* & F_j \end{bmatrix}$$

are positive block matrices. By Theorem 3.4.9, there exists a positive completion of $A_0[\varpi_j]$. In conclusion, a stepwise procedure to construct positive completions of A_0 has been obtained, thereby concluding the proof.   □

Some of the notation introduced above will be frequently used further on and it is convenient to stick to this terminology. Thus, a sequence of chordal graphs $G = G_0, G_1, \ldots, G_s = G'$ as in Theorem 1.6 will be called a *chordal sequence from G to G'*. When $G' = K_n$ we will simply call it a *chordal sequence of G*. Moreover, we will keep the notation (u_j, w_j) to denote the unique edge added to G_{j-1} in order to obtain G_j and the notation ϖ_j for the unique maximal clique in G_j which is not a clique in G_{j-1}.

We may note at this point that the choice of an arbitrary chordal sequence of G could appear as a rather complex operation. However, the existence of perfect schemes enables us to indicate a fairly simple procedure to obtain chordal sequences for a chordal graph G.

2.4 Procedure

Let G be a chordal graph and $\sigma = [v_1, v_2, \ldots, v_n]$ be a perfect scheme of G. Define for $k = 1, 2 \ldots, n$, the sets $\alpha_k = \{v_k, v_{k+1}, \ldots, v_n\}$ and let $R = \min\{k \mid \alpha_k$ is a clique $\}$. Then, we write

$$\alpha_{R-1} = \{v_{R-1}\} \cup X_{R-1} \cup \beta_{R-1},$$

where β_{R-1} is the complement of $\{v_{R-1}\} \cup X_{R-1}$ in α_{R-1} and X_{R-1} is defined by (1.1). Consider the positive integer $r_{R-1} = n - R + 2$ and the ordering $\sigma(r_{R-1})$ of α_{R-1} given by

$$\sigma(r_{R-1}) = [\beta_{R-1}, X_{R-1}, v_{R-1}],$$

where the ordering in β_{R-1} and X_{R-1} is that in σ. Consequently, the positive partial block matrix $A_0[\alpha_{R-1}]$, written with respect to the ordering $\sigma(r_{R-1})$, has the matrix representation

$$A_0[\alpha_{R-1}] = \begin{bmatrix} B_{R-1} & H_{R-1} & ? \\ H_{R-1}^* & D_{R-1} & E_{R-1} \\ ? & E_{R-1}^* & F_{R-1} \end{bmatrix},$$

where

$$\begin{bmatrix} B_{R-1} & H_{R-1} \\ H_{R-1}^* & D_{R-1} \end{bmatrix} \quad \text{and} \quad \begin{bmatrix} D_{R-1} & E_{R-1} \\ E_{R-1}^* & F_{R-1} \end{bmatrix}$$

are positive matrices with specified entries and the unspecified block of $A_0[\alpha_{R-1}]$ is a $|\beta_{R-1}| \times 1$ column. If $\beta_{R-1} = [x_1, x_2, \ldots, x_S]$, then we define $E_1 = E \cup \{(v_{R-1}, x_S)\}$, $E_2 = E_1 \cup \{(v_{R-1}, x_{S-1})\}$, $\ldots$, $E_S = E_{S-1} \cup \{(v_{R-1}, x_1)\}$.

For $i = 2, 3, \ldots, R - 1$, we write

$$\alpha_{R-i} = \{v_{R-i}\} \cup X_{R-i} \cup \beta_{R-i},$$

where β_{R-i} is the complement of $\{v_{R-i}\} \cup X_{R-i}$ in α_{R-i}. Define $r_{R-i} = r_{R-i+1} + 1$ and consider the ordering $\sigma(r_{R-i})$ of α_{R-i} given by

$$\sigma(r_{R-i}) = [\beta_{R-i}, X_{R-i}, v_{R-i}],$$

where the ordering in β_{R-i} and X_{R-i} is that in σ. Consequently, the partial matrix $A_0[\alpha_{R-i}]$, written with respect to the ordering in $\sigma(r_{R-i})$, has the matrix representation

$$A_0[\alpha_{R-i}] = \begin{bmatrix} B_{R-i} & H_{R-i} & ? \\ H_{R-i}^* & D_{R-i} & E_{R-i} \\ ? & E_{R-i}^* & F_{R-i} \end{bmatrix},$$

where

$$\begin{bmatrix} D_{R-i} & E_{R-i} \\ E_{R-i}^* & F_{R-i} \end{bmatrix}$$

is a positive matrix with specified entries, while

$$\begin{bmatrix} B_{R-i} & H_{R-i} \\ H_{R-i}^* & D_{R-i} \end{bmatrix}$$

coincides with $A_0[\alpha_{R-i+1}]$ after a reordering of vertices and the unspecified block of $A_0[\alpha_{R-i}]$ is a $|\beta_{R-i}| \times 1$ column.

We continue to define the chordal sequence of G by successively connecting each v_{R-i} with the vertices in β_{R-i}, taken in the reverse order. Since σ is a perfect scheme, X_{R-i} is a clique for each $i = 1, 2, \ldots, R-1$, and this assures that a chordal sequence of G is obtained by the above procedure. This chordal sequence will be referred to as the *chordal sequence of G associated to σ*. □

Let us illustrate this procedure with the following example.

2.5 Example   Let $G = (V, E)$ be the graph in Figure 7.3 given as follows: $V = \{1, 2, 3, 4, 5, 6\}$ and $E = \{(1, 2), (2, 3), (3, 4), (4, 5), (5, 6), (1, 6), (2, 6), (2, 4), (4, 6)\}$. A perfect scheme of G is $\sigma = [1, 3, 5, 2, 4, 6]$. We see that $X_1 = \{2, 6\}$, $X_2 = \{2, 4\}$, $X_3 = \{4, 6\}$, $X_4 = \{4, 6\}$, $X_5 = \{6\}$ and $\alpha_1 = V$, $\alpha_2 = \{2, 3, 4, 5, 6\}$, $\alpha_3 = \{2, 4, 5, 6\}$, $\alpha_4 = \{2, 4, 6\}$, $\alpha_5 = \{4, 6\}$, $\alpha_6 = \{6\}$. Consequently, $R = 4$.

Let A_0 be a positive partial matrix subjacent to G_2. We can illustrate the completion process in the proof of Theorem 2.3 along the chordal sequence associated to the above perfect scheme σ. We have $E_1 = E \cup \{(2, 5)\}$ and

$$A_0[\varpi_1] = \begin{bmatrix} \circ & \circ & \circ & ? \\ \circ & \circ & \circ & \circ \\ \circ & \circ & \circ & \circ \\ ? & \circ & \circ & \circ \end{bmatrix}$$

with respect to the ordering $\sigma(r_3) = [2, 4, 6, 5]$. Consequently, a positive completion of $A_0[\varpi_1]$ is obtained using Theorem 3.4.9.

Then, we define $E_2 = E_1 \cup \{(3, 6)\}$, $E_3 = E_2 \cup \{(3, 5)\}$ and the partial matrix which must be completed is:

$$\begin{bmatrix} \circ & \circ & \bullet & \circ & ? \\ \circ & \circ & \circ & \circ & ? \\ \bullet & \circ & \circ & \circ & \circ \\ \circ & \circ & \circ & \circ & \circ \\ ? & ? & \circ & \circ & \circ \end{bmatrix}.$$

The entries denoted by $\bullet$ were determined at the previous step and the ordering according to which this partial matrix is written is $\sigma(r_2) = [5, 6, 2, 4, 3]$. The above partial matrix is positive because it consists of two blocks, one of which coincides with the previously obtained positive completion of $A_0[\varpi_1]$ and the second one consists of specified entries. Consequently, the existence of a positive completion of this positive partial matrix is guaranteed by Theorem 3.4.9.

Finally, we define $E_4 = E_3 \cup \{(1,4)\}$, $E_5 = E_4 \cup \{(1,5)\}$ and $E_6 = E_5 \cup \{(1,3)\}$. In this case, we have to complete to a positive matrix the following partial matrix:

$$\begin{bmatrix} \circ & \bullet & \circ & \circ & \bullet & ? \\ \bullet & \circ & \circ & \bullet & \circ & ? \\ \circ & \circ & \circ & \circ & \circ & ? \\ \circ & \bullet & \circ & \circ & \circ & \circ \\ \bullet & \circ & \circ & \circ & \circ & \circ \\ ? & ? & ? & \circ & \circ & \circ \end{bmatrix},$$

where the ordering is $\sigma(r_1) = [3,5,4,2,6,1]$. The same reasoning shows that this is a positive partial matrix and another application of Theorem 3.4.9 leads to a positive completion of this matrix which, in fact, turns into a positive completion of the initial partial matrix.   □

It is worth mentioning that there exist chordal sequences which are not associated to perfect schemes through the Procedure 2.4.

2.6 Example   Let $G = (V, E)$ be a graph with $V = \{1,2,3,4\}$ and $E = \emptyset$. Let $\sigma = [v_1, v_2, v_3, v_4]$ be an arbitrary perfect scheme of G. Then, $\alpha_4 = \{v_4\}$, $\alpha_3 = \{v_3, v_4\}$, $\alpha_2 = \{v_2, v_3, v_4\}$. Further, $\beta_3 = \{v_4\}$, $\beta_2 = \{v_3, v_4\}$ and this shows that the associated chordal sequence is given by $E_1 = E \cup \{(v_3, v_4)\}$, $E_2 = E_1 \cup \{(v_2, v_3)\}$, $E_3 = E_2 \cup \{(v_2, v_4)\}$ and so on. But it is possible to start a chordal sequence of G with $E_1 = E \cup \{(v_3, v_4)\}$, $E_2 = E_1 \cup \{(v_2, v_3)\}$, $E_3 = E_2 \cup \{(v_1, v_2)\}$ and this chordal sequence is not associated to a perfect scheme of G.   □

We have seen that all the constructions presented here in connection with chordal graphs essentially depend on the existence of a perfect scheme. Although a detailed analysis is beyond our purposes, let us mention that there exist fast (*i.e.* linear time) algorithms for constructing perfect schemes of chordal graphs.

We conclude this section by indicating a method to construct positive partial matrices subjacent to chordal graphs and by parametrizing all the strictly positive completions of a strictly positive partial matrix subjacent to a chordal graph–a partial matrix is called *strictly positive* if all the principal submatrices consisting of specified elements are invertible. We have the following result.

2.7 Theorem   (a) *Let A_0 be a strictly positive partial block matrix subjacent to a given chordal graph $G = (V, E)$ and suppose that the diagonal entries of A_0 are identity matrices. Then A_0 is uniquely determined by a set $\{\Gamma(u_j, w_j) \mid j = 1, 2, \ldots, s\}$ of strict contractions associated to a chordal sequence $G_\emptyset = G_0$, G_1, ..., G_s from $G_\emptyset = (V, \emptyset)$ to G.*

(b) *Every strictly positive completion of A_0 is uniquely determined by a set $\{\Gamma(u_j, w_j) \mid j = 1, 2, \ldots, t\}$ of strict contractions associated to a chordal sequence $G = G'_0, G'_1, \ldots, G'_t = K_{|V|}$ of G.*

Proof (a) By a remark made in the proof of Theorem 2.3, there exists exactly one maximal clique ϖ_j in G_j which is not a clique in G_{j-1}. The edge added to the graph G_{j-1} in order to obtain G_j was denoted by (u_j, w_j). Then, $A_0[\varpi_1]$ has the matrix representation

$$A_0[\varpi_1] = \begin{bmatrix} I & A_{(u_1,w_1)} \\ A^*_{(u_1,w_1)} & I \end{bmatrix}$$

and by Lemma 1.2.1, $A[\varpi_1]$ is strictly positive if and only if $A_{(u_1,w_1)} = \Gamma(u_1, w_1)$, with $\Gamma(u_1, w_1)$ a strict contraction. Further, suppose that there are defined the strict contractions $\Gamma(u_j, w_j)$, $j = 1, 2, \ldots, k-1$ and $k-1 < s$, uniquely determining the entries $A(u_1, w_1)$, $A(u_2, w_2)$, $\ldots$, $A(u_{k-1}, w_{k-1})$ of A_0. After a reordering of ϖ_k if necessary, $A_0[\varpi_k]$ has the matrix representation

$$A_0[\varpi_k] = \begin{bmatrix} B_k & H_k & A_{(u_k,w_k)} \\ H^*_k & D_k & E_k \\ A^*_{(u_k,w_k)} & E^*_k & F_k \end{bmatrix},$$

where

$$\begin{bmatrix} B_k & H_k \\ H^*_k & D_k \end{bmatrix} \quad \text{and} \quad \begin{bmatrix} D_k & E_k \\ E^*_k & F_k \end{bmatrix}$$

are strictly positive matrices whose entries are among the elements of the set $\{A_{(u_1,w_1)}, A_{(u_2,w_2)}, \ldots, A_{(u_{k-1},w_{k-1})}, I\}$. By Theorem 1.5.3, $A_{(u_k,w_k)}$ is uniquely determined by the strict contractions already defined and by a new one, denoted by $\Gamma(u_k, w_k)$. The required description of A_0 is concluded.

(b) The proof is essentially the same as for (a) and can be omitted.   □

In accordance to our terminology in Chapter 1, we call the elements of the set $\{\Gamma(u_j, w_j) \mid j = 1, 2, \ldots, s\}$ associated to a strictly positive partial matrix A_0 by Theorem 2.7(i), the *Schur parameters* of A_0 along the considered chordal sequence from $G_\emptyset$ to G. Similarly, when A_0 is given and a chordal sequence of G is taken into account, the elements of the set $\{\Gamma(u_j, w_j) \mid j = 1, 2, \ldots, t\}$ associated to a strictly positive completion A of A_0 by Theorem 2.7(ii) are called the *Schur parameters* of A along the considered chordal sequence of G.

7.3 Completing Positive Partial Matrices. II

In this section we address another generalization of the band completion problem. Two results will be discussed in more details. Thus, we obtain a reformulation of this new problem in terms of Schur products and then, a solution of the problem is presented for n-cycles, $n \geq 4$. The statement of this new generalization of the band completion problem is the following.

3.1 Problem *Characterize those partial matrices which admit positive completions.*

One solution to this problem can be obtained using the notion of Schur product of matrices. Consider two matrices $A = [A_{ij}]^n_{i,j=1}$ and $B = [B_{ij}]^n_{i,j=1}$, then their *Schur product* is defined by:

$$A \odot B = [A_{ij}B_{ij}]^n_{i,j=1}.$$

We will need the following result.

3.2 Lemma   *The Schur product of two positive matrices is a positive matrix.*

Proof The spectral theorem for positive matrices shows that

$$A = \sum_{k=1}^{n} \lambda_k P_k \quad \text{and} \quad B = \sum_{k=1}^{n} \mu_k Q_k,$$

where $\{P_k\}_{k=1}^{n}$ and $\{Q_k\}_{k=1}^{n}$ are two sets of 1-dimensional orthogonal projections and $\{\lambda_k\}_{k=1}^{n}$, $\{\mu_k\}_{k=1}^{n}$ are sets of positive numbers. It is readily checked that

$$A \odot B = \sum_{k,j=1}^{n} \lambda_k \mu_j P_k \odot Q_j.$$

We claim that if P and Q are 1-dimensional orthogonal projections, then $P \odot Q$ is a positive matrix. For suppose $P = uu^*$ and $Q = vv^*$, where u and v are column vectors

$$u = [u_1 \quad u_2 \quad \ldots \quad u_n]^\top \quad \text{and} \quad v = [v_1 \quad v_2 \quad \ldots \quad v_n]^\top.$$

Then

$$P \odot Q = \begin{bmatrix} u_1 u_1^* v_1 v_1^* & u_1 u_2^* v_1 v_2^* & & u_1 u_n^* v_1 v_n^* \\ u_2 u_1^* v_2 v_1^* & u_2 u_2^* v_2 v_2^* & \ddots & \\ & \ddots & \ddots & \\ u_n u_1^* v_n v_1^* & & & u_n u_n^* v_n v_n^* \end{bmatrix}$$

$$= [u_1^* v_1^* \quad u_2^* v_2^* \quad \ldots \quad u_n^* v_n^*]^* [u_1^* v_1^* \quad u_2^* v_2^* \quad \ldots \quad u_n^* v_n^*] \geq 0,$$

thereby concluding the proof. $\square$

Consider $G = (V, E)$ a graph with n vertices and let $\mathcal{S}_G$ be the set of $n \times n$ matrices S with the property that $S_{ij} = 0$ if $i \neq j$ and $(i,j) \notin E$. Remark that $\mathcal{S}_G$ is an operator system of $\mathcal{M}_n$, *i.e.* it is a selfadjoint subspace of $\mathcal{M}_n$ containing the identity matrix. Let A_0 be a partial matrix subjacent to G, $A_0 = [A_{ij} \mid (i,j) \in E]$, and define the *Schur product* map by:

$$\Phi_{A_0} : \mathcal{S}_G \longrightarrow \mathcal{S}_G \tag{3.1}$$

$$\Phi_{A_0}(S) = \begin{cases} S_{ij} A_{ij} & \text{if } (i,j) \in E \\ 0 & \text{if } i \neq j \text{ and } (i,j) \notin E. \end{cases}$$

Moreover, define the functional

$$\Psi_{A_0} : \mathcal{S}_G \longrightarrow \mathbb{C} \tag{3.2}$$

$$\Psi_{A_0}(S) = \sum_{(i,j) \in E} S_{ij} A_{ij}.$$

3.3 Theorem   *Let A_0 be a positive partial matrix subjacent to G. The following are equivalent:*

(a)   *A_0 has a positive completion.*

(b)   *Φ_{A_0} is a positive map.*

(c)   *Ψ_{A_0} is a positive functional.*

Proof $(a) \Rightarrow (b)$ Let A be a positive completion of A_0. For S in $\mathcal{S}_G$, $\Phi_{A_0}(S) = A \odot S$. It follows by Lemma 3.2 that if S is a positive matrix, then so is $\Phi_{A_0}(S)$, i.e. Φ_{A_0} is a positive map.

$(b) \Rightarrow (c)$ Remark that $\Psi_{A_0} = \delta \circ \Phi_{A_0}$, where

$$\delta : \mathcal{S}_G \longrightarrow \mathbb{C}$$

$$\delta(S) = \sum_{(i,j)\in E} S_{ij} = \langle Sf, f \rangle$$

and $f = [1, 1, \ldots, 1]^\top$. Consequently, δ is a positive functional on $\mathcal{S}_G$ and Ψ_{A_0} is positive if Φ_{A_0} is positive.

$(c) \Rightarrow (a)$ If Ψ_{A_0} is a positive functional then, by a result of M.G.Krein, there exists a positive functional $\Psi : \mathcal{M}_n \longrightarrow \mathbb{C}$ extending Ψ_{A_0}. Defining $A = [\Psi(E_{ij})]_{i,j=1}^n$, where E_{ij}, $i, j = 1, 2, \ldots, n$ are the matrix units of $\mathcal{M}_n$, then A is a completion of A_0. It is easily verified that A is a positive matrix: if $\lambda_1, \lambda_2, \ldots, \lambda_n$ belong to $\mathbb{C}$, then

$$\sum_{i,j=1}^n A_{ij}\lambda_j\lambda_i^* = \sum_{i,j=1}^n \Psi(E_{ij})\lambda_j\lambda_i^* = \Psi([\lambda_1, \lambda_2, \ldots, \lambda_n]^*[\lambda_1, \lambda_2, \ldots, \lambda_n]) \geq 0. \quad \square$$

3.4 Remark   We may note that the functional Ψ_{A_0} is also connected with the trace functional. Thus, we see that if we formally define the product SA_0 of an element S in $\mathcal{S}_G$ with a partial matrix according to the usual rule of multiplication of matrices and the additional rules

$$0\times? = 0, \quad \lambda\times? =?, \quad \lambda+? =?,$$

for λ in $\mathbb{C}$, then the trace of $SA_0^\top$ is well defined and for any S in $\mathcal{S}_G$, $\Psi_{A_0}(S) = \mathrm{tr}(SA_0^\top)$.
$\square$

Theorem 3.3 provides a certain answer to Problem 3.1. However, one feels that a more effective solution to Problem 3.1 is desirable. Such a solution can be obtained, for instance, in the case of partial matrices subjacent to an n-cycle, $n \geq 4$.

3.5 Lemma   *Let $n \geq 4$ and $0 \leq \vartheta_1, \vartheta_2, \ldots, \vartheta_n \leq \pi$ be such that at most one of*

$\vartheta_1, \vartheta_2, \ldots, \vartheta_n$ is greater than $\pi/2$. Then the positive partial matrix

$$
A_0 = \begin{bmatrix}
1 & \cos\vartheta_1 & ? & & \cos\vartheta_n \\
\cos\vartheta_1 & 1 & \cos\vartheta_2 & ? & \\
? & \cos\vartheta_2 & 1 & \ddots & \\
& ? & \ddots & \ddots & \cos\vartheta_{n-1} \\
\cos\vartheta_n & & & \cos\vartheta_{n-1} & 1
\end{bmatrix}
\tag{3.3}
$$

has positive real completions if and only if

$$
2 \max_{1 \le k \le n} \vartheta_k \le \sum_{k=1}^{n} \vartheta_k.
$$

Proof This result is proven by induction on n. For $n = 4$, remark that A_0 has a positive real completion if and only if there exists $\vartheta \in [0, \pi]$ such that

$$
\begin{bmatrix}
1 & \cos\vartheta_1 & \cos\vartheta \\
\cos\vartheta_1 & 1 & \cos\vartheta_2 \\
\cos\vartheta & \cos\vartheta_2 & 1
\end{bmatrix}
\quad \text{and} \quad
\begin{bmatrix}
1 & \cos\vartheta_4 & \cos\vartheta \\
\cos\vartheta_4 & 1 & \cos\vartheta_3 \\
\cos\vartheta & \cos\vartheta_3 & 1
\end{bmatrix}
\tag{3.4}
$$

are positive matrices. By Theorem 1.5.3, the first matrix in (3.4) is positive if and only if

$$
\cos\vartheta = \cos\vartheta_1 \cos\vartheta_2 + g \sin\vartheta_1 \sin\vartheta_2
$$

where $|g| \le 1$. Equivalently, ϑ must satisfy the inequality

$$
\cos(\vartheta_1 + \vartheta_2) \le \cos\vartheta \le \cos(|\vartheta_1 - \vartheta_2|).
\tag{3.5}
$$

Similarly, the second matrix in (3.4) is positive if and only if

$$
\cos(\vartheta_3 + \vartheta_4) \le \cos\vartheta \le \cos(|\vartheta_3 - \vartheta_4|).
\tag{3.6}
$$

Suppose, without loss of generality, that ϑ_1 is the greatest among $\vartheta_1, \vartheta_2, \vartheta_3, \vartheta_4$. Then $\vartheta_2, \vartheta_3, \vartheta_4 \le \pi/2$ and it is easy to see that there exists $\vartheta \in [0, \pi]$ satisfying (3.5) and (3.6) if and only if $\vartheta_1 \le \vartheta_2 + \vartheta_3 + \vartheta_4$, which is exactly the required inequality.

Suppose that the result holds for $n-1$ and consider the numbers $0 \le \vartheta_1, \vartheta_2, \ldots, \vartheta_n \le \pi$ such that at most one of them is greater than $\pi/2$. Assume, without loss of generality, that $\vartheta_n = \max_{k=1,n} \vartheta_k$. We claim that the partial matrix A_0 in (3.3) associated to these numbers has positive real completions if and only if there exists $\vartheta \in [0, \pi]$ such that the positive partial matrix

$$
A_0' = \begin{bmatrix}
1 & \cos\vartheta_1 & ? & & \cos\vartheta \\
\cos\vartheta_1 & 1 & \cos\vartheta_2 & ? & \\
? & \cos\vartheta_2 & 1 & \ddots & \\
& ? & \ddots & \ddots & \cos\vartheta_{n-2} \\
\cos\vartheta & & & \cos\vartheta_{n-2} & 1
\end{bmatrix}
$$

has positive completions and

$$A_0'' = \begin{bmatrix} 1 & \cos\vartheta & \cos\vartheta_n \\ \cos\vartheta & 1 & \cos\vartheta_{n-1} \\ \cos\vartheta_n & \cos\vartheta_{n-1} & 1 \end{bmatrix}$$

is a positive matrix. For suppose A_0'' is positive and A_0' has a positive completion A'. Then the partial matrix

$$\begin{bmatrix} & & & & \cos\vartheta_n \\ & & & & ? \\ & A' & & & \vdots \\ & & & & ? \\ & & & & \cos\vartheta_{n-1} \\ \cos\vartheta_n & ? & \dots & ? & \cos\vartheta_{n-1} & 1 \end{bmatrix}$$

is subjacent to a proper interval graph, hence it has a positive completion which turns into a positive completion of A_0. The converse is obvious.

Using this claim and the induction hypothesis, it follows that if A_0 has a positive completion, then

$$2\vartheta \le 2\max\{\vartheta, \vartheta_1, \dots, \vartheta_{n-2}\} \le \vartheta + \sum_{k=1}^{n-2} \vartheta_k,$$

hence $\vartheta \le \sum_{k=1}^{n-2} \vartheta_k$. Since A_0'' is positive, it follows that $\vartheta_n \le \vartheta_{n-1} + \vartheta$. Consequently,

$$\vartheta_n \le \vartheta_1 + \vartheta_2 + \dots + \vartheta_{n-1}.$$

Conversely, suppose

$$2\max_{k=1,n} \vartheta_k \le \sum_{k=1}^{n} \vartheta_k.$$

Define the number $\vartheta = \max\{\vartheta_n - \vartheta_{n-1}, 2\max_{k=1,n-2}\vartheta_k - \sum_{k=1}^{n-2}\vartheta_k\}$ which belongs to $[0, \pi]$.
We can distinguish two cases.

A. $\vartheta = \vartheta_n - \vartheta_{n-1}$.

For this ϑ, the matrix A_0'' is positive. Moreover,

$$2\max\{\vartheta_1, \dots, \vartheta_{n-2}, \vartheta\} = \max\{2\max_{k=1,n-2} -\vartheta, \vartheta\} + \vartheta \le \sum_{k=1}^{n-2} \vartheta_k + \vartheta$$

and by the induction hypothesis, A_0' has a positive real completion.

B. $\vartheta = 2\max_{k=1,n-2}\vartheta_k - \sum_{k=1}^{n-2}\vartheta_k$.

In this case, $\vartheta \le \max_{k=1,n-2}\vartheta_k \le \vartheta_n$. Since $\vartheta \ge \vartheta_n - \vartheta_{n-1}$, it follows that A_0'' is a positive real matrix. As in the previous case, the numbers $\vartheta_1, \vartheta_2, \dots,$

ϑ_{n-2} and ϑ satisfy the induction hypothesis, therefore the partial matrix A_0' has a positive real completion. The proof is complete.   □

3.6 Lemma   If two or more of the numbers $\vartheta_1, \vartheta_2, \ldots, \vartheta_n$ in $[0, \pi]$ are equal to $\pi/2$, then the partial matrix A_0 defined by (3.3) has positive real completions.

Proof Suppose $\vartheta_n = \pi/2$, and $\vartheta_j = \pi/2$ for a certain $1 \le j \le n-1$. Consider the partial band matrix

$$B_0 = \begin{bmatrix} 1 & \cos \vartheta_1 & ? & & & \\ \cos \vartheta_1 & 1 & \cos \vartheta_2 & ? & & \\ ? & \cos \vartheta_2 & 1 & \ddots & & \\ & ? & \ddots & \ddots & \cos \vartheta_{n-1} \\ & & & \cos \vartheta_{n-1} & 1 \end{bmatrix}$$

and consider the positive completion B of B_0 corresponding to the Schur parameters equal to 0. Since $\vartheta_j = \pi/2$, it is easy to check using the formulae in Theorem 1.5.3 that the $(n, 1)$ entry of B is 0, *i.e.* B is a positive real completion of A_0.   □

The next result gives another answer to Problem 3.1 when the partial matrix A_0 is subjacent to an n-cycle, $n \ge 4$.

3.7 Theorem   Let $c_1, c_2, \ldots c_n$, $n \ge 4$, be real numbers with $|c_i| \le 1$. The partial matrix

$$A_0 = \begin{bmatrix} 1 & c_1 & ? & & c_n \\ c_1 & 1 & c_2 & ? & \\ ? & c_2 & 1 & \ddots & \\ & ? & \ddots & \ddots & c_{n-1} \\ c_n & & c_{n-1} & 1 \end{bmatrix}$$

has positive real completions if and only if

$$2 \max_{k=1,n} \arccos |c_k| \le \sum_{k=1}^{n} \arccos |c_k|, \quad \text{for} \quad c_1 c_2 \ldots c_n > 0$$

and

$$\sum_{k=1}^{n} \arccos |c_k| \ge \pi, \quad \text{for} \quad c_1 c_2 \ldots c_n \le 0.$$

Proof If two or more of the entries of A_0 are 0, then A_0 has positive real completions by Lemma 3.6 and

$$\sum_{k=1}^{n} \arccos |c_k| \ge 2(\pi/2) = \pi.$$

Then, suppose that at most one c_k is 0. Moreover, it may be assumed without loss of generality, that $c_1 c_2 \ldots c_{n-1} \neq 0$. Define the matrix

$$
D = \begin{bmatrix}
1 & 0 & 0 & & & \\
0 & \text{sgn}(c_1) & 0 & & 0 & \\
0 & 0 & \text{sgn}(c_1 c_2) & \ddots & & \\
& 0 & \ddots & \ddots & 0 & \\
& & & 0 & \text{sgn}(c_1 c_2 \ldots c_{n-1})
\end{bmatrix}
$$

and remark that

$$
B_0 = D^* A_0 D = \begin{bmatrix}
1 & |c_1| & ? & & \varepsilon|c_n| \\
|c_1| & 1 & |c_2| & ? & \\
? & |c_2| & 1 & \ddots & \\
& ? & \ddots & \ddots & |c_{n-1}| \\
\varepsilon|c_n| & & & |c_{n-1}| & 1
\end{bmatrix},
$$

where $\varepsilon = \text{sgn}(c_1 c_2 \ldots c_n)$. Notice that A_0 has positive real completions if and only if B_0 has. If $\varepsilon > 0$, then we can use Lemma 3.5 in order to deduce that A_0 has positive real completions if and only if $2 \max_{k=1,n} \arccos |c_k| \leq \sum_{k=1}^{n} \arccos |c_k|$. If $\varepsilon \leq 0$, then Lemma 3.5 is used again to deduce that A_0 has positive real completions if and only if $\arccos(-|c_n|) \leq \sum_{k=1}^{n-1} \arccos |c_k|$, which is exactly $\sum_{k=1}^{n} \arccos |c_k| \geq \pi$. The proof is complete. $\qquad\square$

7.4 Completing Contractive Partial Matrices

In this section we address a completion problem for contractive partial matrices. By *contractive partial matrix* we mean a partial matrix $T_0 = [T_{ij} \mid 1 \leq i \leq n, 1 \leq j \leq m]$ such that each of its submatrices with specified elements are contractions. In this case, it is convenient to associate to T_0 a bipartite graph in the following way: choose disjoint sets $Y = \{y_1, y_2, \ldots, y_n\}$, $Z = \{z_1, z_2, \ldots, z_m\}$ and $E = \{(y_i, z_j) \mid T_{ij} \text{ is specified}\}$. We remark that similar to Problem 2.1 and Problem 3.1, but unlike the band case analyzed in Chapter 3, the condition that all the partial submatrices with specified entries be contractions is no longer sufficient in order for there to exist contractive completions. As an example, we may note the following contractive partial matrix:

$$
T_0 = \begin{bmatrix} ? & \frac{1}{\sqrt{2}} & \frac{1}{\sqrt{2}} \\ \frac{1}{\sqrt{2}} & ? & \frac{1}{\sqrt{2}} \end{bmatrix}. \tag{4.1}
$$

Both of the matrices

$$
\begin{bmatrix} x & \frac{1}{\sqrt{2}} & \frac{1}{\sqrt{2}} \end{bmatrix} \quad \text{and} \quad \begin{bmatrix} x & \frac{1}{\sqrt{2}} \\ \frac{1}{\sqrt{2}} & \frac{1}{\sqrt{2}} \end{bmatrix}
$$

must be contractions. From the first condition it follows that the $(1,1)$ entry (denoted by x) must equal 0, while from the second one, it follows that $x = -1/\sqrt{2}$. Consequently, the contractive partial matrix T_0 given by (4.1) has no contractive completion, even though all its partial submatrices are contractions.

The next remark is that when the problem of finding contractive completions of contractive partial matrices is considered, we can suppose that the associated graph of T_0 is connected. For suppose $G_k = \{Y_k, Z_k, E_k\}$, $k = 1, 2, \ldots, s$, are the connected components of G and let T_0 be a contractive partial matrix subjacent to G. Suppose that the submatrices of T_0 subjacent to G_k, $k = 1, 2, \ldots, s$, have contractive completions. Taking for each $k = 1, 2, \ldots, s$, such a completion $T^{(k)}$, it follows that the contraction $T = \oplus_{k=1}^{s} T^{(k)}$ is a contractive completion of T_0. With these remarks, we can state and prove the main result of this section.

4.1 Theorem   Let $G = (Y, Z, E)$ be a connected bipartite graph. The following are equivalent:

(a)   Every contractive partial matrix subjacent to G has contractive completions.
(b)   For every partial (not necessarily contractive) matrix T_0 subjacent to G, there exist permutation matrices U_1 and U_2 such that

$$
U_1 T_0 U_2 = \begin{bmatrix} B_{11} & \cdots & B_{1,j_1} & ? & & & \\ B_{21} & & & \cdots & B_{2,j_2} & \cdots & ? \\ \vdots & & & & & & \\ B_{n1} & \cdots & & & & & B_{nm} \end{bmatrix}, \qquad (4.2)
$$

where $1 \le j_1 \le j_2 \ldots \le j_n = m$.
(c)   Every partial (not necessarily contractive) matrix subjacent to G has no submatrix of the form

$$
\begin{bmatrix} ? & \circ \\ \circ & ? \end{bmatrix} \quad or \quad \begin{bmatrix} \circ & ? \\ ? & \circ \end{bmatrix}.
$$

(d)   The graph $\hat{G} = (V, F)$ is chordal, where $V = Y \cup Z$ and $F = E \cup (X \times X) \cup (Y \times Y)$.

Proof $(b) \Rightarrow (c)$ This implication is immediate.

$(c) \Rightarrow (d)$ Suppose that the graph $\hat{G}$ is not chordal. By the structure of $\hat{G}$, the only possible k-cycle with $k > 3$ in $\hat{G}$ might be of the form $[y, z, y', z']$. The presence of such a 4-cycle in $\hat{G}$ refflects into the structure of a partial matrix subjacent to G by the existence of a submatrix of the form

$$
\begin{bmatrix} ? & \circ \\ \circ & ? \end{bmatrix} \quad or \quad \begin{bmatrix} \circ & ? \\ ? & \circ \end{bmatrix}.
$$

$(d) \Rightarrow (a)$ Consider T_0 a contractive partial matrix subjacent to G. By definition, the positive partial matrix

$$
A_0 = \begin{bmatrix} I & T_0 \\ T_0^* & I \end{bmatrix}
$$

is subjacent to $\hat{G}$. Consequently, if $\hat{G}$ is a chordal graph, then by Theorem 2.3, the positive partial matrix A_0 has a positive completion

$$A = \begin{bmatrix} I & T \\ T^* & I \end{bmatrix}$$

and T is a contractive completion of T_0.

(a) $\Rightarrow$ (c)   This implication is true because of the example (4.1).

(c) $\Rightarrow$ (b)   Take y, y' in Y, $y \neq y'$. By hypothesis, at least one of the sets $Adj(y) - Adj(y')$ and $Adj(y') - Adj(y)$ must be empty, hence the sets $Adj(y)$ and $Adj(y')$ are either disjoint or one of them is included into the other one. Since G is connected, it follows that there exists an ordering σ of the set $\{1, 2, \ldots, n\}$ such that

$$\emptyset \neq Adj(y_{\sigma(1)}) \subseteq Adj(y_{\sigma(2)}) \subseteq \ldots \subseteq Adj(y_{\sigma(n)}) = \{z_1, z_2, \ldots, z_m\}.$$

Similarly, there exists an ordering τ of the set $\{1, 2, \ldots, m\}$ such that

$$\emptyset \neq Adj(y_{\tau(1)}) \subseteq Adj(y_{\tau(2)}) \subseteq \ldots \subseteq Adj(y_{\tau(m)}) = \{y_1, y_2, \ldots, y_n\}.$$

Permuting the rows of a partial matrix T_0 subjacent to G according to the ordering of σ and its columns according to the inverse of the ordering of τ, a matrix of type (4.2) will be obtained. The proof is complete. $\qquad \square$

7.5   Notes

We have presented in this chapter several results illustrating connections between completion problems and elements of graph theory. Of special interest in this direction appeared to be the class of chordal graphs. These graphs played a critical role in other topics as well, for instance in connection with perfect graphs ([Ber], [HS]) or with the Gaussian elimination for sparse matrices [Di], [FG], [Ro], [RTL]. For the presentation in Section 1 we used the book [Gol]. We presented only the results required in the treatment of the completion problems that was formulated in Section 2. Lemma 1.3 is one of the basic results in the theory of chordal graphs and it was proved in [Di]. Theorem 1.4 is equally important and it was proved in [FG]. Some fast algorithms for constructing perfect schemes of chordal graphs are discussed in [TY]. Corollary 1.5 appears in [RTL] and its usefulness in connection with completion problems is pointed out in [GJSW] for $G' = K_n$ (see also [BC1]). Theorem 1.7 is proved in [Bun], [Gav], [Wa]. The fourth equivalent statement in this theorem is noticed in [BJLu]. The interval graphs are another important class of graphs with many applications that can be found in [Gol]. Section 2 contains the main result, Theorem 2.4, that brought into attention the connection between completion problems (namely, the band completion problem in [DG1]) and elements of the graph theory. This result is proved in the paper [GJSW], which was followed by a number of subsequent developments, some of them mentioned below. Theorem 2.7 was noticed in [BC1]. Section 3 contains only a few partial results in connection with Problem 3.1. Lemma 3.2 is a well-known result of

Schur, Theorem 3.3 is proved in [PPS] and Theorem 3.5 is proved in [Fi2]—for presentation we used the paper [BJT], which contains another alternative of the result in [Fi2]. Other discussions in connection with Problem 3.1 can be found in [Jo1-2]. Theorem 4.1 was proved in [JR1] and we used for its presentation the thesis [Ba]. This work, as well as [Lun], contains an important number of developments of the ideas in [GJSW]. For other recent developments we mention the papers [AHMR], [BJL], [JLN], [JR2-3].

Chapter 8
Determinantal Formulae and Optimization

In this chapter we discuss the role played by the Schur parameters in connection with the computation of the determinants of the positive completions of a positive partial matrix subjacent to a chordal graph. The problem of finding the completion with maximum determinant is also addressed as a generalization of the maximum entropy principle presented in Chapter 5. The behavior of the maximum determinant solution with respect to the operation of compression of matrices is analyzed in the last section.

8.1 Determinantal Formulae

In this section we compute the determinants of the positive completions of a positive partial matrix subjacent to a chordal graph in terms of their Schur parameters. The main tool in our considerations is Theorem 1.5.10 and it is useful to begin with the presentation of a simple consequence of this result.

1.1 Theorem *Let A be a positive $n \times n$ block matrix and let $\alpha = \{1, 2, \ldots, m\}$, $\beta = \{k, k+1, \ldots, n\}$ be two subsets of indices such that $\det A[\alpha \cap \beta] > 0$, where k and m are positive integers, $1 \leq k \leq m+1 \leq n$. Then*

$$\det A = \frac{\det A[\alpha] \det A[\beta]}{\det A[\alpha \cap \beta]} \times \prod_{(i,j) \in \gamma} \det D_{\Gamma_{ij}}^2,$$

where $\{\Gamma_{ij} \mid 1 \leq i \leq j \leq n\}$ is the set of the Schur parameters of A and $\gamma = \{(i,j) \mid 1 \leq i \leq k-1, m+1 \leq j \leq n\}$.

Proof Theorem 1.5.10 gives that

$$\det A[\alpha] = \prod_{i \in \alpha} \det A_{ii} \times \prod_{1 \leq i < j \leq m} \det D_{\Gamma_{ij}}^2,$$

$$\det A[\beta] = \prod_{i \in \beta} \det A_{ii} \times \prod_{k \leq i < j \leq n} \det D_{\Gamma_{ij}}^2$$

203

and respectively,

$$\det A[\alpha \cap \beta] = \prod_{i \in \alpha \cap \beta} \det A_{ii} \times \prod_{k \leq i < j \leq m} \det D^2_{\Gamma_{ij}}.$$

Then the required formula follows directly from these equalities.   □

This result sets the stage for our developments on determinantal formulae. The next remark is that the formula obtained in Theorem 1.1 can be iterated along a chordal sequence associated to a perfect scheme of a chordal graph. Let $G = (V, E)$ be a chordal graph and let $\sigma = [v_1, v_2, \ldots, v_n]$ be a perfect scheme of G. Using Procedure 7.2.4, we associate to σ a chordal sequence G_0, G_1, $\ldots$, G_t of G. The sets X_i, $i = 1, 2, \ldots, n-1$, are introduced by (7.1.1) and since G is chordal, it follows that all these sets are cliques. We also maintain the notation introduced in Chapter 7 to denote by (u_j, w_j) the unique edge added to G_{j-1} in order to obtain G_j and by ϖ_j the unique maximal clique in G_j which is not a clique in G_{j-1}.

1.2 Theorem   *Let $G = (V, E)$ be a chordal graph and let $\sigma = [v_1, v_2, \ldots, v_n]$ be a perfect scheme of G. Let G_0, G_1, $\ldots$, G_t be the chordal sequence of G associated to σ by Procedure 7.2.4. Fix A_0 a strictly positive partial block matrix subjacent to the graph G and suppose that all the diagonal entries of A_0 are identity matrices. Let A be a strictly positive completion of A_0 and let $\{\Gamma(u_j, w_j) \mid j = 1, 2, \ldots, t\}$ be the set of the Schur parameters of A along the chordal sequence G_0, G_1, $\ldots$, G_t. Then*

$$\det A = \prod_{m=1}^{n} \frac{\det A[\{v_m\} \cup X_m]}{\det A[X_m]} \times \prod_{j=1}^{t} \det D^2_{\Gamma(u_j, w_j)}.$$

Proof There are a few other elements associated to the perfect scheme σ by Procedure 7.2.4. Thus, the sets $\alpha_k = \{v_k, v_{k+1}, \ldots, v_n\}$ of vertices were introduced, as well as the positive integer $R = \min\{k \mid \alpha_k$ is a clique $\}$. Every α_{R-i}, $i = 2, 3, \ldots, R - 1$, was partitioned as $\alpha_{R-i} = \{v_{R-i}\} \cup X_{R-i} \cup \beta_{R-i}$. Suppose $\beta_1 = [x_1, x_2, \ldots, x_P]$ and consider the ordering $[[x_P, x_{P-1}, \ldots, x_2, x_1], Adj(v_1), v_1]$ of V, where the ordering of $Adj(v_1)$ may be taken that of σ. Using Theorem 1.1 with respect to the sets $\alpha = \beta_1 \cup Adj(v_1)$ and $\beta = \{v_1\} \cup Adj(v_1)$, gives

$$\det A = \det A[\alpha_2] \times \frac{\det A[V - \{x_P\}]}{\det A[V - \{x_P, v_1\}]} \times \det D^2_{\Gamma(v_1, x_P)}.$$

Using this type of formula for all the elements of β_1, one successively obtains:

$$\det A = \det A[\alpha_2] \times \frac{\det A[V - \{x_P\}]}{\det A[V - \{x_P, v_1\}]} \times \det D^2_{\Gamma(v_1, x_P)}$$

$$= \det A[\alpha_2] \times \frac{\det A[V - \{x_P, v_1\}]}{\det A[V - \{x_P, v_1\}]}$$

$$\times \frac{\det A[V - \{x_P, x_{P-1}\}]}{\det A[V - \{x_P, x_{P-1}, v_1\}]}$$

$$\times \det D^2_{\Gamma(v_1, x_P)} \det D^2_{\Gamma(v_1, x_{P-1})}$$

$$= \ldots = \det A[\alpha_2] \times \frac{\det A[\{v_1\} \cup Adj(v_1)]}{\det A[Adj(v_1)]} \times \prod_{x \in \beta_1} \det D^2_{\Gamma(v_1, x)}.$$

The induced graph G_{α_2} is also chordal and $[v_2, v_3, \ldots, v_n]$ is a perfect scheme of it. Using this remark, one proves by induction that

$$\det A = \det A[\alpha_R] \times \prod_{m=1}^{R-1} \frac{\det A[\{v_m\} \cup X_m]}{\det A[X_m]}$$

$$\times \prod_{w \in \beta_1} \det D^2_{\Gamma(v_1, w)} \times \ldots \times \prod_{w \in \beta_{R-1}} \det D^2_{\Gamma(v_{R-1}, w)}.$$

Finally, note that G_{α_R} is the complete graph by the definition of R and the formula

$$\det A[\alpha_R] = \prod_{m=R}^{n} \frac{\det A[\{v_m\} \cup X_m]}{\det A[X_m]}$$

is obvious. Since $\{(v_m, w) \mid m = 1, 2, \ldots, R-1 \text{ and } w \in \cup_{k=1}^{R-1} \beta_k\} = \{(u_j, w_j) \mid j = 1, 2, \ldots, t\}$, the proof is complete. $\square$

Going back to Theorem 1.1, we note that it implies both of the following results, usually referred to as Hadamard's inequality and, respectively, Fisher-Hadamard's inequality.

1.3 Corollary   *Let A be a positive $n \times n$ block matrix and let $\alpha = \{1, 2, \ldots, m\}$, $\beta = \{k, k+1, \ldots, n\}$ be two subsets of indices such that $\det A[\alpha \cap \beta] > 0$, where k and m are positive integers, $1 \le k \le m+1 \le n$. Then*

$$\det A \le \frac{\det A[\alpha] \det A[\beta]}{\det A[\alpha \cap \beta]}.$$
$\square$

1.4 Corollary   *Let A be a positive $n \times n$ block matrix. Then*

$$\det A \le \prod_{i=1}^{n} \det A_{ii}.$$
$\square$

It is useful to note that Theorem 1.1 tells us exactly when the equality occurs in both inequalities in Theorem 1.3 and Theorem 1.4. All these results can be extended to the setting in Theorem 1.2.

1.5 Theorem   *Let A_0 be a strictly positive partial block matrix subjacent to a chordal graph. Then, among all the positive completions of A_0 there exists a unique one with maximum determinant.*

Proof Suppose, without loss of generality, that the diagonal entries of A_0 are identity matrices. Let $\sigma = [v_1, v_2, \ldots, v_n]$ be a perfect scheme of G and let G_0, G_1, ..., G_t be the chordal sequence of G associated to σ by Procedure 7.2.4. By Theorem 7.2.7, there exists a one-to-one correspondence between strictly positive completions of A_0 and Schur parameters. Using this result and Theorem 1.2, one obtains that

$$\det A \le \prod_{m=1}^{n} \frac{\det A[\{v_m\} \cup X_m]}{\det A[X_m]} \tag{1.1}$$

for every positive completion A of A_0. Moreover, exactly one strictly positive completion of A_0, denoted by $(A_0)^0$, attains this bound. This $(A_0)^0$ is the positive completion corresponding to the Schur parameters $\Gamma^0(u_j, w_j) = 0$ for all $j = 1, 2, \ldots, t$. □

An additional result clarifies the independence of the right side of (1.1) on σ and A.

1.6 Proposition *Let A_0 be a strictly positive partial block matrix subjacent to a chordal graph G. Then the product*

$$D(A, \sigma; A_0) = \prod_{m=1}^{n} \frac{\det A[\{v_m\} \cup X_m]}{\det A[X_m]}$$

is the same for every perfect scheme $\sigma = [v_1, v_2, \ldots, v_n]$ of G and every strictly positive completion A of A_0.

Proof Let $\sigma = [v_1, v_2, \ldots, v_n]$ be a perfect scheme of G and let A be a strictly positive completion of A_0. Since σ is a perfect scheme of G, it follows that X_m and $\{v_m\} \cup X_m$ are cliques of G for all $m = 1, 2, \ldots, n$. Consequently, $A[X_m] = A_0[X_m]$ and $A[\{v_m\} \cup X_m] = A_0[\{v_m\} \cup X_m]$ for all $m = 1, 2, \ldots, n$, which shows that $D(A, \sigma; A_0)$ does not depend on A (and we can denote it by $D(\sigma; A_0)$).

By Theorem 1.2, it follows that for every strictly positive completion A of A_0, the following identity holds:

$$\det A = D(\sigma; A_0) \prod_{j=1}^{t} \det D^2_{\Gamma(u_j, w_j)}.$$

By Theorem 1.5, $D(\sigma; A_0)$ represents the maximum determinant over the determinants of all the strictly positive completions of A_0. Therefore, $D(\sigma; A_0)$ does not depend on σ. □

From now on, the product

$$\prod_{m=1}^{n} \frac{\det A_0[\{v_m\} \cup X_m]}{\det A_0[X_m]}$$

will be denoted by $D(A_0)$.

We know from Example 7.2.6 that not every chordal sequence is obtained as in Procedure 7.2.4. However, the determinantal formula in Theorem 1.2 can be extended to any chordal sequence of a chordal graph. Let us first analyze a particular case.

1.7 Example Consider the graph $G_\emptyset = (V, E)$, $V = \{1, 2, 3, 4\}$, $E = \emptyset$. Define the chordal sequence $G_1, \ldots, G_6$ of $G_\emptyset$ by $E_1 = \{(3,4)\}$, $E_2 = E_1 \cup \{(2,4)\}$, $E_3 = E_2 \cup \{(1,2)\}$, $E_4 = E_3 \cup \{(2,3)\}$, $E_5 = E_4 \cup \{(1,4)\}$, $E_6 = E_5 \cup \{(1,3)\}$. Let A_0 be a strictly positive partial block matrix subjacent to G whose diagonal entries are identity matrices and let A be a strictly positive completion of A_0. If we take into account the Schur parameters of A along the considered chordal sequence, then we use a splitting process based on Theorem 1.1 and compute:

$$
\det A = \frac{\det A[\{1,2,4\}] \det A[\{2,3,4\}]}{\det A[\{2,4\}]} \times \det D^2_{\Gamma(1,3)}
$$

$$
= \frac{\det A[\{1,2\}] \det A[\{2,4\}] \det A[\{2,3,4\}]}{\det A[\{2\}] \det A[\{2,4\}]}
$$

$$
\times \det D^2_{\Gamma(1,3)} \det D^2_{\Gamma(1,4)}
$$

$$
= \frac{\det A[\{1,2\}] \det A[\{2,4\}] \det A[\{3,4\}]}{\det A[\{2\}] \det A[\{4\}]}
$$

$$
\times \det D^2_{\Gamma(1,3)} \det D^2_{\Gamma(1,4)} \det D^2_{\Gamma(2,3)}
$$

$$
= \frac{\det A[\{1\}] \det A[\{2\}] \det A[\{2,4\}] \det A[\{3,4\}]}{\det A[\{4\}]}
$$

$$
\times \det D^2_{\Gamma(1,3)} \det D^2_{\Gamma(1,4)} \det D^2_{\Gamma(2,3)} \det D^2_{\Gamma(1,2)}
$$

$$
= \det D^2_{\Gamma(1,3)} \det D^2_{\Gamma(1,4)} \det D^2_{\Gamma(2,3)}
$$

$$
\times \det D^2_{\Gamma(1,2)} \det D^2_{\Gamma(2,4)} \det D^2_{\Gamma(3,4)}. \qquad \square
$$

We show that this splitting process based on Theorem 1.1 may be used to handle the general case as well.

1.8 Theorem *Let $G = (V, E)$ be a chordal graph and let G_0, G_1, $\ldots$, G_t be a chordal sequence of G. Fix A_0 a strictly positive partial block matrix subjacent to the graph G and suppose that all the diagonal entries of A_0 are identity matrices. Let A be a strictly positive completion of A_0 and let $\{\Gamma(u_j, w_j) \mid j = 1, 2, \ldots, t\}$ be the set of the Schur parameters of A along the chordal sequence G_0, G_1, $\ldots$, G_t. Then*

$$
\det A = D(A_0) \times \prod_{j=1}^{t} \det D^2_{\Gamma(u_j, w_j)}.
$$

Proof This result is proved by induction on the number $|V|$ of vertices of G. For $n \leq 3$, this is easy to be checked out and assume the formula is true for any chordal graph with at most $n - 1$ vertices. Consider a strictly positive partial block matrix A_0 subjacent to a chordal graph $G = (V, E)$ with $|V| = n$. Moreover, consider a chordal sequence $G_0, G_1, \ldots, G_t$ of G and a strictly positive completion $A = [A_{ij}]_{1 \leq i, j \leq n}$ of A_0. Let $\{\Gamma(u_j, w_j) \mid j = 1, 2, \ldots, t\}$ be the set of the Schur parameters of A along the fixed chordal sequence of G. Define for $0 \leq m \leq t$ the strictly positive partial matrices $A_0(G_m) = [A_{ij}(G_m) \mid 1 \leq i, j \leq n]$ by

$$
A_{ij}(G_m) = \begin{cases} A_{ij} & \text{if} \quad (i, j) \in E_m \\ \text{unspecified} & \text{otherwise.} \end{cases} \tag{1.2}
$$

Remark that $A(G_0) = A_0$, $A_0(G_t) = A$ and A is a strictly positive invertible completion of any $A_0(G_m)$ for $0 \leq t$. Moreover, G_m is the associated graph of $A_0(G_m)$ and $G_m, G_{m+1}, \ldots, G_t$ is a chordal sequence of G_m. The set of the Schur parameters of A (viewed as a strictly positive completion of $A_0(G_m)$) along this chordal sequence of G_m is exactly $\{\Gamma(u_j, w_j) \mid j = m+1, \ldots, t\}$. By Theorem 1.1 we have

$$
\det A = \frac{\det A[V - \{u_t\}] \det A[V - \{w_t\}]}{\det A[V - \{u_t, w_t\}]} \times \det D^2_{\Gamma(u_t, w_t)} \tag{1.3}
$$

and we claim that

$$
\frac{\det A[V - \{u_t\}] \det A[V - \{w_t\}]}{\det A[V - \{u_t, w_t\}]} = D(A_0(G_{t-1})). \tag{1.4}
$$

Indeed, $[u_t, w_t, (V - \{u_t, w_t\})]$ is a perfect scheme of G_{t-1}, where the ordering in $V - \{u_t, w_t\}$ is chosen at random, and then (1.4) is a consequence of Proposition 1.6. From (1.3) and (1.4), it results that

$$
\det A = D(A_0(G_{t-1})) \det D^2_{\Gamma(u_t, w_t)}. \tag{1.5}
$$

Suppose now that the formula

$$
\det A = D(A_0(G_k)) \det D^2_{\Gamma(u_t, w_t)} \cdots \det D^2_{\Gamma(u_{k+1}, w_{k+1})} \tag{1.6}
$$

was proven for every k, $m \leq k \leq t - 1$, where $0 < m < t$. We claim that the same formula holds for $m - 1$. Let $\sigma = [v_1, v_2, \ldots, v_n]$ be a perfect scheme of G_{m-1}. We distinguish two cases.

A. $v_1 \neq u_m$ and $v_1 \neq w_m$.

Since v_1 is simplicial in G_{m-1}, it follows that u_m and w_m are not simultaneously adjacent to v_1 in G_{m-1}. Consequently, v_1 remains a simplicial vertex in G_m and by Theorem 7.1.4, a perfect scheme $[v_1, x_2, \ldots, x_n]$ of G_m may be found. Now, denote by $\hat{G}_m$, $0 \leq m \leq t$, the induced graphs $(G_m)_{\{v_2, v_3, \ldots, v_n\}}$. In particular, $[v_2, v_3, \ldots, v_n]$ is a perfect scheme of $\hat{G}_{m-1}$ and $[x_2, x_3, \ldots, x_n]$ is a perfect scheme of $\hat{G}_m$. Some of these graphs may appear several times. However, taking into account those consecutive coinciding graphs only once, a chordal sequence $\hat{G} = \hat{G}_0, \hat{G}_1, \ldots, \hat{G}_{t'} = K_{n-1}$ of $\hat{G}$ is obtained, where $\hat{G}_{m-1}$ and $\hat{G}_m$ remain consecutive, but possible on other positions in the sequence. Further on, consider $\hat{A}_0 = A_0[\{v_2, v_3, \ldots, v_n\}]$, $\hat{A} = A[\{v_2, v_3, \ldots, v_n\}]$ and remark that $\hat{A}$ is a strictly positive completion of the strictly positive partial block matrix $\hat{A}_0$. Let $\{\hat{\Gamma}(u_j, w_j) \mid j = 1, 2, \ldots, t'\}$ be the set of the Schur parameters of $\hat{A}$ along the chordal sequence $\hat{G}_0$, $\hat{G}_1$, ..., $\hat{G}_{t'}$ of $\hat{G}$. We may note that if $\hat{A}$ is viewed as a strictly positive completion of $\hat{A}_0(\hat{G}_{m-1})$, then $\{\hat{\Gamma}(u_j, w_j) \mid j = m, m+1, \ldots, t'\}$ is the set of the Schur parameters of $\hat{A}$ along the chordal sequence $\hat{G}_{m-1}$, $\hat{G}_m$, ..., $\hat{G}_{t'}$ of $\hat{G}_{m-1}$. Similarly, $\{\hat{\Gamma}(u_j, w_j) \mid j = m+1, m+2, \ldots, t'\}$ is the set of the Schur parameters of $\hat{A}$ along the chordal sequence $\hat{G}_m$, $\hat{G}_{m+1}$, ..., $\hat{G}_{t'}$ of $\hat{G}_m$, this time $\hat{A}$ being viewed as a strictly positive completion of $\hat{A}_0(\hat{G}_m)$. By the induction hypothesis,

$$\det \hat{A} = D(\hat{A}_0(\hat{G}_{m-1})) \times \prod_{j=m}^{t'} \det D^2_{\hat{\Gamma}(u_j, w_j)}$$

and

$$\det \hat{A} = D(\hat{A}_0(\hat{G}_m)) \times \prod_{j=m+1}^{t'} \det D^2_{\hat{\Gamma}(u_j, w_j)}.$$

Since $\| \hat{\Gamma}(u_j, w_j) \| < 1$ for $j = 1, 2, \ldots, t'$, it follows that

$$D(\hat{A}_0(\hat{G}_m)) = D(\hat{A}_0(\hat{G}_{m-1}))) \det D^2_{\hat{\Gamma}(u_m, w_m)}.$$

Now, the main point is the following: as another consequence of the fact that u_m and w_m are not simultaneously adjacent in G_{m-1}, it follows that the unique maximal clique in G_m which is not a clique in G_{m-1} is also the unique maximal clique in $\hat{G}_m$ which is not a clique in $\hat{G}_{m-1}$. In view of the dependence on Schur parameters in Theorem 7.2.7, this means that

$$\hat{\Gamma}(u_m, w_m) = \Gamma(u_m, w_m)$$

and the formula

$$D(\hat{A}_0(\hat{G}_m)) = D(\hat{A}_0(\hat{G}_{m-1}))) \det D^2_{\Gamma(u_m, w_m)} \tag{1.7}$$

is deduced. By Proposition 1.6, the formula (1.7) can be rewritten as

$$\prod_{s=2}^{n} \frac{\det A[\{x_s\} \cup X_s]}{\det A[X_s]} = \prod_{s=2}^{n} \frac{\det A[\{v_s\} \cup X_s]}{\det A[X_s]} \times \det D^2_{\Gamma(u_m, w_m)}. \tag{1.8}$$

Multiply both sides of (1.8) by the number

$$\frac{\det A[\{v_1\} \cup Adj(v_1)]}{\det A[Adj(v_1)]}.$$

Now $\sigma = [v_1, v_2, \ldots, v_n]$ and $[v_1, x_2, \ldots, x_n]$ are perfect schemes of the chordal graphs G_{m-1} and G_m, respectively. It follows by Proposition 1.6 that

$$D(A_0(G_m)) = D(A_0(G_{m-1})) \det D^2_{\Gamma(u_m, w_m)}. \tag{1.9}$$

Equation (1.6) was supposed to be true for $k = m$. Using (1.9), the same formula is obtained for the required case $k = m - 1$.

B. u_m and w_m are the only two simplicial vertices of G_{m-1}.

The fact that this is the only remaining possibility is a consequence of Lemma 7.1.3. We show that in this case $G_m = K_n$, so that this situation can occur only for $m = t$, a case already covered by (1.5). By a remark in the proof of Theorem 7.2.3, G_m has one more maximal clique which is not a clique in G_{m-1} and which contains both u_m and w_m. Since u_m and w_m are simplicial vertices, this clique is exactly $\omega = \{u_m, w_m\} \cup Adj_{G_{m-1}}(u_m) \cap Adj_{G_{m-1}}(w_m)$. Excepting the cliques $\omega_1 = \{u_m\} \cup Adj_{G_{m-1}}(u_m)$ and $\omega_2 = \{w_m\} \cup Adj_{G_{m-1}}(w_m)$, all the other maximal cliques in G_{m-1} remain maximal cliques in G_m. Now, we prove that it is not possible for both ω_1 and ω_2 to remain maximal cliques in G_m. Otherwise, using the remark that a vertex in a chordal graph is simplicial if and only if it is contained into exactly one maximal clique, we get that G_m has no simplicial vertex and this is a contradiction with the fact that G_m is chordal.

Suppose that ω_2 is not maximal in G_m, so w_m is a simplicial vertex in G_m. Since the unique clique in G_m which may contain ω_2 is ω, it follows that $Adj_{G_{m-1}}(w_m) \subseteq Adj_{G_{m-1}}(u_m)$. Suppose $G_m \neq K_n$. As a consequence of Lemma 7.1.3, it follows that there exists a second simplicial vertex v of G_m which is non-adjacent to w_m. Let ω_0 be the unique maximal clique in G_m containing v. Since v is not adjacent to v_m, it follows that $\omega \neq \omega_0$ and then v is also simplicial in G_{m-1}. This means that $v = u_m$, which is not possible because u_n and w_m are adjacent in G_m. Consequently, $G_m = K_n$.

From the analysis of the two cases it follows that the formula (1.6) holds for any k in $\{0, 1, \ldots, t - 1\}$. In particular, for $k = 0$, this is exactly the required formula.   □

8.2   Maximum Determinant Formulae

In this section we present several formulae for the determinant of the positive completion $(A_0)^0$. We begin with a characterization of the maximum determinant completion in terms of its Schur parameters.

2.1 Theorem   *Let $G = (V, E)$ be a chordal graph. Let A_0 be a strictly positive partial block matrix subjacent to G and suppose that all the diagonal entries of A_0 are identity matrices. Let $G_0, G_1, \ldots, G_t$ be a chordal sequence of G. Then the Schur parameters of the maximum determinant completion $(A_0)^0$ of A_0 along this chordal sequence of G are exactly $\Gamma^0(u_j, w_j) = 0$ for $j = 1, 2, \ldots, t$.*

Proof For chordal sequences associated to perfect schemes the result was already noticed in the proof of Theorem 1.5. The general case of an arbitrary chordal sequence of G is a direct consequence of Theorem 1.8 and Theorem 7.2.7.   □

We have seen that a formula for $\det(A_0)^0 (= D(A_0))$ in terms of a perfect scheme of G may be obtained as a combination of Theorem 1.5 and Proposition 1.6:

$$D(A_0) = \prod_{m=1}^{n} \frac{\det A_0[\{v_m\} \cup X_m]}{\det A_0[X_m]}, \tag{2.1}$$

where $\sigma = [v_1, v_2, \ldots, v_n]$ is a perfect scheme of G and X_m, $m = 1, 2, \ldots, n$, are the sets associated by (7.1.1) to this perfect scheme of G. A simple remark here is that a cancellation process may take place in the formula (2.1).

2.2 Example Let G be the graph introduced in Example 7.2.5. Let A_0 be a strictly positive partial matrix subjacent to G. By (2.1),

$$
\begin{aligned}
D(A_0) &= \frac{\det A_0[\{1,2,6\}] \det A_0[\{2,3,4\}] \det A_0[\{4,5,6\}]}{\det A_0[\{2,6\}] \det A_0[\{2,4\}] \det A_0[\{4,6\}]} \\
&\quad \times \frac{\det A_0[\{2,4,6\}] \det A_0[\{4,6\}] \det A_0[\{6\}]}{\det A_0[\{4,6\}] \det A_0[\{6\}]} \\
&= \frac{\det A_0[\{1,2,6\}] \det A_0[\{2,3,4\}] \det A_0[\{4,5,6\}] \det A_0[\{2,4,6\}]}{\det A_0[\{2,6\}] \det A_0[\{2,4\}] \det A_0[\{4,6\}]}. \ \square
\end{aligned}
$$

The general result explaining the cancellation process in (2.1) expresses $D(A_0)$ in terms of a tree of G satisfying the intersection property. The existence of such a tree is guaranteed by Theorem 7.1.7.

2.3 Theorem *Let G be a connected chordal graph and let $T = (\mathcal{C}, \mathcal{E})$ be a tree of G satisfying the intersection property. Let A_0 be a strictly positive partial block matrix subjacent to G. Then*

$$D(A_0) = \frac{\prod_{C \in \mathcal{C}} \det A_0[C]}{\prod_{(C_1, C_2) \in \mathcal{E}} \det A_0[C_1 \cap C_2]}. \tag{2.2}$$

Proof This result is proven by induction on the number $n = |V|$ of vertices of G. For small values of n ($n = 1, 2, 3$), this is easy to be checked out and suppose that the formula is true for every chordal graph with at most $n - 1$ vertices.

Consider a strictly positive partial block matrix A_0 subjacent to a chordal graph $G = (V, E)$ with $|V| = n$ and suppose, without loss of generality, that all the diagonal entries of A_0 are identity matrices. Moreover, let $T = (\mathcal{C}, \mathcal{E})$ be a tree of G which satisfies the intersection property. Let C_0 in $\mathcal{C}$ be an endpoint of T and pick some v in $C_0 - C_1$, where C_1 belongs to $Adj_T(C_0)$. Since v is contained into exactly one maximal clique (*i.e.* C_0) of G, it follows that v is a simplicial vertex of G. Consider $G' = G_{V - \{v\}}$ and let $[w_1, w_2, \ldots, w_{n-1}]$ be a perfect scheme of G'. Then $\sigma = [v, w_1, w_2, \ldots w_{n-1}]$ is a perfect scheme of G. We distinguish two cases.

A. The clique $X = Adj_G(v)$ is maximal in G'.

In this case one obtains the tree $T' = (\mathcal{C}', \mathcal{E}')$ by renaming C_0 with X and taking $\mathcal{E}' = \mathcal{E}$. Then T' is a tree of G' satisfying the intersection property. By the induction hypothesis and (2.1),

$$\prod_{m=1}^{n-1} \frac{\det A_0[\{w_m\} \cup X_m']}{\det A_0[X_m']} = \frac{\prod_{C' \in \mathcal{C}'} \det A_0[C']}{\prod_{(C_1', C_2') \in \mathcal{E}'} \det A_0[C_1' \cap C_2']}, \qquad (2.3)$$

where X_m', $m = 1, 2, \ldots, n-1$, are the sets of type (7.1.1) associated to the perfect scheme $[w_1, w_2, \ldots, w_{n-1}]$ of G'. Multiplication of both sides of (2.3) by the number

$$\frac{\det A_0[\{v\} \cup X]}{\det A_0[X]}$$

gives in the left side exactly $D(A_0)$. But $\{v\} \cup X = C_0$ and $\det A_0[X]$ cancels with the term in $\prod_{C' \in \mathcal{C}'} \det A_0[C']$ corresponding to the maximal clique X obtained by renaming C_0 with X. Thus, the right side is exactly

$$\frac{\prod_{C \in \mathcal{C}} \det A_0[C]}{\prod_{(C_1, C_2) \in \mathcal{E}} \det A_0[C_1 \cap C_2]},$$

as it was required.

B.   The clique $X = Adj_G(v)$ is not maximal in G'.

In this case, consider the tree $T' = (\mathcal{C}', \mathcal{E}') = T_{\mathcal{C} - \{C_0\}}$. T' is a tree of G' which satisfies the intersection property. By the induction hypothesis,

$$\prod_{m=1}^{n-1} \frac{\det A_0[\{w_m\} \cup X_m']}{\det A_0[X_m']} = \frac{\prod_{C' \in \mathcal{C}'} \det A_0[C']}{\prod_{(C_1', C_2') \in \mathcal{E}'} \det A_0[C_1' \cap C_2']}$$

and this time, multiplication in both sides by

$$\frac{\det A_0[\{v\} \cup X]}{\det A_0[X]} = \frac{\det A_0[C_0]}{\det A_0[C_0 \cap C_1]}$$

leads directly to the required formula.   □

We can explain the meaning of the sets appearing in the denominator of the formula in Theorem 2.3. To this end, let G be a connected chordal graph and let $T = (\mathcal{C}, \mathcal{E})$ be a tree of G satisfying the intersection property. Define

$$\mathcal{M}_T = \{C_1 \cap C_2 \mid (C_1, C_2) \in \mathcal{E}\}.$$

We note the following result.

2.4 Theorem S belongs to $\mathcal{M}_T$ if and only if S is a minimal vertex separator of G.

Proof   Consider a perfect scheme $\sigma = [v_1, v_2, \ldots, v_n]$ of G. By the proof of Theorem 2.3, $\mathcal{M}_T$ is the set of the cliques X_i which are not maximal cliques in $G_{\{v_{i+1},v_{i+2},\ldots,v_n\}}$. It is proven by induction on $n = |V|$ that these sets are exactly the minimal vertex separators of G. For small values of n ($n \le 3$), this is easy to be checked out and suppose the affirmation is true for every chordal graph with at most $n-1$ vertices. Let G be a chordal graph with n vertices and let $\sigma = [v_1, v_2, \ldots, v_n]$ be a perfect scheme of G. Consider S a minimal vertex separator in G. We can distinguish two cases.

A. S is a $v_k - v_m$ separator for $k, m \ge 2$.

Since v_1 is a simplicial vertex, S is a $v_k - v_m$ minimal separator in $G_{\{v_2,v_3,\ldots,v_n\}}$ as well, and it follows by the induction hypothesis that S has the required structure.

B. S is a $v_1 - v_m$ minimal separator.

Let G_α be the connected component of $G - S$ containing v_1 and let G_β be the connected component of $G - S$ containing v_m. By Remark 7.1.2, there exists a in α adjacent to all the vertices in S. If $a \ne v_1$, then we are reduced to the case **A**, since S is also an $a - v_m$ separator. If v_1 is the only vertex of α adjacent to all the vertices in S, then it follows that $S = Adj(v_1) = X_1$. By the same Remark 7.1.2, there exists b in β adjacent to all the vertices of S, so that X_1 is not a maximal clique in $G_{\{v_2,v_3,\ldots,v_n\}}$.

For the converse implication we do not use the induction. If X_i is not a maximal clique in $G_{\{v_{i+1},v_{i+2},\ldots,v_n\}}$, then consider ω' the maximal clique in $G_{\{v_{i+1},v_{i+2},\ldots,v_n\}}$ containing X_i and pick some w in $\omega' - X_i$. Consequently, the set X_i is a minimal $v_i - w$ separator. $\square$

Finally, here is another formula for $D(A_0)$ in terms of the Schur parameters of A_0.

2.5 Theorem   *Let $G = (V, E)$ be a chordal graph. Let A_0 be a strictly positive partial block matrix subjacent to G and suppose that all the diagonal entries of A_0 are identity matrices. Let $G_\emptyset = G_0$, G_1, ..., $G_s = G$ be a chordal sequence from $G_\emptyset = (V, \emptyset)$ to G. Let $\{\Gamma(u_j, w_j) \mid j = 1, 2, \ldots, s\}$ be the Schur parameters of A_0 along this chordal sequence. Then*

$$D(A_0) = \prod_{j=1}^{s} \det D^2_{\Gamma(u_j,w_j)}. \qquad (2.4)$$

Proof Let $G = G_{s+1}$, G_{s+2}, ..., $G_t = K_n$ be a chordal sequence of G, then G_0, G_1, ..., G_t is a chordal sequence from $G_\emptyset$ to K_n. Let $\{\Gamma(u_j, w_j) \mid j = 1, 2, \ldots, t\}$ be the set of the Schur parameters of a strictly positive completion A of A_0 along this chordal sequence. It follows that the elements of the set $\{\Gamma(u_j, w_j) \mid j = 1, 2, \ldots, s\}$ are the Schur parameters of A_0 along the chordal sequence G_0, G_1, ..., G_s from $G_\emptyset$ to G. By Theorem 1.8,

$$\det A = \prod_{j=1}^{s} \det D^2_{\Gamma(u_j,w_j)} \times \prod_{j=s+1}^{t} \det D^2_{\Gamma(u_j,w_j)} \le \prod_{j=1}^{s} \det D^2_{\Gamma(u_j,w_j)}$$

with equality only for $\Gamma(u_j, w_j) = 0$, $s < j \le t$, *i.e.* for $A = (A_0)^0$. $\square$

8.3   Maximum Determinant for Nonchordal Graphs

We address an optimization problem which contains the maximum determinant problem solved by Theorem 1.5 as a particular case. We begin our discussion with a new characterization of the maximum determinant completion and we emphasize that Schur parameters still may be used to this end. For a graph $G = (V, E)$, we denote by E^c the subset $\{(i, j) \mid i \neq j, (i, j) \notin E\}$ of $V \times V$.

3.1 Theorem   *Let $G = (V, E)$ be a chordal graph. Let A_0 be a strictly positive partial matrix subjacent to G. Then the maximum determinant completion $(A_0)^0$ of A_0 is the unique strictly positive completion of A_0 with the property:*

$$(((A_0)^0)^{-1})_{ij} = 0 \quad \text{for} \quad (i, j) \in E^c.$$

Proof It may be useful to illustrate this phenomenon for the generic case of a strictly positive partial matrix of the form:

$$A_0 = \begin{bmatrix} A_{11} & A_{12} & ? \\ A_{12}^* & A_{22} & A_{23} \\ ? & A_{23}^* & A_{33} \end{bmatrix}.$$

Let A be a strictly positive completion of A_0 and consider the Cholesky factorization $A = FF^*$ of A. In terms of the Schur parameters Γ_{12}, Γ_{23} and Γ_{13} of A,

$$F = \begin{bmatrix} A_{11}^{1/2} & 0 & 0 \\ 0 & A_{22}^{1/2} & 0 \\ 0 & 0 & A_{33}^{1/2} \end{bmatrix} \begin{bmatrix} I & \Gamma_{12} & \Gamma_{12}\Gamma_{23} + D_{\Gamma_{12}^*}\Gamma_{13}D_{\Gamma_{23}} \\ 0 & D_{\Gamma_{12}} & D_{\Gamma_{12}}\Gamma_{23} - \Gamma_{12}^*\Gamma_{13}D_{\Gamma_{23}} \\ 0 & 0 & D_{\Gamma_{13}}D_{\Gamma_{23}} \end{bmatrix}.$$

Remark that the $(1, 3)$ entry of F^{-1} is $-D_{\Gamma_{12}^*}^{-1}D_{\Gamma_{13}^*}^{-1}\Gamma_{13}A_{33}^{-1/2}$. Since $A^{-1} = F^{-1}(F^{-1})^*$ and F is upper triangular, it follows that $(A^{-1})_{13} = 0$ if and only if $\Gamma_{13} = 0$. In other words, $(A^{-1})_{13} = 0$ if and only if $A = (A_0)^0$, as required.

The proof of the general case is based on this remark and uses Procedure 7.2.4. Let G be a chordal graph and let A_0 be a strictly positive partial matrix subjacent to G. Without loss of generality, it is supposed that all the diagonal entries of A_0 are identity matrices. If $\sigma = [v_1, v_2, \ldots, v_n]$ is a perfect scheme of G, then $G_0 = G$, G_1, ..., $G_t = K_n$ is the chordal sequence of G associated to σ by Procedure 7.2.4. By Theorem 2.1, the Schur parameters of $(A_0)^0$ along this chordal sequence are $\Gamma^0(u_j, w_j) = 0$ for $j = 1, 2, \ldots, t$. The remark at the beginning of this proof assures that $(((A_0)^0[\varpi_1])^{-1})_{(u_1, w_1)} = 0$. Let R be the number introduced by Procedure 7.2.4 as the least positive integer with the property that $\alpha_R = \{v_R, v_{R+1}, \ldots, v_n\}$ is a clique.

We now claim that if $(((A_0)^0[\alpha_{R-i+1}])^{-1})_{pq} = 0$ for a certain $(p, q) \in (\alpha_{R-i+1} \times \alpha_{R-i+1}) - E$ with $1 \leq i \leq k$ and $1 \leq k \leq R - 1$, then $(((A_0)^0[\alpha_{R-i}])^{-1})_{pq} = 0$. It is known (see Procedure 7.2.4) that the partial matrix $A_0[\alpha_{R-i}]$ written with respect to the ordering $\sigma(r_{R-i})$ has the block matrix representation

$$A_0[\alpha_{R-i}] = \begin{bmatrix} B_{R-i} & H_{R-i} & ? \\ H_{R-i}^* & D_{R-i} & E_{R-i} \\ ? & E_{R-i}^* & F_{R-i} \end{bmatrix},$$

where
$$\begin{bmatrix} D_{R-i} & E_{R-i} \\ E_{R-i}^* & F_{R-i} \end{bmatrix}$$
is a strictly positive matrix with specified elements, while
$$A_0' = \begin{bmatrix} B_{R-i} & H_{R-i} \\ H_{R-i}^* & D_{R-i} \end{bmatrix}$$
is a reordering of $A_0[\alpha_{R-i+1}]$. Similarly, consider the matrix representation
$$(A_0)^0[\alpha_{R-i}] = \begin{bmatrix} a & b \\ b^* & c \end{bmatrix},$$
where $a = (A_0)^0[\alpha_{R-i+1}]$, $b = [X \quad E_{R-i}]^\top$ and $c = F_{R-i}$. Using the formula (1.6.14) and the equality $a^{-1}bs^{-1} = [0 \quad Y]^\top$, where $s - c - B^*a^{-1}b$, it follows that
$$a^{-1} + a^{-1}bs^{-1}b^*a^{-1} = a^{-1} + \begin{bmatrix} 0 & 0 \\ 0 & Z \end{bmatrix}.$$

This formula proves our claim and now we can easily conclude the proof of the theorem. ☐

Hereafter we restrict our discussion to matrices with scalar entries and we show that both Theorem 1.5 and Theorem 3.1 may be deduced as consequences of the solution of an optimization problem with objective function
$$f_B(X) = \log\det X - \mathrm{tr}\,BX \tag{3.1}$$
on a certain set $\mathcal{R}$ of positive matrices. This function is also motivated by the consideration of the relative entropy defined in Section 5.4. Thus, if f and g are probability densities of two Gaussian distributions defined by covariance matrices P and R respectively, then $D(f\|g) = -\frac{1}{2}[\log\det PR^{-1} + \mathrm{tr}(I - PR^{-1})]$. This shows that in connection with the completion problem 7.3.1, the maximum determinant solution is the right analogue of the maximum entropy solution in Section 5.4. The following result clarifies the nature of the considered optimization problem. It is convenient to introduce the notation $\mathcal{M}_n^+$ ($\mathcal{M}_n^{\geq}$) to denote the set of (strictly) positive matrices in $\mathcal{M}_n$.

3.2 Lemma   *The function f_B defined by (3.1) is strictly concave on the set $\mathcal{M}_n^+$.*

Proof We show that the function $f_0(X) = \log\det X$ is strictly concave on $\mathcal{M}_n^+$. For every $0 \leq \lambda \leq 1$, A in $\mathcal{M}_n^{\geq}$ and C in $\mathcal{M}_n^+$,
$$f_0(\lambda A + (1-\lambda)C) = f_0(A(\lambda + (1-\lambda)A^{-1/2}CA^{-1/2}))$$
and the spectral theorem for positive matrices shows that
$$A^{-1/2}CA^{-1/2} = U \begin{bmatrix} d_1 & 0 & \cdots & 0 \\ 0 & d_2 & & \\ \vdots & & \ddots & \\ 0 & & & d_n \end{bmatrix} U^*$$

with a unitary matrix U. Consequently,

$$
\begin{aligned}
f_0(\lambda A + (1-\lambda)C) &= f_0(A) + \textstyle\sum_{j=1}^n \log(\lambda + (1-\lambda)d_j) \\
&\geq f_0(A) + (1-\lambda)\textstyle\sum_{j=1}^n \log d_j \\
&= \lambda f_0(A) + (1-\lambda)\log\det(AA^{-1/2}CA^{-1/2}) \\
&= \lambda f_0(A) + (1-\lambda)f_0(C).
\end{aligned}
$$

A simple approximation argument shows that f_0 is strictly concave on $\mathcal{M}_n^+$. Since $\mathrm{tr}(BX)$ is linear in X, it follows that f_B is strictly concave. $\square$

Domains $\mathcal{R}$ which are of interest for our purposes are described as follows. Let $\mathcal{W} \subset \mathcal{M}_n$ be a linear subspace with the property that $\mathcal{W} \cap \mathcal{M}_n^+ = \{0\}$ and for A in $\mathcal{M}_n^>$, define

$$
\mathcal{R} = (A + \mathcal{W}) \cap \mathcal{M}_n^>. \tag{3.2}
$$

3.3 Theorem *If A and B belong to $\mathcal{M}_n^>$, and $\mathcal{R}$ is a set as described by (3.2), then the optimization problem*

$$
\max_{X \in \mathcal{R}} f_B(X)
$$

has a unique solution A^0 in $\mathcal{R}$. Moreover, for all $W \in \mathcal{W}$,

$$
\mathrm{tr}(((A^0)^{-1} - B)W) = 0.
$$

Proof It is easily seen that $(A + \mathcal{W}) \cap \mathcal{M}_n^+$ is a closed convex subset of $\mathcal{M}_n^+$. Since $\mathcal{W} \cap \mathcal{M}_n^+ = \{0\}$, it follows that $(A + \mathcal{W}) \cap \mathcal{M}_n^+$ is bounded. By Lemma 3.2, f_B is strictly concave, therefore f_B has a unique maximum, denoted by A^0, in $(A + \mathcal{W}) \cap \mathcal{M}_n^+$. Near the boundary f_B tends to $-\infty$, hence A^0 belongs to $\mathcal{R}$.

Then, consider an arbitrary matrix W in $\mathcal{W}$ and define

$$
g_W(x) = f_B(A^0 + xW)
$$

in a neighborhood of 0 on $\mathbb{R}$. In this neighborhood, $g_W(x) \leq g_W(0)$ by the maximum property of A^0. Moreover, g_W is differentiable and the necessary condition of extremum is $g_W'(0) = 0$. But,

$$
\begin{aligned}
g_W'(x) &= \frac{(\det(I + x(A^0)^{-1}W))'}{\det(I + x(A^0)^{-1}W)} - \mathrm{tr}(B(A^0 + xW))' \\
&= \frac{1}{\det(I + x(A^0)^{-1}W)} \\
&\quad \times (1 + x\mathrm{tr}(A^0)^{-1}W + \text{terms of higher powers of } x)' - \mathrm{tr}BW,
\end{aligned}
$$

consequently

$$
g_W'(0) = \mathrm{tr}((A^0)^{-1}W) - \mathrm{tr}(BW) = \mathrm{tr}(((A^0)^{-1} - B)W).
$$

The proof is complete. □

Let us note several particular cases of this result.

3.4 Corollary Let $G = (V, E)$ be an undirected graph and let A, B be given matrices in $\mathcal{M}_n^>$. Then there exists a unique matrix $A^0 \in \mathcal{M}_n^>$ such that

(a) $A_{ij}^0 = A_{ij}$ for $(i, j) \in E$ or $i = j$.

(b) $((A^0)^{-1})_{ij} = B_{ij}$ for $(i, j) \in E^c$.

Proof Consider the set

$$\mathcal{W} = \{W \in \mathcal{M}_n \mid W \text{ is selfadjoint and } W_{ij} = 0 \text{ for } (i, j) \in E \text{ or } i = j\}.$$

Then $\mathcal{W} \cap \mathcal{M}_n^+ = \{0\}$, since the only positive matrix with zero diagonal is the zero matrix. By Theorem 3.3, there exists A^0 in $(A + \mathcal{W}) \cap \mathcal{M}_n^>$ which is the unique solution in $\mathcal{R} \cap \mathcal{M}_n^>$ of the optimization problem

$$\max_{X \in \mathcal{R}} f_B(X).$$

Since A^0 belongs to $(A + \mathcal{W}) \cap \mathcal{M}_n^>$, it follows that $A_{ij}^0 = A_{ij}$ for (i, j) in E or $i = j$.

Furthermore, $\operatorname{tr}(((A^0)^{-1} - B)W) = 0$ for all W in $\mathcal{W}$. If (j, k) belongs to E^c, then define

$$(W_R^{(j,k)})_{lm} = \begin{cases} 1 & \text{if } (l, m) = (j, k) \text{ or } (l, m) = (k, j) \\ 0 & \text{otherwise} \end{cases}$$

and

$$(W_I^{(j,k)})_{lm} = \begin{cases} \sqrt{-1} & \text{if } (l, m) = (j, k) \\ -\sqrt{-1} & \text{if } (l, m) = (k, j) \\ 0 & \text{otherwise.} \end{cases}$$

Since the matrices $W_R^{(j,k)}$ and $W_I^{(j,k)}$ belong to $\mathcal{W}_G$, it is deduced from the equalities $\operatorname{tr}(((A^0)^{-1} - B)W_R^{(j,k)}) = \operatorname{tr}(((A^0)^{-1} - B)W_I^{(j,k)})$ that $((A^0)^{-1})_{jk} = B_{jk}$. □

We show how to deduce Theorem 1.5 and Theorem 3.1 from the previous corollary of Theorem 3.3.

3.5 Corollary Let A_0 be a strictly positive partial block matrix subjacent to a chordal graph. Then, among all the strictly positive completions of A_0 there exists a unique one, denoted by A^0, with maximum determinant. Moreover,

$$(((A_0)^0)^{-1})_{ij} = 0 \quad \text{for } (i, j) \in E^c.$$

Proof Since G is chordal, Theorem 7.2.3 assures that there exists a strictly positive completion A of A_0. The proof is concluded by using Corollary 3.4 for this A and $B = 0$. □

3.6 Corollary *Let A_0 be a symmetric partial Toeplitz matrix with a prescribed main diagonal and which admits strictly positive Toeplitz completions. Then there exists a unique positive Toeplitz completion A^0 of A_0 such that*

(a) A^0 is the maximum determinant positive Toeplitz completion of A_0.

(b)   The sum of the entries of $(A^0)^{-1}$ on each of the diagonals corresponding to unspecified diagonals of A_0, equals 0.

Proof Consider $B = 0$ and $\mathcal{W}$ is the linear space generated by the Toeplitz matrices

$$
W_R^{(j)} = \begin{bmatrix}
0 & 0 & & 1 & 0 & & 0 \\
0 & 0 & & & 1 & & 0 \\
& & \ddots & & & \ddots & \\
& & & 0 & & & 1 \\
1 & & & & & & \\
0 & 1 & & & \ddots & & \\
& & \ddots & & & & \\
0 & & & 1 & & & 0
\end{bmatrix}
$$

and

$$
W_I^{(j)} = \begin{bmatrix}
0 & 0 & & \sqrt{-1} & 0 & & 0 \\
0 & 0 & & & \sqrt{-1} & & 0 \\
& & \ddots & & & \ddots & \\
& & & 0 & & & \sqrt{-1} \\
-\sqrt{-1} & & & & & & \\
0 & -\sqrt{-1} & & & \ddots & & \\
& & \ddots & & & & \\
0 & & & -\sqrt{-1} & & & 0
\end{bmatrix} ;
$$

the nonzero elements are on the j-th and $-j$-th diagonals, where j denotes an index with the property that the j-th and $-j$-th diagonals of A_0 are unspecified. Take A a strictly positive Toeplitz completion of A_0 and then the proof may be concluded by applying Theorem 3.3.   □

8.4 Inheritance Principles

We have seen in the previous sections that the problem of maximizing the determinant over the set of strictly positive completions is interesting by itself and has several important applications. The purpose of this section is to study the behavior of the maximum determinant completion under the operation of compression of matrices. This will permit the replacement of the global problem of finding the maximum determinant completion with a sequence of similar problems, but of smaller complexity.

     Let us state the problem. Fix $G = (V, E)$ a chordal graph. For a strictly positive partial block matrix A_0 subjacent to G, we denote by $(A_0)^0$ its maximum determinant completion. We are interested to characterize those subsets V' of V with the property that

$$(A_0[V'])^0 = (A_0)^0[V'] \tag{4.1}$$

for every strictly positive partial block matrix A_0 subjacent to G. We use chordal sequences with some specified properties in order to characterize the subsets V' of V which satisfy (4.1). More precisely, consider a chordal sequence $G = G_0, G_1, \ldots, G_t = K_n$ of G, $n = |V|$. Recall that we denoted by ϖ_j the unique maximal clique in G_j which is not a clique in G_{j-1} and by (u_j, w_j) the edge added to G_{j-1} in order to obtain G_j. Now, we say that the considered chordal sequence has the property $(P(V'))$ if there exists $r \leq t$ such that $\varpi_j \subseteq V'$ for all $j = 1, 2, \ldots, r$ and $(G_r)_{V'}$ is the complete graph with vertex set V'. Then, we say that V' has the property (P) if there exists a chordal sequence of G with the property $(P(V'))$.

4.1 Theorem *Let $G = (V, E)$ be a chordal graph. A subset V' of V satisfies (4.1) for every strictly positive partial block matrix subjacent to G if and only if V' has the property (P).*

Proof Suppose that V' satisfies the property (P). Consider the induced subgraph $G' = G_{V'}$ of G and let $G_0, G_1, \ldots, G_t$ be a chordal sequence of G satisfying the property $(P(V'))$. This implies that $\{G'_k = (G_k)_{V'}\}_{k=0}^r$ is a chordal sequence from G' to K_p, $p = |V'|$. Moreover, for $k = 1, 2, \ldots, r$ the unique maximal clique in G'_k which is not a clique in G'_{k-1} is exactly ϖ_k, i.e. the unique maximal clique in G_k which is not a clique in G_{k-1}. Let A_0 be a strictly positive partial block matrix subjacent to G. By Theorem 7.2.7, any strictly positive completion A of A_0 is uniquely determined by its Schur parameters $\Gamma(u_j, w_j)$, $j = 1, 2, \ldots, t$, along the fixed chordal sequence $G_0, G_1, \ldots, G_t$. It follows that $\{\Gamma(u_j, w_j) \mid j = 1, 2, \ldots, r\}$ is the set of the Schur parameters of $A[V']$ along the chordal sequence $G'_0, G'_1, \ldots, G'_r$. By Theorem 2.1, the Schur parameters of $(A_0)^0$ are $\Gamma^0(u_j, w_j) = 0$ for $j = 1, 2, \ldots, t$. Consequently, (4.1) holds for V' and A_0.

Conversely, suppose that V' is a subset of V such that (4.1) holds for every strictly positive partial block matrix subjacent to G. It is useful to introduce another simple property of a chordal sequence. Thus, a chordal sequence $G_0, G_1, \ldots, G_t$ of G has the property $(P(V'))_0$ if there exists $r' \leq t$ such that $(V', E_{V'} \cup \{(u_j, w_j) \mid j = 1, 2, \ldots, r'\})$ is the complete graph with vertex set V'. Then, V' is said to have the property $(P)_0$ if there exists a chordal sequence of G with the property $(P(V'))_0$.

We claim that V' has the property $(P)_0$ if and only if the graph $K(V') = (V, F)$ is chordal, where $F = E \cup \{(u, w) \mid u, w \in V'\}$, and this is easy to be seen using Theorem 7.1.6.

Now, suppose that V' has not the property (P) and we can distinguish two cases:

A. V' does not have $(P)_0$.

Then $K(V')$ is not chordal by the previous claim, so it contains a k-cycle $O_k = [v_1, v_2, \ldots, v_k]$ for some k, $4 \leq k \leq n$. This k-cycle has exactly two vertices in V', say v_1 and v_k. Since G is chordal, all edges of O_k except for (v_1, v_k) belong to E. Now we indicate a way to construct a strictly positive partial matrix A_0 subjacent to G for which (4.1) does not hold. First, construct a special chordal sequence between $G_\emptyset = (V, \emptyset)$ and G. Employ Theorem 7.1.6 to get a chordal sequence $G_\emptyset = G_0, G_1, \ldots, G_s = G_{V'}$ from $G_\emptyset$ to $G_{V'}$. Let $G_{s+1}, G_{s+2}, \ldots, G_{s+k-2}$ be the (chordal) graphs obtained by successively adding to G_s the edges $(v_1, v_2), \ldots, (v_{k-2}, v_{k-1})$. Using Theorem 7.1.6, we can continue the chordal sequence from G_{s+k-2} to $G_{V' \cup \{v_2, \ldots v_{k-1}\}}$ and then further on to G. We now construct a strictly positive partial matrix A_0 subjacent to G by using Theorem 7.2.7 along this chordal sequence. Take the first s Schur parameters zero. This would uniquely determine the positive partial matrix $A_0[V']$. The entry of index (v_1, v_k) in $(A_0[V'])^0$ is $z = 0$. We further choose the Schur parameters of A_0 corresponding to the positions $(v_1, v_2), (v_2, v_3), \ldots, (v_{k-1}, v_k)$ to be $a_2, a_3, \ldots, a_k$ as in Lemma 7.2.2. Set zero all the remaining Schur parameters. Since the maximal cliques obtained for G_{s+1}, $\ldots, G_{s+k-2}$ are respectively $\varpi_{s+1} = \{v_1, v_2\}, \ldots, \varpi_{s+k-2} = \{v_{k-2}, v_{k-1}\}$ and no maximal clique along the considered chordal sequence may contain two of the edges of C_k, we see that all the specified entries of A_0 equal the corresponding Schur parameter. If (4.1) holds for A_0, then $(A_0)^0[\{v_1, v_2, \ldots v_k\}]$ should be positive, which contradicts Lemma 7.2.2.

B. V' has $(P)_0$ but not (P).

Consider a chordal sequence of G having the property $(P(V'))_0$. Let s be the smallest index such that the maximal clique ϖ_s is not contained in V'. (u_s, w_s) is the edge added to get the next graph in the chordal sequence. Pick some v in $\varpi_s - V'$. Again we construct a chordal sequence between $G_\emptyset$ and G. Choose an arbitrary chordal sequence $G_0, G_1, \ldots, G_r$ from $G_\emptyset$ to $G_{V'}$. Furthermore, let G_{r+1} be the (chordal) graph obtained by adding to G_r the edge (v, u_s) and let G_{r+1}, $G_{r+2}, \ldots, G_{V' \cup \{v\}}$ be a chordal sequence from G_{r+1} to $G_{r+p} = G_{V' \cup \{v\}}$. Using once again Theorem 7.1.6, we can find a chordal sequence from $G_{V' \cup \{v\}}$ to G. The concatenation of these sequences produces a chordal sequence from $G_\emptyset$ to G. We construct along this chordal sequence a strictly positive partial matrix A_0 in the following way: its first r Schur parameters are equal to zero; the Schur parameter corresponding to the edge (v, u_s) is arbitrarily chosen a in the open interval $(0, 1)$. We now set the remaining Schur parameters zero except for b in the open interval $(0, 1)$, the one corresponding to the edge (v, w_s). Since u_s, w_s and v cannot enter simultaneously in a maximal clique along the sequence $G_{r+1}, \ldots, G_{r+p}$, the entries in A_0 coincide with their corresponding Schur parameters. Suppose (4.1) holds. Since the entry of $(A_0[V'])^0$ corresponding to (u_s, w_s) is zero, the same holds for $(A_0)^0[V']$, which is a contradiction. Consequently, (4.1) does not hold for these V' and A_0 and the theorem is proved.   □

The introduction of the property $(P)_0$ makes it easy to obtain a simple characterization of the property (P) directly in terms of V'. For this purpose, define

$$Adj(V') = \{v \in V - V' \mid (v, w) \in E \quad \text{for some} \quad w \in V'\}$$

and we note the following result.

4.2 Theorem   Let $G = (V, E)$ be a chordal graph. Then, a subset $V' \subset V$ has the property that (4.1) holds for every strictly positive partial block matrix subjacent to G, if and only if $K(V')$ is chordal and for each $v \in Adj(V')$, the set $Adj(v) \cap V'$ is a clique.

Proof Suppose there exists a chordal sequence of G with the property $(P(V'))$. In particular, it has the property $(P(V'))_0$, so $K(V')$ is chordal. Suppose there is some v in $Adj(V')$ such that $Adj(v) \cap V'$ is not a clique. Hence there exist $u, w \in Adj(v) \cap V'$ such that $(u, w) \notin E_{V'}$. Since $(G_s)_{V'}$ is the complete graph with vertex set V' by definition, it follows that there exists $1 \le s' \le s$ with $(u, w) \in E_{s'}$ and $(u, w) \notin E_{s'-1}$. Since $\varpi_{s'}$ is a maximal clique, it follows that v belongs to $\varpi_{s'}$, contradicting the assumption that $\varpi_{s'} \subseteq V'$.

Conversely, $K(V')$ being chordal, there exists a chordal sequence G_0, G_1, ..., G_t of G with the property that for a certain $r \le t$, $(V', E_{V'} \cup \{(u_j, w_j) \mid j = 1, 2, \ldots, r\})$ is the complete graph with vertex set V'. Consider an index $s \le r$ such that ϖ_s is not a subset of V' and take v in $\varpi_s - V'$. Consequently, v belongs to $Adj(u_s) \cap Adj(w_s)$ and then $\{u_s, w_s\} \subset Adj(v) \cap V'$. Since $(u_s, w_s) \notin E$, $Adj(v) \cap V'$ is not a clique of G. □

Let us note several particular cases of this result. This type of results are referred to as *inheritance*, or *permanence principles*.

4.3 Corollary Let $G = (V, E)$ be a chordal graph. Then, a subset $V' \subset V$ has the property that (4.1) holds for every strictly positive partial matrix subjacent to G, if and only if we can't find two vertices $a, b \in V'$ such that $(a, b) \notin E$ and which can be connected through a simple path in G having all the vertices but a and b in $V - V'$. □

Recall that a partial positive matrix A_0 is called block band partial $n \times n$ matrix if it has the property: if the (i, j) entry is specified then the (k, m) entry is specified whenever $i \le k \le m \le j$.

4.4 Corollary Let A_0 be a block band strictly positive partial $n \times n$ matrix. Then (4.1) holds for any $V' = \{r, r+1, \ldots, s-1, s\}$, where $1 \le r \le s \le n$. □

Another interesting version of an inheritance principle is described by the following result.

4.5 Corollary Let $G = (V, E)$ be a chordal graph and v a simplicial vertex. Then $V' = V - \{v\}$ has the property that (4.1) holds for every strictly positive partial block matrix subjacent to G. □

Finally, we note the following result which emphasizes once again the meaning of the inheritance principles: the several-variable optimization problem of finding the maximum determinant completion of a strictly positive partial block matrix subjacent to a chordal graph is reduced to a sequence of one-variable optimization problems of the same kind.

4.6 Corollary For a chordal graph $G = (V, E)$, every chordal sequence G_0, G_1, ..., G_t of G has the following property. For every strictly positive partial block matrix A_0 subjacent to G, construct a sequence of strictly positive partial block matrices as follows: A_j is obtained from A_{j-1} by completing the (u_j, w_j) entry in such a way that its principal submatrix subjacent to ϖ_j is the maximum determinant completion of $A_{j-1}[\varpi_j]$. Then the matrix A_t coincides with the maximum determinant completion $(A_0)^0$ of A_0. □

8.5 Notes

Various presentations of the Hadamard type inequalities can be found in [BJ], [Helt5], [Jo1-2]. The maximum determinant principle is also developed in many places. The present formulation was noticed in [DG1] for proper interval graphs and in [GJSW] for the general case. For various formulations of Theorem 3.3, see [BW2], [BLW], [De], [Fi1], [LPK], [Lu], [NN], [SpK]. Corollary 3.4 is a result in [De] – see [SpK] for a recent presentation. Corollary 3.6 was noted in [LJ]. Here we used [BW2]. Theorem 2.3 and Theorem 2.4, as well as other formulae for $D(A_0)$, appear in [BF], [BJ], [JB], [BJLu]. The results involving Schur parameters (Theorem 1.2, Theorem 1.8, Theorem 2.5) were proved in [BC1]. The method used in the proof of Theorem 4.1 follows [BC1] and we used here some notes written together with O. Farcasanu. The first result of this type (Corollary 4.4) was proved in [EGL]. Corollary 4.3 was proved in [BJOD] (see also [JL]), Corollary 4.5 appears in [JR2] and Corollary 4.6 was proved in [BJLu].

References

[AA] Adamjan, V.M. and D.Z. Arov,

[1] A class of scattering operators and of characteristic operator-functions of contractions, *Soviet Math. Doklady*, **160**(1965), 1–5.

[2] On the unitary coupling of isometric operators, *Mat. Issled. Kisinev*, 1(1966), 3–66.

[AAK] Adamjan, V.M., Arov, D.Z. and M.G. Krein,

[1] Infinite Hankel matrices and generalized problems of Carathéodory-Fejér and I. Schur, *Functional Analysis and its Applications*, **2**(1968), 1–19.

[2] Analytic properties of Schmidt pairs for a Hankel operator and the generalized Schur-Takagi problem, *Math. USSR Sbornik*, **15**(1971), 31–73.

[3] Infinite Hankel block matrices and related extension problems, *Izv. Akad. Nauk. Armjan SSR, Matematika*, **6**(1971), 87–112.

[AHMR] Agler, J., Helton, J.W., McCullough, S. and L. Rodman,

[1] Positive semidefinite matrices with a given sparsity pattern, *Linear Algebra Appl.*, **107**(1988), 101–149.

[Ak] Akhiezer, N.I.,

[1] *The Classical Moment Problem*, Olivier and Boyd, Edinburgh, Scotland, 1965.

[AK] Akhiezer, N.I., and M.G. Krein,

[1] *Some Questions in the Theory of Moments*, Amer. Math. Soc. Transl. Math. Monographs, **2**, Providence, RI, 1962.

[ABDS] Alpay, D., Bruinsma, P., Dijksma, A., and H.S.V. de Snoo,

[1] Interpolation problems, extensions of symmetric operators and reproducing kernel spaces I, in *Operator Theory: Advances and Applications*, **50**, Birkhäuser Verlag, pp. 35–82, 1991.

[ADD] Alpay, D., Dewilde, P., and H. Dym,

[1]   On the existence and construction of solutions to the partial lossless inverse scattering problem with applications to estimation theory, *IEEE Trans. Information Theory*, **35**(1989), 1184–1205.

[2]   Lossless inverse scattering and reproducing kernels for upper triangular operators, in *Operator Theory: Advances and Applications*, **47**, Birkhäuser Verlag, pp. 61–135, 1990.

[AD] Alpay, D., and H. Dym,

[1]   On applications of reproducing kernel spaces to the Schur algorithm and rational *J* unitary factorization, in *Schur Methods in Operator Theory and Signal Processing, Operator Theory: Advances and Applications*, **18**, Birkhäuser Verlag, pp. 89–159, 1986.

[An] Ando, T.,

[1]   On a pair of commutative contractions, *Acta Sci. Math.*, **24**(1963), 88–90.

[AnCF] Ando, T., Ceausescu, Z., and C. Foias,

[1]   On intertwining dilations II., *Acta Sci. Math.*, **31**(1977), 3–14.

[Ar] Arocena, R.,

[1]   Generalized Toeplitz kernels and dilations of intertwining operators, *Integral Equations and Operator Theory*, **6**(1983), 759–778.

[2]   On the extension problem for a class of translation invariant forms, *J. Operator Theory*, **21**(1989), 323–347.

[Aro] Aronszajn, N.,

[1]   The theory of reproducing kernels, *Trans. Amer. Math. Soc.*, **68**(1950), 337–404.

[Arov] Arov, D.Z.,

[1]   Darlington's method for dissipative systems, *Sov. Phys. Dokl.*, **16**(1972), 954–956.

[AG] Arov, D.Z., and Z. Grossman,

[1]   Scattering matrices in the theory of dilation of isometric operators, *Soviet Math. Doklady*, **27**(1983), 848–854.

[2]   Scattering matrices in the theory of unitary extension of isometric operators, *Math. Nach.*, **157**(1992), 105–123.

[ArK] Arov, D.Z., and M.G. Krein,

[1]   Problem of search of the minimum entropy in indeterminate extension problems, *Functional Analysis and its Applications*, **15**(1981), 123–126.

[2]   On the evaluation of entropy functionals and their minima in generalized extension problems, *Acta Sci. Math.*, **45**(1983), 33–50.

[ACC] Arsene, Gr., Ceausescu, Z., and T. Constantinescu,

[1] Schur analysis of some completion problems, *Linear Algebra Appl.*, **109** (1988), 1–36.

[ACF] Arsene, Gr., Ceausescu, Z., and C. Foias,

[1] On intertwining dilations VIII., *J. Operator Theory*, **4**(1980), 55–91.

[AC] Arsene, Gr., and T. Constantinescu,

[1] Structure of positive block matrices and nonstationary prediction, *J. Functional Analysis*, **70**(1987), 402–425.

[ACG] Arsene, Gr., Constantinescu, T., and A. Gheondea,

[1] Lifting of operators and prescribed number of negative squares, *Michigan Math. J.*, **34**(1987), 201–216.

[AGh] Arsene, Gr., and A. Gheondea,

[1] Completing matrix contractions, *J. Operator Theory*, **7**(1982), 179–189.

[At] Atkinson, F.,

[1] *Discrete and Continuous Boundary Value Problems*, New York, Academic Press, 1964.

[Arv] Arveson, W.B.,

[1] Interpolation problems in nest algebras, *J. Functional Analysis*, **3**(1975), 208–233.

[Ba] Bakonyi, M.,

[1] *Completion of Partial Operator Matrices*, Thesis, The College of William and Mary, Williamsburg, VA, Aug. 1992.

[BC] Bakonyi, M., and T. Constantinescu,

[1] Inheritance principles for chordal graphs, *Linear Algebra Appl.*, **148**(1991), 125–143.

[2] *Schur's Algorithm and Several Applications*, Longman, London, 1992.

[BW] Bakonyi, M., and H.J. Woerdeman,

[1] On the strong Parrott completion problem, *Proc. Amer. Math. Soc.*, **117**(1993), 429–433.

[2] Maximum entropy elements in the intersection of an affine space and the cone of positive definite matrices, *SIAM J. Matrix Analysis Appl.*, 1995.

[BaC] Ball, J.A., and N. Cohen,

[1] De Branges-Rovnyak operator models and system theory: a survey, in *Operator Theory: Advances and Applications*, **50**, Birkhäuser Verlag, pp. 93–136, 1991.

[BFHT] Ball, J.A., Foias, C., Helton, J.W., and A. Tannenbaum,

[1]   On a local nonlinear commutant lifting theorem, *Indiana Univ. Math. J.*, **36**(1987), 693–707.

[BG] Ball, J.A., and I. Gohberg,

[1]   A commutant lifting theorem for triangular matrices with diverse applications, *Integral Equations and Operator Theory*, **8**(1985), 205–267.

[2]   Classification of shift invariant subspaces of matrices with Hermitian form and completion of matrices, in *Operator Theory: Advances and Applications*, **45**, Birkhäuser Verlag, pp. 23–85, 1990.

[BaGK] Ball, J.A., Gohberg, I., and M.A. Kaashoek,

[1]   Nevanlinna-Pick interpolation problem for time-varying input-output maps: The discrete case, in *Operator Theory: Advances and Applications*, **56**, Birkhäuser Verlag, pp. 1–51, 1992.

[2]   Nevanlinna-Pick interpolation problem for time-varying input-output maps: The continuous case, in *Operator Theory: Advances and Applications*, **56**, Birkhäuser Verlag, pp. 52–89, 1992.

[3]   Bitangential interpolation for input-output operators of time-varying systems: the discrete time case, in *Operator Theory: Advances and Applications*, **64**, Birkhäuser Verlag, pp. 33–72, 1993.

[4]   Bitangential interpolation for input-output operators of time-varying systems: the continuous time case, *Integral Equations and Operator Theory*, to appear.

[5]   Input-output operators of J-unitary time-varying continuous time systems, preprint 1994.

[BGR] Ball, J.A., Gohberg, I., and L. Rodman,

[1]   *Interpolation of Rational Matrix Functions*, Birkhäuser Verlag, 1990.

[BH] Ball, J.A., and J.W. Helton,

[1]   Lie groups over the field of rational functions, signed spectral factorization, signed interpolation, and amplifier design, *J. Operator Theory*, **8**(1982), 19–64.

[2]   A Beurling-Lax theorem for the Lie group $U(m,n)$ which contains most classical interpolation theory, *J. Operator Theory*, **9**(1983), 107–142.

[BF] Barrett, W.W., and P. Feinsilver,

[1]   Inverses of banded matrices, *Linear Algebra Appl.*, **41**(1981), 111–130.

[BJ] Barrett, W.W., and C.R. Johnson,

[1]   Determinantal formulae for matrices with sparse inverses, *Linear Algebra Appl.*, **56**(1984), 73–88.

[BJL] Barrett, W.W., Johnson, C.R., and R. Loewy,

[1] The real positive definite completion problem: Cycle compatibility, preprint 1993

[BJLu] Barrett, W.W., Johnson, C.R., and M. Lundquist,

[1] Determinantal formulae for matrix completions associated with chordal graphs, *Linear Algebra Appl.*, **121**(1989), 265–289.

[BJOD] Barrett, W.W., Johnson, C.R., Olesky, D., and P. van den Driessche,

[1] Inherited matrix entries: principal submatrices of the inverse, *SIAM J. Alg. Disc. Meth.*, **8**(1987), 313–322.

[BJT] Barrett, W.W., Johnson, C.R., and P. Tarazaga,

[1] The real positive definite completion problem for a simple cycle, *Linear Algebra Appl.*, **192**(1993), 3–31.

[BGK] Bart, H., Gohberg, I., and M.A. Kaashoek,

[1] *Minimal Factorization of Matrix and Operator Functions,* Birkhäuser Verlag, 1979.

[Bax] Baxter, G.,

[1] A convergence equivalence related to polynomials orthogonal on the unit circle, *Trans. Amer. Math. Soc.*, **99**(1961), 478–487.

[BAG] Ben-Artzi, A., and I. Gohberg,

[1] Band matrices and dichotomy, in *Operator Theory: Advances and Applications*, **50**, Birkhäuser Verlag, pp. 137–170, 1991.

[2] Dichotomy, discrete Bohl exponents, and spectrum of block weighted shifts, *Integral Equations and Operator Theory*, **14**(1991), 613–677.

[3] Dichotomies of perturbed time varying systems and the power method, *Indiana Univ. Math. J.*, **42**(1993), 699–720.

[Be] Bercovici, H.,

[1] *Operator Theory and Aritmetic in H^∞,* Amer. Math. Soc. Providence, Rhode Island, 1988.

[Ber] Berge, C.,

[1] *Graphs and Hypergraphs*, Chapter 16, North-Holland, Amsterdam, 1973.

[Bev] Bevensee, R.M.,

[1] *Maximum Entropy Solutions to Scientific Problems*, Prentice Hall, Englewood Cliffs, New Jersey, 1993.

[BoL] Borwein, J.M., and A.S. Lewis,

[1]   Convergence of best entropy estimate, *SIAM J. Optimization*, **1**(1991), 191–205.

[2]   Duality relationships for entropy-like minimization problems, *SIAM J. Control and Optimization*, **29**(1991), 325–338.

[Bo] Boyd, D.W.,

[1]   Schur's algorithm for bounded holomorphic functions, *Bull. London Math. Soc.*, **11**(1979), 145–150.

[BS] Böttcher, A., and B. Silbermann,

[1]   *Invertibility and Asymptotics of Toeplitz Matrices*, Academie-Verlag, Berlin, 1983.

[2]   *Analysis of Toeplitz Operators*, Academie-Verlag, Berlin, 1990.

[dB] de Branges, L.,

[1]   *Hilbert Spaces of Entire Functions*, Prentice-Hall Englewood Cliffs, New Jersey, 1968.

[dBR] de Branges, L., and J. Rovnyak,

[1]   Canonical models in quantum scattering theory, in *Perturbation Theory and its Applications in Quantum Mechanics*, Ed. C.H. Wilcox, New York-London-Sidney, pp. 295–395, 1966.

[2]   *Square Summable Power Series*, Holt Richard and Winson, New York, 1966.

[Br] Brodskii, M.S.,

[1]   *Triangular and Jordan Representations of Linear Operators and Intermediate Systems*, Amer. Math. Soc. Transl. Math. Monographs, **32**, Providence, RI, 1969.

[2]   Unitary operator colligations and their characteristic functions, *Russian Math. Surveys*, **22**(1978), 159–191.

[BL] Brodskii, M.S., and M.S. Livsic,

[1]   Spectral analysis of non-selfadjoint operator and intermediate systems, *Uspehi Mat. Nauk* **13**(1958), 3–85; English translation: Amer. Math. Soc. Transl. **13**(1960), 265–346.

[BK] Bruckstein, A.M., and T. Kailath,

[1]   Inverse scattering for discrete transmission-line models, *SIAM Review*, **29** (1987), 359–389.

[Bu] Bultheel, A.,

[1]   Error analysis of incoming and outcoming schemes for trigonometric moment problems, in *Proceedings of the Conference on Padé Approximation and its Applications*, Lecture Notes in Mathematics, Springer, **888**, pp. 100–109, 1980.

[2]   *Laurent Series and their Padé Approximation*, Birkhäuser, Boston, 1987.

[Bun] Buneman, P.,

[1]   The recovery of trees from measures of dissimilarity, in *Mathematics in the Archaeological and Historical Sciences*, Edinburgh Univ.Press, pp. 387–395, 1972.

[Bur] Burg, J.P.,

[1]   *Maximum Entropy Spectral Analysis*, Thesis, Department of Geophysics, Stanford University, California, 1975.

[BLW] Burg, J.P., Luenberger, D.G., and D.L. Wenger,

[1]   Estimation of structured covariance matrices, *Proc. IEEE*, **70**(1982), 963–974.

[ByL] Byrnes, C.I., and A. Lindquist,

[1]   On the geometry of the Kimura-Georgiou parametrization of modelling filters, *Int. J. Control*, **50**(1989), 99–105.

[BLM] Byrnes, C.I., Lindquist, A., and T. McGregor,

[1]   Predictability and unpredictability in Kalman filtering, *IEEE Trans. Automatic Control*, **36**(1991), 563–579.

[Ca] Carathéodory, C.,

[1]   Über den Variabilitätsbereich der Koeffizienten von Potenzreihen, die gegebene Werte nicht annehmen, *Math. Ann.*, **64**(1907), 95–115.

[2]   Über den Variabilitätsbereich der Fourierschen Konstanten von positiven harmonischen Funktionen, *Rend. Circ. Mat. Palermo*, **32**(1911), 193–217.

[CF] Carathéodory, C., and L. Fejér,

[1]   Über den Zusammenhang der Extremen von harmonischen Funktionen mit ihren Koeffizienten und über den Picard-Landauschen Satz, *Rend. Circ. Mat. Palermo*, **32**(1911), 218–239.

[CeF] Ceausescu, Z., and C. Foias,

[1]   On intertwining dilations V., *Acta Sci. Math.*, **40**(1978), 9–32.

[2]   On intertwining dilations VI., *Rev. Roumaine Math. Pures Appl.*, **23**(1978), 1471–1482.

[CP] Chang, B.C., and J.B. Pearson,

[1]   Optimal disturbance reduction on linear multivariable systems, *IEEE Trans. Automatic Control*, **29**(1984), 880–887.

[Ch] Chover, J.,

[1]   On normalized entropy and the extensions of a positive definite function, *J. Math. Mech.*, **10**(1961), 927–945.

[Cl] Claerbout, J.F.,

[1]   *Fundamentals of Geophysical Data Processing*, McGraw-Hill, New York, 1976.

[Co] Constantinescu, T.,

[1]   On the structure of the Naimark dilation, *J. Operator Theory*, **12**(1984), 159–175.

[2]   Schur analysis of positive block matrices, in *Schur Methods in Operator Theory and Signal Processing, Operator Theory: Advances and Applications*, **18**, Birkhäuser Verlag, pp. 191–206, 1986.

[3]   On a general extrapolation problem, *Rev. Roumaine Math. Pures Appl.*, **32**(1987), 509–521.

[4]   Some aspects of nonstationarity II, *Mathematica Balkanica*, **4**(1990), 211–235.

[5]   Factorization of positive-definite kernels, in *Hellinger Volume, Operator Theory: Advances and Applications*, **48**, Birkhäuser Verlag, pp. 245–260, 1990.

[CG] Constantinescu, T., and A. Gheondea,

[1]   On unitary dilations and characteristic functions in indefinite product spaces, in *Operator Theory: Advances and Applications*, **43**, Birkhäuser Verlag, pp. 87–102, 1987.

[2]   Minimal signature in lifting of operators, II, *J. Functional Analysis* **103**(1992), 317–351.

[CSK] Constantinescu, T., Sayed, A.H., and T. Kailath,

[1]   Displacement structure and completion problems, *SIAM J. Matrix Analysis Appl.*, **16**(1995), 58–78.

[2]   Displacement structure and maximum entropy, 1994, preprint.

[CS] Cotlar, M., and C. Sadosky,

[1]   Weakly positive matrix measures, generalized Toeplitz forms, and their applications to Hankel and Hilbert transform operators, in *Operator Theory: Advances and Applications*, **58**, Birkhäuser Verlag, pp. 93–120, 1992.

[Cy] Cybenko, G.,

[1]   The numerical stability of the Levinson-Durbin algorithm for Toeplitz systems of equations, *SIAM J. Sci. Statist. Comp.*, **1**(1980), 303–319.

[Da] Davidson, K.,

[1]   *Nest Algebras*, Longman, London, 1988.

[Dav] Davis, C.,

[1]   A factorization of an arbitrary $m \times n$ contractive operator, in *Proceeding of the Toeplitz Centennial Conference, Tel Aviv*, Birkhäuser Verlag, pp. 217–232, 1982.

[DKW] Davis, C., Kahan, W.M., and H.F. Weinberger,

[1]   Norm preserving dilations and their applications to optimal error bounds, *SIAM J. Numer. Anal.*, **19**(1982), 444–469.

[DM] Delosme, J.-M., and M. Morf,

[1]   Mixed and minimal representations for Toeplitz and related systems, In *Proceedings 14th Asilomar Conference on Circuits, Systems and Computers*, Nov. 1980.

[DI] Delosme, J.-M., and I.C.F. Ipsen,

[1]   Parallel solution of symmetric positive definite systems with hyperbolic rotations, *Linear Algebra Appl.*, **77**(1986), 75–111.

[DGK] Delsarte, Ph., Genin, Y., and Y. Kamp,

[1]   Schur parametrization of positive definite block, Toeplitz systems, *SIAM J. Appl. Math.*, **36**(1979), 34–46.

[2]   The Nevanlinna-Pick problem for matrix-valued functions, *SIAM J. Appl. Math.*, **36**(1979), 47–61.

[3]   On the Toeplitz embedding of arbitrary matrices, *Linear Algebra Appl.*, **51**(1983), 97–119.

[De] Dempster, A.P.,

[1]   Covariance selections, *Biometrics*, **28**(1972), 157–175.

[Dep] Deprettere, E.F.,

[1]   Mixed form time-variant lattice recursions, in *Outils et Modèles Mathématique pour l'Automatique, l'Analyse de Systèmes et le Traitement du Signal*, CNRS, Paris, 1981.

[Dev] Devinatz, A.,

[1]   The factorization of the operator valued functions, *Ann. of Math.*, **73**(1961), 458–495.

[Dew] Dewilde, P.,

[1]   New algebraic methods for modelling large-scale integrated circuits, *International J. of Circuit Theory and Appl.*, **16**(1988), 473–503.

[2]   A course on the algebraic Schur and Nevanlinna-Pick interpolation problems, in *Algorithms and Parallel VLSI Architectures*, Eds. E.F. Deprettere and A.J. van der Veen, Elsevier Sciences Publications, pp. 13–69, 1991.

[DD] Dewilde, P., and E.F. Deprettere,

[1]   The generalized Schur algorithm: approximation and hierarchy, in *Operator Theory: Advances and Applications*, **29**, Birkhäuser Verlag, pp. 97–116, 1988.

[DeD] Dewilde, P., and H. Dym,

[1]   Lossless chain scattering matrices and optimum linear prediction: The vector case, *Circuit Theory and Appl.*, **9**(1981), 135–175.

[2]   Lossless inverse scattering, digital filters, and estimation theory, *IEEE Trans. on Information Theory*, **30**(1984), 644–662.

[3]   Interpolation for upper triangular operators, in *Operator Theory: Advances and Applications*, **56**, Birkhäuser Verlag, pp. 153–260, 1992.

[DVK] Dewilde, P., Vieira, A.C., and T. Kailath,

[1]   On a generalized Szegö-Levinson realization algorithm for optimal predictors based on a network synthesis approach, *IEEE Trans. on Circuit and Systems*, **25**(1978), 663–675.

[DKV] Dewilde, P., Kaashoek, M.A., and Verhaegen, M., Editors.,

[1]   *Challenges of a Generalized System Theory*, North-Holland, Amsterdam, 1993.

[DV] Dewilde, P., and A.-J. van de Veen,

[1]   On the Hankel-norm approximation of upper triangular operators and matrices, *Integral Equations and Operator Theory*, **17**(1993), 1–45.

[Di] Dirac, G.A.,

[1]   On Rigid Circuit Graphs, *Abh. Math. Sem. Univ. Hamburg*, **25**(1961), 71–76.

[Dor] Dorato, P., Editor,

[1]   *Robust Control*, IEEE Press, 1987.

[Do] Douglas, R.G.,

[1] On majoration, factorization and range inclusion on Hilbert space, *Proc. Amer. Math. Soc.*, **17**(1966), 413–415.

[2] On factoring positive operator functions, *J. Math. and Mech.*, **232**(1966), 119–126.

[3] *Banach Algebra Techniques in Operator Theory*, Academic Press, New York, 1972.

[4] Canonical models, in *Topics in Operator Theory*, Ed. C. Pearcy, Mathematical Surveys, **13**, Amer. Math. Soc., Providence, pp. 161–218, 1974.

[DH] Douglas, R.G., and J. W. Helton,

[1] Inner dilations of analytic matrix functions and Darlington synthesis, *Acta Sci. Math.*, **34**(1973), 301–310.

[DMP] Douglas, R.G., Muhly, P.S., and C. Pearcy,

[1] Lifting commuting operators, *Michigan Math. J.*, **15**(1968), 385–395.

[DP] Douglas, R.G., and V.I. Paulsen,

[1] *Hilbert Modules over Function Algebras*, Longman, London, 1989.

[DFT] Doyle, J.C., Francis, B.A., and A.R. Tannenbaum,

[1] *Feedback Control Theory*, Macmillan, 1992.

[Dr] Dritschel, M.A.,

[1] *Extension theorems for operators on Krein spaces*, Thesis, University of Virginia, 1989.

[2] A lifting theorem for bicontractions in Krein spaces, *J. Functional Analysis*, **88**(1990), 61–89.

[DR] Dritschel, M.A., and J. Rovnyak,

[1] Extension theorems for contraction operators on Krein spaces, in *Operator Theory: Advances and Applications*, **47**, Ed. I. Gohberg, Birkhäuser Verlag, pp. 221–305, 1990.

[DFK] Dubovoi, V., Fritzsche, B., and B. Kirstein,

[1] *Matricial Version of the Classical Schur Problem*, Teubner, Stuttgart, 1992.

[Dy] Dym, H.,

[1] *J Contractive Matrix Functions, Reproducing Kernel Hilbert Spaces and Interpolation*, CBMS Regional Conferences series, **71**, Amer. Math. Soc., Providence, Rhode Island, 1989.

[2] On reproducing kernel spaces, *J*-unitary matrix functions, interpolation and displacement rank, in *The Gohberg Anniversary Collection, Operator Theory: Advances and Applications*, **41**, Birkhäuser Verlag, pp. 173–239, 1989.

[DG] Dym, H., and I. Gohberg,

[1]   Extensions of band matrices with band inverses, *Linear Algebra Appl.*, **36** (1981), 1–24.

[2]   Extension of kernels of Fredholm operators, *J. d Analyse Math.*, **42**(1982), 83–125.

[3]   A maximum entropy principle for contractive interpolants, *J. Functional Analysis*, **65**(1986), 83–125.

[DM] Dym, H., and H.P. McKean, Jr.,

[1]   *Gaussian Processes, Function Theory and the Inverse Spectral Problem*, Academic Press, 1976.

[EGL] Ellis, R.L., Gohberg, I., and D. Lay,

[1]   Band extensions, maximum entropy and the permanence principle, in *Maximum Entropy and Bayesian Methods in Applied Statistics*, Ed. J. Justice, Cambridge, U. P., 1986.

[ET] Erdös, P., and P. Túran,

[1]   On interpolation, III, *Ann. of Math.*, **41**(1940), 510–553.

[EL] Evans, D.E., and J.T. Lewis,

[1]   *Dilations of Irreversible Evolutions in Algebraic Quantum Theory*, Communications of the Dublin Institute of Advaced Studies, Series A(Theoretical Physics), **24**, 1977.

[Fe] Fedcina, P.I.,

[1]   A criterion for the solvability of the Nevanlinna-Pick tangent problem, *Mat. Issled.*, **7**(1972), 213–227.

[2]   The tangential Nevanlinna-Pick problem with multiple points, *Akad. Nauk Armjan. SSR Dokl.*, **61**(1975), 214–218.

[FF] Feintuch, A., and B.A. Francis,

[1]   Uniformly optimal control of linear feedback systems, *Automatica*, **21**(1985), 563–574.

[2]   Distance formulas for operator algebras arising in optimal control problems, in *Operator Theory: Advances and Applications*, **29**, Birkhäuser Verlag, pp. 151–170, 1988.

[Fi] Fiedler, M.,

[1]   *Czech. Math. J.*, **13**(1963), 524–586.

[2]   Matrix inequalities, *Numerische Mathematik*, **9**(1966), 109–119.

[Fo] Foias, C.,

[1]   Contractive intertwining dilations and waves in layered media, *Proceedings of the International Congress of Mathematicians*, Helsinki, pp. 605–613, 1978.

[FoF] Foias, C., and A. Frazho,

[1]   *The Commutant Lifting Approach to Interpolation Problems*, Birkhäuser Verlag, Basel, 1990.

[2]   Simultaneous H^∞ and L^2 suboptimization, *SIAM J. Math. Analysis*, **23** (1992), 984–994.

[FFG] Foias, C., Frazho, A, and I. Gohberg,

[1]   Central intertwining lifting, maximum entropy and their permanence, *Integral Equations and Operator Theory*, **18**(1994), 166–201.

[FT] Foias, C., and A. Tannenbaum,

[1]   A strong Parrott theorem, *Proc. Amer. Math. Soc.* **106**(1989), 777–784.

[2]   On the four block problem I, *Operator Theory: Advances and Applications*, **32**, Birkhäuser Verlag, pp. 93–112, 1988; II, The singular system, *Integral Equations and Operator Theory*, **11**(1988), 727–767.

[FR] Frazho, A.E., and M.A. Rotea,

[1]   A remark on mixed L^2/L^∞ bounds, *Integral Equations and Operator Theory*, **15**(1992), 343–348.

[Fr] Francis, B.A.,

[1]   *A course in H^∞ Control Theory*, Springer Verlag, New York, 1987.

[FHZ] Francis, B.A., Helton, J.W., and G. Zames,

[1] H_∞-optimal feedback controllers for linear multivariable systems, *IEEE Trans. Automatic Control*, **29**(1984), 888–900.

[FG] Fulkerson, D.R., and O.A. Gross,

[1]   Incidence matrices and interval graphs, *Pacific J. Math.*, **15**(1965), 835–855

[Fu] Fuhrmann, P.A.,

[1]   *Linear Systems and Operators in Hilbert Spaces*, McGraw-Hill, New York, 1981.

[Ga] Garnett, J.B.,

[1]   *Bounded Analytic Functions*, Academic Press, NY, 1981.

[Gav] Gavril, F.,

[1]   The intersection graphs of subtrees in trees are exactly the chordal graphs, *J. Combin. Theory B*, **16**(1974), 47–56.

[GDKDM] Genin, Y., Van Dooren, P., Kailath, T., Delosme, J., and M. Morf,

[1]   On Σ-lossless transfer functions and related questions, *Linear Algebra Appl.*, **50**(1983), 251–275.

[Ge] Geronimus, Ya. L.,

[1]   *Orthogonal Polynomials*, Consultant Bureau, New York, 1961.

[Geo] Georgiou, T.T.,

[1]   Realization of power spectra from partial covariance sequences, *IEEE Trans. Acoustics, Speech and Signal Processing*, **35**(1987), 438–449.

[GeK] Georgiou, T.T., and P.P. Khargonekar,

[1]   Linear fractional transformations and spectral factorization, *IEEE Trans. Automatic Control*, **31**(1986), 345–347.

[Gl] Glover, K.,

[1]   All optimal Hankel-norm approximations of linear multivariable systems and their L_∞-error bounds, *Int. J. Control*, **39**(1984), 1115–1193.

[2]   Robust stabilization of linear multivariable systems: relations to approximation, *Int. J. Control*, **43**(1986), 741–766.

[GLDKS] Glover, K., Limebeer, D.J.N., Doyle, J.C., Kasenally, E.M., and M.G. Safonov,

[1]   A characterization of all solutions to the four block general distance problem, *SIAM J. Control and Optimization*, **29**(1991), 283–324.

[Go] Gohberg, I., Editor,

[1]   *I. Schur Methods in Operator Theory and Signal Processing, Operator Theory: Advances and Applications*, **18** Birkhäuser Verlag, 1986.

[2]   *Topics in Operator Theory and Interpolation, Operator Theory: Advances and Applications*, **29** Birkhäuser Verlag, 1988.

[3]   *Time-Variant Systems and Interpolation, Operator Theory: Advances and Applications*, **56** Birkhäuser Verlag, 1992.

[4]   *New Aspects in Interpolation and Completion Theories, Operator Theory: Advances and Applications*, **64** Birkhäuser Verlag, 1994.

[GGK] Gohberg, I., Goldberg, S., and M.A. Kaashoek,

[1]   *Classes of linear operators, Operator Theory: Advances and Applications*, **63**, Birkhäuser Verlag, 1993.

[GoK] Gohberg, I., and M.G. Krein,

[1]   *Introduction to the Theory of Linear Nonselfadjoint Operators*, Amer. Math. Soc. Transl. Math. Monographs, Providence, RI, 1969.

[2]   *Theory and Applications of Voltera Operators in Hilbert Space*, Amer. Math. Soc. Transl. Math. Monographs, Providence, RI, 1970.

[GKL] Gohberg, I., Kaashoek, M.A., and L. Lerer,

[1] Minimality and realization of discrete time-varying systems, in *Operator Theory: Advances and Applications*, **56**, Ed. I. Gohberg, Birkhäuser Verlag, Basel, 1992, pp. 261–296

[GKW] Gohberg, I., Kaashoek, M.A., and H.J. Woerdeman,

[1] The band method for positive and contractive extension problems, *J. Operator Theory*, **22**(1989), 109–155

[2] A maximum entropy principle in the general framework of the band method, *J. Functional Analysis*, **95**(1991), 231–254.

[GKKL] Gohberg, I., Kailath, T., Koltracht, I., P. Lancaster,

[1] Linear complexity parallel algorithms for linear sysytems of equations with recursive structure, *Linear Algebra Appl.*, **88/89**(1987), 271–315.

[GKO] Gohberg, I., Kailath, T., and V. Olshevsky,

[1] Fast Gaussian elimination with partial pivoting for matrices with displacement structure, preprint 1994.

[GS] Gohberg, I., and A.A. Semencul,

[1] On the inversion of finite Toeplitz matrices and their continuous analogs, *Mat. Issled.*, **2**(1972), 201–233.

[Gol] Golumbic, M.C.,

[1] *Algorithmic Graph Theory and the Perfect Graph*, Academic Press, New York, 1980

[Gr] Gragg, W. B.,

[1] Positive definite Toeplitz matrices, the Arnoldi process for isometric operators, and gaussian quadrature on the unit circle, in *Numerical Methods in Linear Algebra*, Ed. E. S. Nikolaev, Moscow Univ. Press, pp. 16–32, 1982.

[GrS] Grenander, U., and G. Szegö,

[1] *Toeplitz Forms and Their Applications*, Univ. of California Press, Berkeley, California, 1958.

[GJSW] Grone, R., Johnson, C.R., Sa, E.M., and H. Wolkowitz,

[1] Positive definite completions of partial Hermitian matrices, *Linear Algebra and Appl.*, **58**(1984), 102–124

[GL] Grossmann, M., and H. Langer,

[1] Über indexerhaltende Erweiterungen eines hermiteschen Operators im Pontrjaginraum, *Math. Nach.*, **64**(1974), 289–317.

[HS] Hajnal, A., and J. Surányi,

[1]   Über die Auflösung von Graphen in vollständige Teilgraphen, *Ann. Univ. Sci. Budapest Eötvös. Sect. Math.* **1**(1944), 113–121.

[Ha] Halmos, P.R.,

[1]   *A Hilbert Space Problem Book,* Springer Verlag, New York, 1982.

[Hay] Haykin, S.,

[1]   *Adaptive Filters Theory,* Prentice Hall, Englewood Cliffs, New Jersey, 1991.

[He] Helson, H.,

[1]   *Lectures on Invariant Subspaces,* Academic Press, New York, 1964.

[HL] Helson, H., and D. Lowdenslager,

[1]   Prediction theory and Fourier series in several variables, I. *Acta Math.,* **99**(1958), 165–202; II. *Acta Math.,* **106**(1961), 175–213.

[Helt] Helton, J.W.,

[1]   Discrete time systems, operator models and scattering theory, *J. Functional Analysis,* **16**(1974), 15–38.

[2]   The distance of a function to H^∞ in the Poincaré metric; electrical power transfer, *J. Functional Analysis,* **38**(1980), 273–314.

[3]   Broadbanding: gain equalization directly from data, *IEEE Trans. Circuit and Systems,* **28**(1981), 1125–1137.

[4]   Non-Euclidean functional analysis and electronics, *Bull. Amer. Math. Soc.,* **7**(1982), 1–64.

[5]   *Operator Theory, Analytic Functions, Matrices and Electrical Engineering,* CBMS Regional Conferences Series in Math., **68**, Amer. Math. Soc., Providence, Rhode Island, 1987.

[Her] Herglotz, G.,

[1]   Über Potenzreihen mit positivem, reellem Teil im Einheitskreis, *Sächs. Akad. Wiss. Leizig, Math.-Phys. Kl.,* **63**(1911), 501–511.

[Herg] Herglotz, G., et al.,

[1]   *Ausgewählte Arbeiten zu den Ursprüngen der Schur-Analysis,* Teubner, Stuttgart, 1991.

[Ho] Hoffman, K.,

[1]   *Banach Spaces of Analytic Functions,* Prentice-Hall Englewood Cliffs, New Jersey, 1962.

[HM] Honig, M.L., and D.G. Messerschmitt,

[1]   *Adaptive Filters,* Kluwer Academic Publishers, Boston, 1984.

[IR] Ibragimov, I.A., and Yu.A. Rozanov,

[1]   *Gaussian Stochastic Processes*, Springer Verlag, 1978.

[IS] Itakura, F., and S. Saito,

[1]   Digital filtering techniques for speech analysis and synthesis, in *Proc. 7 th Int. Cong. Acoust.*, Budapest, pp. 261–264, 1971.

[Jo] Johnson, C.R.,

[1]   Combinatorial matrix analysis: an overview, *Linear Algebra Appl.*, **107** (1988), 3–15.

[2]   Matrix completion problems: A survey, in *Matrix Theory and its Applications*, Proc. of Symposia in Appl. Math., **40**(1989), AMS, Providence, RI.

[JB] Johnson, C.R., and W.W. Barrett,

[1]   Spanning tree extensions of the Hadamard-Fischer inequalities, *Linear Algebra Appl.*, **66**(1985), 177–194

[JL] Johnson, C.R., and M. Lundquist,

[1]   Matrices with chordal inverse zero-patterns, *Linear and Multiliniar Algebra*, **36**(1993), 1–17.

[JLN] Johnson, C.R., Lundquist, M., and G. Nevdal,

[1]   Positive definite Toeplitz completions, preprint 1993.

[JR] Johnson, C.R., and L. Rodman,

[1]   Completion of partial matrices to contractions, *J. Functional Analysis*, **69** (1986), 260–267.

[2]   Chordal inheritance principles and positive definite completions of partial matrices over function rings, in *Operator Theory: Advances and Applications*, **35** Birkhäuser Verlag, 1988.

[3]   Completion of Toeplitz partial contractions, *SIAM J. Matrix Analysis Appl.*, **9**(1988), 159–167.

[JNT] Jones, W.B., Njastad, O., and W.J. Thron,

[1]   Moment theory, orthogonal polynomials, quadrature and continuous fractions associated with the unit circle, *Bull. London Math. Soc.*, **21**(1989), 113–152.

[KK] Kaashoek, M.A., and J. Kos,

[1]   The Nehari-Takagi problem for input-output operators of time-varying continuous time systems, *Integral Equations and Operator Theory*, to appear.

[KLW] Kaftal, V.G., Larson, D.R., and G. Weiss,

[1]   Quasitriangular subalgebras of semifinite von Neumann algebras ae closed, *J. Functional Analysis*, **107**(1992), 387–401.

[Ka] Kailath, T.,

[1]   *Linear Systems*, Prentice-Hall Englewood Cliffs, New Jersey, 1980.

[2]   Time-variant and time-invariant lattice filters for nonstationary processes, in *Outils et Modèles Mathématique pour l'Automatique, l'Analyse de Systèmes et le Traitement du Signal*, CNRS, Paris, 1981.

[3]   A theorem of I. Schur and its impact on modern signal processing, in *Schur Methods in Operator Theory and Signal Processing, Operator Theory: Advances and Applications*, **18**, Ed. I. Gohberg, Birkhäuser Verlag, pp. 9–30, 1986.

[4]   Signal processing applications of some moment problems, in *Moments in mathematics*, Ed. H. Landau, Amer. Math. Soc. **37**, pp. 71–109, 1987.

[5]   Remarks on the origin of the displacement-rank concept, *Applied Mathematics and Computation*, **45**(1991), 193–206.

[KB] Kailath, T., and A.M. Bruckstein,

[1]   Naimark dilations, state-space generators and transmission lines, in *Operator Theory: Advances and Applications*, Birkhäuser Verlag, pp. 173–186, 1984.

[KKM] Kailath, T., Kung, S.-Y., and M. Morf,

[1]   Displacement ranks of matrices and linear equations, *J. Math. Anal. Appl.*, **68**(1979), 395–407.

[KL] Kailath, T., and H. Lev-Ari,

[1]   Generalized Schur parametrizations of nonstationary second-order processes, in *Proceeding of the Toeplitz Centennial Conference, Tel Aviv*, Birkhäuser Verlag, 1982.

[KO] Kailath, T., and V. Olshevsky,

[1]   Displacement structure approach to Chebyshev-Vandermonde and related matrices, *Integral Equations and Operator Theory*, **22**(1995), 65–92.

[KP] Kailath, T., and B. Porat,

[1]   State-space generators for orthogonal polynomials, in *Prediction Theory and Harmonic Analysis*, North-Holland, Amsterdam, 1983.

[KFA] Kalman, R.E., Falb, P.L., and B. M. Arbib,

[1]   *Topics in Mathematical System Theory*, McGraw-Hill, New York, 1969.

[Kam] Kamen, E.W.,

[1]   The poles and zeros of a linear time-varying system, *Linear Algebra Appl.*, **98**(1988), 263–289.

[KKP] Kamen, E.W., Khargonekar, P.P., and K.R. Poola,

[1]   A transfer function approach to linear time-varying discrete-time systems, *SIAM J. Control and Optimization*, **23**(1985), 550–565.

[K] Katsnelson, V.E.,

[1] Methods of J theory in continuous problems of analysis, I–IV, *Amer. Math. Soc. Transl.*, **136**(1987), 49–108.

[KaR] Khargonekar, P.P., and M.A. Rotea,

[1] Coprime factorization for linear time-varying systems, *Proc. IEEE American Control Conference*, pp. 848–851, Atlanta, GA, 1988.

[Ki] Kimura, H.,

[1] Robust stabilizability for a class of transfer functions, *IEEE Trans. Automatic Control*, **29**(1984), 788–793.

[2] Directional interpolation approach to H^∞-optimization and robust stabilization, *IEEE Trans. Automatic Control*, **32**(1987), 1085–1093.

[3] Conjugation, interpolation and model matching in H^∞, *Int. J. Control*, **49**(1989), 269–307.

[Ko] Kolmogorov, A.N.,

[1] Sur l'interpolation et l'extrapolation des suites stationaire, *C. R. Acad. Sci. (Paris)*, **208**(1939), 2043–2045.

[2] Stationary sequences in Hilbert spaces, *Bull. Math Univ., Moscow*, **2**(1941).

[Ko] Kos, J.,

[1] Higher order time-varying Nevanlinna-Pick interpolation, in *Challenges of a Generalized System Theory*, North-Holland, Amsterdam, pp. 59–71, 1993.

[KoP] Kovalishina, I.V., and V.P. Potapov,

[1] *Integral Representation of Hermitian Positive Functions*, Private Translation by T. Ando, Division of Applied Math., Research Institute of Applied Electricity, Hokkaido University, Sapporo, Japan.

[Kr] Krein, M.G.,

[1] Sur le probleme du prolongement des functions hermitiens positives et continues, *C. R. (Dokl.) Akad. Sci. URSS*, **26**(1940), 17–22.

[2] On Hermitian operators whose deficiency indices are 1, *Dokl. Akad. Nauk. SSSR*, **43**(1944), 323–326.

[3] The theory of self-adjoint extensions of semi-bounded Hermitian transformations and applications, *Math. Sb.*, **20** (1947), 431–495; 21, No. 63(1947), 365–404.

[KrL] Krein, M.G., and H. Langer,

[1] Über die verallgemeinerten Resolventen und die charakteristische Funktion eines isometrischen Operators im Raume Π_κ, in: *Colloquia Math. Soc. János Bolyai*, **5**, Hilbert Space Operators and Operator Algebras, pp. 353–399, North-Holland, Amsterdam - London 1972.

[2] Über einige Fortsetzungsprobleme, die eng mit der Theorie hermitescher Operatoren im Raume Π_κ zusammenhängen. I. Einige Funktionenklassen und ihre Darstellungen, *Math. Nachr.*, **77**(1977), 187–236. II. Verallgemeinerte Resolventen, u-Resolventen und ganze Operatoren, *J. Functional Analysis*, **30**(1978), 390–447.

[3] On some extension problems which are closely connected with the theory of Hermitian operators in a space Π_κ. III. Indefinite analogues of the Hamburger and Stieltjes moment problem, *Beiträge zur Analysis*, Part (I), **14** (1979), 25–40. Part (II), **15**(1981), 27–45.

[4] On some propositions on analytic matrix functions related to the theory of operators in the spaces Π_κ, *Acta Sci. Math.*, **43**(1981), 181–205.

[5] On some continuation problems which are closely related to the theory of operator spaces Π_κ, IV, *J. Operator Theory*, **13**(1985), 299–417.

[KN] Krein, M.G., and A.A. Nudelman,

[1] *The Markov Moment Problem and Extremal Problems,* Amer. Math. Soc. Transl. Math. Monographs, Providence , RI, 1977.

[KR] Krein, M.G., and P.G. Rehtman,

[1] On the problem of Nevanlinna-Pick, *Trudi Odeskogo Derz. Univ. Mat.*, **2**(1938), 63–68.

[La] Landau, H.J.,

[1] Maximum entropy and the moment problem, *Bull. Amer. Math. Soc.*, **16**(1987), 44–77.

[Lan] Langer, H.,

[1] On measurable Hermitian indefinite functions with a finite number of negative squares, *Acta Sci. Math.*, **45**(1983), 281–292.

[LP] Lax, P., and R.S. Phillips,

[1] *Scattering Theory*, New York, 1967.

[Le] Leech, R.B.,

[1] Factorization of analytic functions and operator inequalities.

[Lev] Lev-Ari, H.,

[1] *Lattice Modelling of Nonstationary Processes*, Thesis, Stanford University, 1983.

[LBK] Lev-Ari, H., Bistritz, Yu., and T. Kailath,

[1] Generalized Bezoutians and families of efficient zero-location procedures, *IEEE Trans. Circuits and Systems*, **38**(1991), 170–186.

[LAK] Lev-Ari, H., and T. Kailath,

[1] Schur and Levinson algorithms for nonstationary processes, Proceedings ICASSP, 1981.

[2] Lattice filter parametrization and modeling of nonstationary processes, *IEEE Trans. Information Theory*, **30**(1984), 2–16.

[3] Triangular factorization of structured Hermitian matrices, in *Schur Methods in Operator Theory and Signal Processing, Operator Theory: Advances and Applications*, **18**, Ed. I. Gohberg, Birkhäuser Verlag, pp. 301–324, 1986.

[4] State-space approach to factorization of lossless transfer functions and structured matrices, *Linear Algebra Appl.*, **162–164**(1992), 273–295.

[LPK] Lev-Ari, H., Parker, S.R., and T. Kailath,

[1] Multidimensional maximum-entropy covariance extension, *IEEE Trans. Information Theory*, **35**(1989), 497–508.

[Levi] Levinson, N.,

[1] The Wiener RMS (root mean square) error criterion in filter design and prediction, *J. Math. Phys.*, **25**(1947), 261–278.

[LA] Limebeer, D.J.N., and B.D.O. Anderson,

[1] An interpolation theory approach to H^∞ controller degree bounds, *Linear Algebra Appl.*, **98**(1988), 347–386.

[LG] Limebeer, D.J.N., and M.Green,

[1] Parametric interpolation, H^∞-control and model reduction, *Int. J. Control*, **52**(1990), 293–318.

[LY] Livsic, M.S., and A.A. Yantsevich,

[1] *Operator Colligations in Hilbert Spaces*, John Wiley and Sons, New York, 1979.

[Lo] Lowdenslager, D.B.,

[1] On factoring matrix valued functions, Ann. of Math., **78**(1963), 450–454.

[Lu] Luenberger, D.G.,

[1] *Introduction to Linear and Nonlinear Programming*, Addison-Wesley, Reading, MA, 1973.

[Lun] Lundquist, M.,

[1] *Zero Patterns, Chordal Graphs and Matrix Completion*, Thesis, Clemson University, South Carolina, 1990.

[LJ] Lundquist, M., and C.R. Johnson,

[1] Linearly constrained positive definite completions, *Linear Algebra Appl.*, **150**(1991), 195–207.

[MG] Markel, J.D., and A.H. Grey, Jr.,

[1] *Linear Prediction of Speach*, Springer Verlag, New York, 1976.

[Ma] Masani, P.,

[1] The prediction theory of multivariable stochastic processes III, *Acta Math.*, **104**(1960), 141–162.

[2] On isometric flows on Hilbert spaces, *Bull. Amer. Math. Soc.*, **68**(1962), 624–632.

[3] On the representation theorem of scattering, *Bull. Amer. Math. Soc.*, **74**(1968), 219–234.

[MNT] Maté, A., Nevai, P., and V. Totik,

[1] Asymptotics for the ratio of leading coefficients of orthogonal polynomials on the unit circle, *Constructive Approximation*, **1**(1985), 63–69.

[Mur] Murnaghan, F.D.,

[1] *The unitary and rotation groups*, Spartan Books, Washington, D.C., 1962.

[Na] Naimark, M.A.,

[1] Selfadjoint extensions of the second kind of a symmetric operator, *Bulletin Acad. Sci. URSS* (Ser. Math.) **4**(1940), 53–104.

[2] Positive definite operators on a commutative group, *Bulletin Acad. Sci. URSS* (Ser. Math.) **7**(1943), 237–244.

[Ne] Nehari, Z.,

[1] On bounded bilinear forms, *Ann. of Math.*, **65**(1957), 153–162.

[Nel] Nellis, H.,

[1] *Sparse approximations of inverse matrices*, Thesis, Delft University of Technology, October, 1989.

[NN] Nesterov, Yu., and A. Nemirovsky,

[1] *Interior Point Polynomial Methods in Convex Programming: Theorey and Applications*, Lecture Notes in Mathematics, Springer Verlag, 1992.

[Neu] von Neumann, J.,

[1] Allgemeine Eigenwerttheorie Hermitescher Funktionaloperatoren, *Math. Ann.*, **102**(1929), 49–131.

[Nev] Nevai, P.,

[1]   Weakly convergent sequences of functions and orthogonal polynomials, *J. Approximation Theory*, **65**(1991), 322–340.

[2]   Orthogonal polynomials, recurrences, Jacobi matrices and measures, in *Progress in Approximation Theory*, Eds. A.A. Gonchar and E.B. Saff, Springer Verlag, pp. 79–104, 1992.

[Nevan] Nevanlinna, R.,

[1]   Über beschränkte Funktionen, die in gegebenen Punkten vorgeschriebene Werte annehmen, *Ann. Acad. Sci. Fenn.*, **13**(1919).

[2]   Über beschränkte analytische Funktionen, *Ann. Acad. Sci. Fenn.*, **32**(1929).

[Ni] Nikolskii, N.K.,

[1]   *Treatise on the Shift Operator*, Springer Verlag, New York, 1986.

[NV] Nikolskii, N.K., and V.I. Vasyunin,

[1]   A unified approach to functional models, and the transcription problem, in *The Gohberg Anniversary Collection, Operator Theory: Advances and Applications*, **41**, Birkhäuser Verlag, pp. 405–434, 1989.

[Nu] Nudelman, A.A.,

[1]   On a generalization of classical interpolation problems, *Soviet Math. Doklady*, **23**(1981), 125–128.

[Pa] Parrott, S.,

[1]   On a quotient norm and the Sz. Nagy-Foias lifting theorem, *J. Functional Analysis*, **30**(1978), 311–328.

[Pau] Paulsen, V.I.,

[1]   *Completely Bounded Maps and Dilations*, Pitman, New York, 1986.

[PP] Paulsen, V.I., and S.C., Power,

[1]   Lifting theorems for nest algebras, *J. Operator Theory*, **20**(1988), 311–327.

[PPS] Paulsen, V.I., Power, S.C., and R.R. Smith,

[1]   Schur products and matrix completions, *J. Functional Analysis*, **85**(1989), 151–179.

[Ph] Phillips, R.S.,

[1]   The Extension of Dual Subspaces Invariant under an Algebra, in: *Proc. Internat. Sympos. Linear Spaces*, Jerusalem Academic Press and Pergamon, Jerusalem and Oxford, pp. 366–398, 1961.

[Pi] Pick, G.,

[1]   Über die Beschränkungen analytischer Funktionen, welche durch vorgegebene Funktionswerte bewirkt sind, *Math. Ann.*, **77**(1916), 7–23.

[PKTKN] Poola, K., Khargonekar, P., Tikku, A., Krause, J., and K. Nagpal,

[1]   A time-domain approach to model validation, *IEEE Trans. Automatic Control*, **39**(1994), 951–959.

[Pop] Popescu, G.,

[1]   On intertwining dilations for sequences of noncommuting operators, *J. Math. Anal. Appl.*, **167**(1992), 382–402.

[Po] Potapov, V.P.,

[1]   Multiplicative structure of J-expansive matrix functions, Amer. Math. Soc. Transl., **15**(1960), 131–244.

[Pow] Power, S.C.,

[1]   *Hankel Operators on Hilbert Spaces*, Pitman, London, 1982.

[Ra] Rahmanov, E.A.,

[1]   On the asymptotics of the ratio of orthogonal polynomials, *Math. USSR. Sbornik*, **46**(1983), 105–117.

[Re] Redheffer, R.M.,

[1]   On the relation of transmission-line theory to scattering and transfer, *J. Math. Phys.*, **41**(1962), 1–41.

[Ri] Riesz, F.,

[1]   Über Potenzreihen mit vorgeschriebenen Anfangsgliedern, *Acta Math.*, **42** (1920), 147–171.

[RT] Robinson, E.A., and S. Treitel,

[1]   *Geophysical Signal Analysis*, Prentice-Hall, Englewood Cliffs, New Jersey, 1980.

[Ro] Rose, D.J.,

[1]   Triangulated graphs and the elimination process, *J. Math. Anal. Appl.*, **32**(1970), 597–609.

[RTL] Rose, D.J., Tarjan, R.E., and G. S. Leuker,

[1]   Algorithmic aspects of vertex eliminations on graphs, *SIAM J. Comput.*, **5**(1976), 266–283.

[Ros] Rosenblum, M.,

[1]   A corona theorem for countable many functions, *Integral Equations and Operator Theory*, **3**(1980), 125–137.

[RR] Rosenblum, M., and J. Rovnyak,

[1]   *Hardy Classes and Operator Theory*, Oxford Univ. Press, New York, 1985.

[2]   *Topics in Hardy Classes and Univalent Functions*, Birkhäuser Verlag, 1994.

[Rot] Rota, G.C.,

[1]   On models for linear operators, *Comm. Pure Appl. Math.*, **13**(1960), 468–472.

[Sa] Sakhnovich, L.A.,

[1]   Equations with a difference kernel on a finite interval, *Russian Math. Surveys*, **35**(1980), 81–152.

[2]   Interpolation problems, inverse spectral problems and nonlinear equations, in *OperatorTheory: Advances and Applications*, **59**, Birkhäuser Verlag, pp. 292–304, 1992.

[Sar] Sarason, D.,

[1]   Generalized interpolation in H^∞, *Trans. Amer. Math. Soc.*, **127**(1967), 179–203.

[2]   Moment problems and operators in Hilbert spaces, in *Proc. of Symposia in Applied Mathematics*, **37**(1987), 54–70.

[Say] Sayed, A.H.,

[1]   *Displacement Structure in Signal processing and Mathematics*, Thesis, Stanford University, 1992.

[SCK] Sayed, A.H., Constantinescu, T., and T. Kailath,

[1]   Time-variant displacement structure and interpolation problems, *IEEE Trans. Automatic Control*, **39**(1994), 960–976.

[SLK] Sayed, A.H., Lev-Ari, H., and T. Kailath,

[1]   Time-variant displacement structure and triangular array, *IEEE Trans. Signal Processing*, **42**(1994), 1052–1062.

[SKLC] Sayed, A.H., Kailath, T., Lev-Ari, H., and T. Constantinescu,

[1]   Efficient recursive solutions of rational interpolation problems, *Integral Equations Operator Theory*, **20**(1994), 84–118.

[SK] Sayed, A.H., and T. Kailath,

[1]   Extended Chandrasekhar recursions, *IEEE Trans. Automatic Control*, **39** (1994), 960–976.

[2]   A state-space approach to adaptive RLS filtering, *IEEE Signal Processing Magazine*, **11**(1994), 18–60.

[Sc] Schur, I.,

[1]   On power series which are bounded in the interior of the unit circle I, *J. für die Reine und Angewandte Mathematik*, 147(1917), 205–232, English translation in *I. Schur Methods in Operator Theory and Signal Processing, Operator Theory: Advances and Applications*, **18**, pp. 31–59, 1986.

[Sch] Schwartz, L.,

[1]   Sous-espaces Hilbertiens d'espaces vectoriels topologiques et noyaux associés (noyaux reproduisants), *J. Analyse Math.*, **13**(1973), 115–256.

[SD] Smith, R.S., and J.C. Doyle,

[1]   Model validation: A connection between robust control and identification, *IEEE Trans. Automatic Control*, **37**(1992), 942–952.

[SY] Smuljan, Ju.L., and R.N. Yanovskaia,

[1]   On matrices whose entries are contractions, *Izo. Visch. Uchab. Zaved Matematica*, **7**(230), 1981.

[SpK] Speed, T.P., and H.T. Kiiveri,

[1]   Gaussian Markov distributions over finite graphs, *The Annals of Statistics*, **14**(1986), 138–150.

[St] Stinespring, W.,

[1]   Positive functions on C^*-algebras, *Proc. Amer. Math. Soc.*, **6**(1955), 211–216.

[SV] Suciu, I., and I. Valusescu,

[1]   Factorization theorems and prediction theory, *Rev. Roumaine Math. Pures Appl.*, **23**(1978), 1393–1423.

[Sz] Szegö, G.,

[1]   *Orthogonal Polynomials*, Colloquium Publications, **23**, Amer. Math. Soc., Providence, Rhode Island, 1939.

[Sz.-NF] Sz.-Nagy, B., and C. Foias,

[1]   Dilatation des commutants d'opérateurs, *C.R. Acad. Sci. Paris, Serie A*, **266**(1968), 493–495.

[2]   *Harmonic Analysis of Operators on Hilbert Space*, North Holland, Amsterdam-Budapest, 1970.

[Sz.-NK] Sz.-Nagy, B., and A. Koranyi,

[1] Operatortheoretische Behandlung and Verallgemeinerung eines Problemkreises in der komplexen Funktiontheorie, *Acta Math.*, **100**(1958), 171–202.

[T] Tadmor, G.,

[1] A time-varying Beurling-Lax theorem and a related interpolation problem, *Math. Control Signals Systems*, **7**(1994), 148–166.

[Ta] Tannenbaum, A.,

[1] Feedback stabilization of linear dynamical plants with uncertainty in the gain factor, *Int. J. Control*, **32**(1980), 1–16.

[2] Modified Nevanlinna-Pick interpolation and feedback stabilization of linear plants with uncertainty in the gain factor, *Int. J. Control*, **36**(1982), 331–336.

[TY] Tarjan, R.E., and M. Yannakakis,

[1] Simple linear-time algorithms to test chordality of graphs, test acyclicity of hypergraph, and selectively reduce acyclic hypergraphs, *SIAM J. Comput.*, **13**(1984), 566–579.

[TV] Treil, S., and A. Volberg,

[1] A fixed point approach to Nehari's problem and its applications, in *Operator Theory: Advances and Applications*, **71**, Birkhäuser Verlag, 1994, 165–186.

[Ti] Timotin, D.,

[1] Prediction theory and choice sequences: an alternate approach, in *Operator Theory: Advances and Applications*, **17**, Birkhäuser Verlag, 1986, 341–352.

[Va] Vaidyanathan, P.R.,

[1] *Multirate Systems and Filter Banks*, Prentice Hall, Englewood Cliffs, NJ 1992.

[Ve] van der Veen, A.-J.,

[1] *Time-varying system theory and computational modeling*, Thesis, Delft University of Technology, 1993

[Vi] Vidyasagar, M.,

[1] *Control Systems Synthesis: a Factorization Approach*, The MIT Press, Cambridge, MA, 1985.

[Wa] Walter, J.R.,

[1] Representation of chordal graphs as subtrees of a tree, *J. Graph Theory*, **2**(1978), 265–267.

[Wiener] Wiener, N.,

[1]   *Smoothing of Stationary Time Series*, Wiley, New York, 1949.

[WM] Wiener, N., and P. Masani,

[1]   The prediction theory of multivariate stochastic processes, I. *Acta Math.*, **98**(1957), 111–150; II. *Acta Math.*, **99**(1958), 93–139.

[Wo] Wohlers, M.R.,

[1]   *Lumped and Distributed Passive Networks*, Academic Press, 1969.

[Wol] Wold, H.,

[1]   *A Study in the Analysis of Stationary Time Series*, Stockholm, 1938.

[YS] Youla, D.C., and M. Saito,

[1]   Interpolation with positive real functions, *J. Franklin Institute*, **284**(1967), 77–108.

[YJN] Youlal, H., Janati-Idrissi, M., and M. Najim,

[1]   *Modélisation Paramétrique en Traitement d'Images*, Masson, 1994.

[Yo] Young, N.J.,

[1]   *An Introduction to Hilbert space*, Cambridge Univ. Press, 1989.

[Za] Zames, G.,

[1]   Optimal sensitivity and feedback: weighted seminorms, approximate inverses, and plant invariant schemes, *Proc. Allerton Conf.*, 1979.

[2]   Feedback and optimal sensitivity: model reference transformations, multiplicative seminorms, and approximate inverses, *IEEE Trans. Auto. Control*, **26**(1981), 301–320.

[ZF] Zames, G., and B.A. Francis,

[1]   Feedback, minimax sensitivity, and optimal robustness, *IEEE Trans. Auto. Control*, **28**(1983), 585–601.

Index